PHYSICS

History of the ZGS
(Argonne, 1979)

AIP Conference Proceedings
Series Editor: Hugh C. Wolfe
Number 60

History of the ZGS
(Argonne, 1979)

Editors
Joanne S. Day
Argonne National Laboratory
Alan D. Krisch
University of Michigan
Lazarus G. Ratner
Argonne National Laboratory

American Institute of Physics
New York   1980

L.C. Catalog Card No. 80-67694
ISBN 0-88318-159-2
DOE CONF- 7909111

Argonne National Laboratory

1957	1963	1979
CONCEPTION	**TURN-ON**	**SHUTDOWN**

SYMPOSIUM
on the
HISTORY of the ZGS

SEPTEMBER 13-14, 1979
ARGONNE NATIONAL LABORATORY

Preliminary List of Speakers

J. J. Livingood	W. D. Walker	R. J. Prien	E. F. Parker
L. C. Teng	K. M. Terwilliger	C. W. Potts	J. B. Roberts
A. V. Crewe	E. G. Pewitt	K. C. Wali	M. Derrick
E. L. Goldwasser	F. F. Bacci	L. G. Ratner	R. L. Martin
R. H. Hildebrand	R. Bouie	A. Abashian	H. A. Neal
R. G. Sachs	J. A. Bywater	A. Yokosawa	W. A. Wallenmeyer
B. Cork	R. L. Kustom	D. S. Ayres	R. B. Duffield
	T. H. Fields		

Sponsored by:
ARGONNE UNIVERSITIES ASSOCIATION
ARGONNE NATIONAL LABORATORY
ZGS USERS GROUP

Organizing Committee

A. D. Krisch	Michigan (Chairman)	J. S. Day	Argonne (Secretary)
U. Camerini	Wisconsin	T. A. Romanowski	Ohio State
F. J. Loeffler	Purdue	R. G. Sachs	Chicago
H. A. Neal	Indiana	A. Wattenberg	Illinois
L. G. Ratner	Argonne		

FOR FURTHER INFORMATION CONTACT:

Mrs. Beverly Marzec, Bldg. 360, Argonne National Laboratory, Argonne, IL 60439, (312) 972-6555/6181

TABLE OF CONTENTS

INTRODUCTORY COMMENTS

A.D. Krisch
Randall Laboratory of Physics
The University of Michigan,Ann Arbor MI 48109

I am very pleased to see so many familiar faces in the audience, including many that I have not seen in a number of years. On behalf of the Organizing Committee, Argonne National Laboratory, and AUA, I would like to welcome all of you to the Symposium on the History of the ZGS. We have a very distinguished group of speakers who are well known to most of us. Their lectures on the life of the ZGS will certainly give rise to memories and will hopefully give us some new insights on constructing and operating major technological facilities.

I would like to remind you that this evening there will be a tour of the ZGS followed by cocktails at the staff club. Friday night there will be a banquet at the Drake Oakbrook Hotel followed by several hours of anecdotes about interesting and amusing experiences at the Zero Gradient Synchrotron.

I would like to say a few words about why we are having this symposium. It is because the ZGS is closing down after 16 years of very successful operation. The ZGS is the largest High Energy Physics facility to ever be shut down. We feel that this symposium is a most fitting way to commemorate the success of the ZGS and the crucial role which it played in fostering High Energy Physics in the Midwest.

Looking over the list of attendees I see that many of you former ZGS staff and users are now in diverse scientific and technological pursuits in the Chicago area and in the nation. I find this very pleasing, because you are spreading and transfering the skills and knowledge that you developed at the ZGS. This direct transfer of frontier research skills is a very important benefit that High Energy Physics gives to our society.

I would like to take this opportunity to thank AUA for its generous support of this symposium. I would also like to thank the AEC, then ERDA and now the Department of Energy for supporting the ZGS through good years and bad, until 0900 on October 1, 1979. There has been a long and very fruitful relationship between the ZGS staff, the Users, AUA, the Universities, and the Government. It thus seems especially appropriate that we all come together to commemorate the retirement of an old friend.

ISSN:0094-243X/80/600001-01$1.50 Copyright 1980 American Institute of Physics

WELCOMING REMARKS

R. E. Diebold

Argonne National Laboratory, Argonne IL   60439

I would like to add a few words of welcome and share a few
thoughts with you. It is good to see so many old friends here, and
the symposium gives us all the opportunity of renewing old acquaint-
ances as well as remembering the technical triumphs of bringing a
machine and its experiments into operation.

While we all rejoice at the remembrance of past achievements,
there is a certain amount of sadness and nostalgia associated with
the retirement of an old friend. Two weeks from now the ZGS will be
shut down, marking the end of an era. As with many retirements, the
melancholy is heightened by the knowledge that the retiree is capable
of additional useful and productive work, that the accomplishments
are being arbitrarily cut off in midstream. Not only does the polarized program of experiments continue to generate surprises and change
our thinking about the importance of spin at both low and high energies, but the conventional unpolarized program was busily generating
interesting results when it was closed down a year and a half ago.

Even though the machine will soon be shut down, its spirit will
live on in many ways. Over the next few years the enormous amount of
data collected by recent experiments will be analyzed and the results
published. These results and those from previous ZGS experiments form
an important part of our knowledge of the elementary particles and
their interactions, and are interwoven into the fabric of the field.

The effects of the ZGS will also continue to be felt in a large
number of university physics departments, especially in the Midwest,
where many high energy physics groups got their start on the ZGS.
Experiments at the ZGS could be done with relatively small groups of
physicists, giving the opportunities for younger faculty to establish
themselves in the field, as well as providing an excellent training
ground for students.

In these days of experimental teams of forty or more, it is not
hard to look back with nostalgia to the good old days and the small
groups at the ZGS. Further, the machine itself has run very smoothly
in the years I have known it. Those of you, who like myself have
done experiments at the ZGS and have since run at other machines,
may well fondly remember your ZGS days with an operating efficiency
in the 90% range. This is a real tribute to the operations crews
and as both a user and an administrator I would personally like to
thank all those concerned with the maintenance of this enviable record. This achievement is all the more remarkable in the closing
years when one considers the attrition of the operations personnel as
people have gone to work on other projects. The transfer of these
people and the maintenance of smooth operation during this period has
been facilitated by the announced plan of an orderly shutdown of the
ZGS by the Department of Energy, which deserves considerable credit
and our thanks.

The world of physics moves on, and many past ZGS users are now working on new projects at other laboratories. The Argonne in-house groups in recent years have also worked on experiments at other laboratories, in order to follow to higher energies the physics studied at the ZGS. In the post-ZGS era the Argonne groups will become involved in the construction of facilities at other laboratories, and we too are looking forward to our new projects and the promise of new exciting physics, while at the same time remembering with nostalgia the physics and friends of the ZGS.

The accelerator people led by Ron Martin have also been turning to new projects inside our multidisciplinary laboratory. Indeed, these projects represent important spinoffs from the national high energy physics program to other parts of the Department of Energy. One of these activities, heavy ion fusion, holds great promise for the long-range future energy needs of this country. Ron and Malcolm Derrick deserve a large amount of credit for their leadership in encouraging and developing new projects such that the talents, built up around the ZGS so painstakingly over the years, will not be lost to the scientific enterprise, but will move on in a most productive manner to attack tomorrow's problems.

It is time to move on to the fascinating program compiled by Alan Krisch and the organizing committee. I would like to take this opportunity to thank the organizers of the symposium, especially Alan and Joanne Day, who have spent an enormous amount of time and effort on this task.

Again, let me welcome you and wish you a pleasant and instructive two days here with old friends and new.

Session I

THE BEGINNING

John J. Livingood
(Retired)
Argonne National Laboratory, Argonne, Illinois 60439

The genesis of high energy physics accelerators occurred in 1947 and 1948 when plans were made for the 1-BeV machine at Birmingham, the 3-BeV Cosmotron at Brookhaven and the 6-BeV Bevatron at Berkeley, these devices coming into operation from 1952 to 1955. Guidance for the protons in these devices was accomplished by the so-called "weak focusing" principle, the magnets all having a constant radial gradient in their fields.

In early 1953 Argonne possessed two Van de Graaff machines of a few million volts and the brand new 21-MeV cyclotron for deuteron acceleration. Several members of the Physics Division (then directed by Louis Turner) felt that the Laboratory should have a more potent armamentum, so in order to arouse enthusiasm, I gave a series of lectures in the fall covering the theory and mathematics of weak-focusing cyclotrons, synchrocyclotrons, synchrotrons, betatrons and linear accelerators. (I learned a lot in teaching others, a fact well known in academic circles.) Morton Hamermesh then continued with a detailed discussion of the "strong focusing" alternating gradient idea then being studied at Brookhaven and at CERN. Ed Crosbie and Mel Ferentz became sufficiently intrigued to join us in forming a group of four that concentrated on understanding and advancing (where possible) the theory of orbits in magnetic fields of various sorts including the Fixed Field Alternating Gradient (FFAG) idea that had just surfaced.

Here we profited immensely by attending the seminars of the midwestern group initially called MAC (Midwest Accelerator Conference) somewhat later to become MURA (Midwest University Research Association). Keith Symon, Kent Terwilliger, Larry Jones, Don Kerst, Bob Haxby and Jackson Laslett were then in the early stages of developing the sophisticated theory of stability of orbits in these specially shaped FFAG fields.

In one of these meetings I encountered John Williams who suggested that Argonne would serve itself well by inducing Lee Teng, then in Kansas, to join us. Fortunately, Lee accepted, in 1955, and soon made a suggestion of inestimable value--the multiunit circular accelerator.

The spread in momenta acceptable by a single magnet of given radius is fixed by the least and the greatest field strengths available. The upper limit was 16 to 20 kilogauss (because of the properties of iron) while the lower limit was perhaps a few hundred gauss, difficult to attain with reliability because of the residual magnetism and inhomogeneities of the core.

Two successive synchrotrons could therefore reach previously unattainable energy provided a means of extraction from one and

ISSN: 0094-243X/80/600004 -06-$1.50 Copyright 1980 American Institute of Physic

insertion into the next could be devised. The extraction scheme
that did this was the regenerative deflector of Tuck, Teng and
LeCouteur, successfully applied at the Liverpool synchrocyclotron
by Albert Crewe and about to be installed in the Chicago machine.
Lee recognized that following extraction by this method (inducing
radial oscillations to cause the ions to enter a deflecting field)
a reverse procedure could be employed to inject the ions into a
second accelerator. Such a transfer of particles from one machine
to another is now standard practice in all high energy machines.

In early 1955, Argonne acquired the services of Martyn Foss and
of John Martin and on July 25, 1955, a formal proposal was made to
the AEC for a 25-GeV proton accelerator consisting of such a two-
stage machine. Stage 1 was planned as a FFAG synchrocyclotron with
a single dee, wherein ions starting from rest would reach 2 BeV at
a radius of 12 feet where the field would be 25 kilogauss. Because
of the field contouring of the FFAG principle, two-dimensional stability would be maintained even though the field was increasing with
rising radius (contrary to the dictates of a weak-focusing machine),
with the result that the radiofrequency accelerating voltage need
fall by only the ratio of 1.45 compared to the figure 3.1 for a conventional weak-focusing machine. (Truly constant frequency would
have required the passage through a large number of beam-destroying
resonances that could be traversed successfully only by an impracti-
cally large dee potential.)
Stage 2 was to be a steady-field FFAG ring magnet of about 150-
feet radius wherein the ions would move radially a few feet into a
rising field of 20 kilogauss final value. Once again, because of
resonances, a constant RF frequency was impossible; it would rise
from 32.7 MHZ to a peak of 33.5 and then fall to 33.0 MHz. The
final energy was to be 25 GeV with an output of 2×10^{10} protons
per second. This ambitious proposal died on the vine, because of
the projected three years of model magnet work, the six years of
construction and, most importantly, because of the as-yet undemon-
strated practicality of the focusing scheme.
The FFAG idea, originally proposed by Thomas in 1938, had not
been understood until 1950, when two small electron cyclotrons embodying it had been built at Berkeley under strict security restrictions. The scheme was independently rediscovered in 1955 by Symon
and by a trio of Russians, with a much more thorough understanding
and analysis. A working model of a sophisticated electron machine
was made in 1957 at MURA.

On January 24, 1956, Laboratory Director Walter Zinn (who was
not one to allow grass to grow under his feet or those of anyone
else) requested the preparation, in 24 hours, of a proposal and cost
estimate for a 10- to 15-BeV proton accelerator to be completed in
four to six years. This time scale and energy range was set by the
AEC.
The instant-design-team percolator was warmed up and a six-
page document was delivered the next day with ten minutes to spare.

It described a 12.5-BeV constant gradient synchrotron (as in the three machines then operating) to be fed from a 30-MeV linac, for a total of $30 million.

The cost estimate was not such a wild guess as might be imagined. We had on hand the cost estimates from the Princeton-Penn, Harvard-MIT, and Brookhaven proposals. At that time the noun "inflation" had not been invented; the verb referred to automobile tires. Later on when the noun became essential, the AEC shunned its use; "contingency" was a more acceptable term. Probably because of its brief time on the stove, this project was not heard of again.

On March 26, 1956, another proposal was made; this was to be a two-stage hybrid consisting of a 50-MeV FFAG cyclotron followed by a 12.5-BeV zero gradient synchrotron. This suggestion did not reach first base.

On April 2, 1956, I was appointed Director of the Particle Accelerator Division with orders to proceed with all deliberate speed on a well-considered plan to outgun the Soviet 10-GeV constant gradient machine then being built and to keep costs low by using a proved design requiring a minimum of experimental work.

The alternating gradient scheme (ultimately of enormous importance to accelerators) was first considered by 1952 at Brookhaven. (It had been suggested in a private letter to Berkeley by Christophilos in 1950, but it either had not been understood or had been thought to be a crack-pot idea.) Because the inevitable oscillations in the orbits are thereby reduced in amplitude, the vacuum chamber and hence the magnets can be much smaller in cross section and so less power is required to drive them than with a constant gradient design. Theoretical studies at Brookhaven and at CERN had unearthed many potential hazards of operation, most of which were laid to rest when a small working model for electrons was completed at BNL employing electrostatic, rather than magnetic fields.

Among these disturbing thoughts was the realization that an AG synchrotron usually displays a critical transition energy at which phase stability momentarily disappears. This was initially very worrysome on paper, but it turned out to be of no consequence in practice. Several years of study and of experimentation with magnetic fields then followed, the CERN 28-BeV and the BNL 33-BeV machines becoming operative in 1959 and 1960.

So when ANL was requested in 1956 to design and build a 12.5-BeV machine, neither the AG nor the FFAG idea had been demonstrated experimentally. With orders to proceed at once, the only alternatives were the constant gradient design (well proved in three laboratories) or the zero gradient device (which had been used in the static fields of mass spectrometers and in beam bending magnets since 1951).

In such a zero gradient synchrotron the orbits are interrupted by straight sections and the magnetic sectors have no radial gradient of field. The magnet ends are set at an angle to the orbits so that a radial component of the fringing field supplies the otherwise missing axial focusing force. The inner top and bottom surfaces of the magnets are then flat and parallel (rather than being sloped with respect to each other), thus leading to easier fabrication and assembly. Model magnet studies should take less time, multiturn injection is possible, and with a window frame core, magnetic poles are eliminated and the field can be driven to considerably higher value than with a gradient.

THUS THE ZGS WAS BORN

During the rest of 1956 the staff grew to 42 warm bodies, buildings 805 and 806 were remodeled as office and laboratory (the larger building, 818, was not acquired as a lab until January 1958) and experiments were started on model magnets under the leadership of Martyn Foss, who was ultimately assisted by Bruce Phillips, Dean McMillan, Bill Siljander, David Orr, Bob Lykken and Joe Mech.

As had been expected, the field was adequately flat across the chamber up to 14 kilogauss, but rose disastrously at 20 kilogauss. This was because at high fields the reluctance of the iron along the long path from a coil to the chamber center is greater than that in the short path from a coil to the nearby edge of the chamber. Foss's solution to this situation was to drill longitudinal holes in the iron in the neighborhood of the coils so that the short path's reluctance is increased to approximate that of the long path. At low fields the reluctance of the gap is much greater than that of the iron, so the effect of the holes almost vanishes. This technique resulted in a satisfactorily flat field up to 21.5 kilogauss—a considerable improvement over the gradient fields of the Cosmotron (13.8 Kg) and of the Bevatron (16 Kg).

Pulsed model magnets then followed. Later, ramps were cut in the end portions of the iron and the coils were flared radially and axially, in order to hold constant the effective length of a magnet and the values of the betatron frequencies throughout the acceleration interval.

The Accelerating Voltage group was headed by John Martin, with help from Henry Kampf, John Dawson, Al Strassenburg, Bob Daniels, Charlie Laverick, Cyril Turner, Lachman Wadhwa and Ken Weberg. The general scheme was to use a variably biased ferrite core as the inductance in a resonant cavity. Five different ferrites were studied in a full-scale mock-up, and circuits were designed to accurately control the frequency during the acceleration interval whether the proton beam was so weak as to be barely detectable or was up to its maximum expected value. I cannot say more about this work because I never did understand it, and still don't.

The problems of first and second step acceleration prior to entry into the synchrotron were dumped into Rolland Perry's lap in the fall of 1956. Phil Livdahl joined this section a year later. Perry spent three months at Brookhaven becoming familiar with the design and manufacture of the 50-MeV Alvarez linac (then being built for the AGS) and BNL very kindly supplied us with drawings of this device.

Because of the much longer planned injection time at Argonne (250 μsec compared with 60 μsec at BNL), the Argonne quadrupoles buried in the drift tubes were redesigned for water cooling and continuous excitation. David Cohen undertook these modifications and, with the aid of search coils designed by Foss and Art Neubauer, obtained magnets in which non-quadrupole error fields were well within tolerances at the boundaries of the beam, that is, at one quarter inch from the axis.

The 750-kilovolt Cockcroft-Walton preinjector voltage supply was purchased from the Haefely Company of Switzerland. Henry Fechter initiated the design of the ion source and acceleration tube.

Perry planned the buncher that precedes the Linac and the debuncher that follows it; this reduces the Linac's energy spread from ± 250 KeV to ± 75 KeV, which is within the acceptance of the synchrotron. Lee Teng calculated the design of the achromatic three-magnet system which steers the Linac beam into the pulsed electrostatic inflector of the ZGS.

The theoretical group of Lee Teng, Ed Crosbie and Mel Ferentz was continually at work, creating a perpetual din with the mechanical desk calculators that had supplanted the slide rule. As I recall, with these machines simple arithmetic simply took time, but extracting a square root required repetitive guess work. When the Wang Company introduced a $2500 typewriter-sized electronic machine, about 1959, there also was a suitcase of apparatus under your desk. To obtain either a sine or a cosine took 15 seconds and you needed both for a tangent. We've come a long way! But the Lab did have the AVIDAC computer and later an IBM machine, so really complicated problems could be handled by these.

Our tripartite brain trust produced a vast quantity of papers on betatron frequencies, damping of these oscillations, resonances, nonlinear effects, gas scattering, chamber size.... . Oh, I cannot begin to list everything.

> The mechanical design of the magnet (and a great many other structures) was in the hands of Bob Lykken.
> Bill Siljander took care of the planning and manufacture of the magnet's coils.
> Gus Calabrese concerned himself with the forces that would be exerted on the coils and the magnet components.
> Calabrese, Ed Frisby and Kostas Burba worried over its power supply, using fly-wheel energy storage in association with a forest of ignitron rectifier tubes.
> Dick Trcka, John Moenich, Ken Weberg, Dave Hafemeister and Finley Markley were responsible for next to nothing--that is, the vacuum.
> Markley also was the resident expert on epoxy resins, plastics and such like.
> Roy Tuma saw to it that pencils, pens, straight edges, beam compasses, drafting machines, and, yes, draftsmen were always in plentiful supply.
> Ken Menefee not only was a superb technician but also was the boss of that most essential gang.
> David Manson was Assistant Director and was in charge of "Operations." I really never knew just what that meant,

but when anything went wrong, the first person to yell for was Dave. (He had been a Seabee during the war.)

Studies of the foundations for the magnets were made by Dick Trcka, Dave Cosgrove, Wayne Nestander, and Professor Donald Deere of the University of Illinois. Initial tests were made on a site (which ultimately was not used for a variety of reasons) by driving 10-inch hollow pipes about 145 feet into the ground and recording their load bearing characteristics. For the final site, it was decided to employ poured concrete caissons of 39-inch diameter. These holes were dug by a Benoto deep caisson excavator which acted as an oversized cork borer and extractor. All wine enthusiasts know that a cork is most readily removed by a pull combined with rotation, so it is not surprising that this monstrous machine was designed in France.

A host of other problems was attacked by the group of seventy-some enthusiasts augmented by Al Crewe, John Marshall, Roger Hildebrand, U. Kruse, S. Wright, and S. Warshaw borrowed from the University of Chicago to advise on experimental areas and beam steering magnets therein. Associate Director John Fitzpatrick concerned himself with the design of buildings.

In August 1957 Congress authorized $27 million for construction. In December $1.5 million was released to acquire an architect-engineer. Fitzpatrick reduced a preliminary list of 65 companies to 21 which were invited to submit proposals. Ten of these were interviewed and on April 9, 1958, the choice was made in favor of Sverdrup and Parcel of St. Louis.

With the basic design now settled and a competent group at work on construction, I resigned as Director in the fall of 1958 in order to write a text book on accelerators. Here the five-year-old mimeographed notes of Hamermesh and myself made an excellent foundation. I felt that if I could write a book that I could understand, then it ought to be intelligible to anyone who could read. Al Crewe took over as Division Director, to be replaced by Lee Teng when Crewe became Laboratory Director. When Lee went to Fermilab, he was replaced by Ronald Martin, who has been running the ship ever since.

THE ZGS - CONCEPTION TO TURN-ON

L. C. Teng

Fermi National Accelerator Laboratory, Batavia, IL 60515

Paraphrasing on Mark Antony I would like to begin by saying: "I come not to bury ZGS, but to praise him." The performance and productivity of the ZGS, especially those of H⁻ injection and polarized beams are something of which everyone who has ever worked on the ZGS can be proud. Although none of us wants to see ZGS operation stopped, I, for one, have no feeling of grief. The ZGS has more than fulfilled its mission and is now being decommissioned.

I am no good at reminiscing, having been taught to always look forward to tomorrow. So I will only give a brief personal account of the 12 years I spent at Argonne from 1955 to 1967. Larry Ratner has my deepest gratitude for having proofread all my dates, especially since this redeems me from all my blunders, inadvertent or otherwise. You may find some overlap between my account and those of the other speakers. As a matter of fact, the morning session may turn out to be an enactment of the movie "Roshomon" - different participants giving their own totally different versions of the same happening.

In 1953 Argonne's management began to seriously consider a project to build a multi-GeV accelerator. I was approached by Jack Livingood in 1954, and arrived on the job one summer day in 1955. There was already a small group of accelerator experts assembled by Jack hard at work in the Physics Building. These included Martyn Foss, John Martin, Ed Crosbie, Mel Ferentz, and Mort Hamermesh. It took me all of one day to get to know the job and the people. By the end of the day I was feeling so much at home that I took on each and every one of the group debating about accelerator design philosophies.

Next day we all travelled to the University of Michigan, Ann Arbor, to a MURA meeting where I engaged in an argument with Keith Symon on the nature of the third-integer resonance. That was my contribution to the MURA-Argonne fight.

Jack had written to me before I came on the job that he was considering seriously the "Symon Plan" as a basis for the design of our accelerator. He was referring to the FFAG invented by Symon and others at MURA in 1954-55. So I concentrated on adapting the FFAG principle to the ANL machine. The FFAG, having a fixed field, can accelerate beam pulses at high repetition rate and hence has the promise of high intensity; but it suffers from needing a rather complicated field shape and hence can accelerate particles over only a relatively narrow momentum range. Even starting with a 50-MeV Linac which was rather novel at the time, an FFAG can comfortably accelerate protons only to a top energy of about 8 GeV. Since we were shooting for an energy close to the 30 GeV of the Brookhaven AGS already in construction, something else had to be added. Beam extraction from a fixed field accelerator was also a problem. The

ISSN: 0094-243X/80/600 0100-08-$1.50 Copyright 1980 American Institute of Physics only efficient scheme available was the regenerative extraction scheme (later known as resonant extraction) which Jim Tuck and I developed for the synchrocyclotron. So I rigged up a scheme of tandem FFAG's. The first one takes the beam to 2 GeV and the second one to 25 GeV. The beam is to be extracted by the regenerative scheme from the first FFAG and injected into the second by a reverse application of the principle to a degenerative injection scheme. Since the regenerative/degenerative beam transfer can be operated dc, the high repetition rate and hence the high intensity feature of the FFAG is retained. I believe we promised an intensity rather daringly high by the standards of that time. This was submitted to the AEC in July 1955 as our first proposal.

In the meantime, Martyn Foss started some daring experimental study for the high current septum required by the beam transfer scheme. In a room in Building 203 he succeeded in passing a current density of more than a half million amperes per square inch through an 1/8-in. copper tubing cooled by water forced through the tubing at tremendous pressure. The spirit of the group was very high indeed. Unhappily, the proposal was rejected by the AEC.

From the very beginning the Argonne project was mired in thick politics. One day in January of 1956 Jack called an emergency meeting of our little but very enthusiastic and devoted group of accelerator physicists to announce what he had just learned about the AEC's "ground rules" for the Argonne accelerator.

1. It shall have an energy of 10 to 15 GeV to just beat the 10-GeV Dubna synchro-phasotron which was under construction at the time.
2. It shall be designed using only well-established principles and conventional technology, which we interpreted as a "weak focusing synchrotron." (Fancy schemes are to be explored only by MURA.)
3. It shall have a "crash" construction schedule to beat the competition of the Dubna machine.

We were all stunned by this new national and international accelerator ballgame. Jack did the only thing he could do by putting the direct question up to the group: "Do you want to build such a machine or do you want to quit?" The ensuing silence was finally broken by Martyn Foss who was more of a magnet man than the rest of us. He allowed that although this was a nasty blow, we could partially salvage our pride by designing and building the world's most efficient magnet using window-frame structure and edge-focusing. Since this was the only sane and positive suggestion expressed at the meeting, it immediately received unanimous support. The proposal was prepared and submitted in "24 hours." Thus was conceived the 12.5-GeV Zero Gradient Synchrotron. Martyn Foss was true to his words. To this day I would challenge anybody to design a more efficient weak focusing magnet or a better performance weak focusing synchrotron. I am sure that the ZGS will go down in history as the highest energy weak focusing synchrotron.

Wally Zinn resigned as Director of Argonne early in 1956 in protest of the AEC's treatment of the Argonne accelerator project.

The construction of a major accelerator at Argonne was apparently
agreed upon between the AEC and Argonne as far back as the formation
of Argonne in 1945. Norman Hilberry then took over first as Acting
Director and later as Director. Whether Zinn's resignation had any
effect I do not know, but the ZGS was approved by the AEC shortly
after in 1956. The authorization by Congress was to come much later
in 1957. In a memo of April 1956 Hilberry asked Jack Livingood to
form the Particle Accelerator Division (PAD) and appointed Jack as
the Division Director.

The people working on the design of the ZGS had been housed in
the Physics Division building. Shortly after PAD was formed the
Division moved into two large wooden frame buildings (805 and 806)
converted from warehouses on the west side of the Argonne site. This
provided space for model magnet measurements (seventh scale and quarter scale) and prototype rf cavity tests. The trouble was that the
old rotten floor planks of the old warehouses kept caving in under
the weight of the very massive equipment. Shoring under the building became a regular operation. The theorists, including myself,
stayed behind in Building 203 and lived with the theorists of the
Physics Division. We shuttled between 203 and 805 a great deal for
discussions and meetings. I was then young and brazen. Someone,
just for fun, kept a statistical record which showed that I talked
almost twice as much in total time at weekly staff meetings as the
second most verbose person.

When the model and prototype program was completed Livingood
resigned as Director of PAD. Albert Crewe took over in the fall
of 1958. At the same time, Roger Hildebrand was appointed Associate
Laboratory Director for High Energy Physics.

It was perhaps the first time in the history of accelerator
building that the architect-engineering firm was 300 miles away
(Sverdrup and Parcell in St. Louis). For well over a year Al Crewe
and several other people led a nomadic life commuting to St. Louis
every week - sometimes twice a week. The ground was broken in June
1959 and the Stonehenge (magnet foundation) went up in 1959 and 1960.
The magnet foundations sit on concrete pillars which are called
caissons and which extend all the way down to the bedrock. The
caisson-making machine was imported from France and was fascinating.
It was like a big drill. A steel cylinder about one meter in diameter with cutting teeth at the front end formed the drill bit. It
was thrust into the earth by a to-and-fro rotary motion. As it sank
out of sight new lengths of steel cylinder were added. After it
reached the bedrock, the dirt in the middle was dug out with a clam
shell. A man with a steam hammer was then lowered some 100 feet to
the bottom to level out the bedrock. Concrete was poured and the
entire length of steel cylinder was pulled out before the concrete
hardened. In this manner they were able to pour one caisson a day.
Director Crewe displayed his bravery one day by riding in a crane
down to the bottom of the steel cylinder to "inspect" the bedrock.

It was an exciting day when the Division moved out of the old
dilapidated warehouse 805 into the brand new, and by comparison,
incredibly plush Lab and Office Building 360. The theorists were
not affected, however, we stayed in 203.

Martyn Foss was promoted to senior physicist in 1960 and I followed in 1961. Hilberry took an early retirement in 1962 and Crewe succeeded him as Laboratory Director. I was then appointed the Director of PAD. Under Crewe the key people were John Fitzpatrick, Associate Director; Martyn Foss, Associate Director and Head of Magnets; Dave Manson, Assistant Director; John Martin, Head of RF; Rolland Perry, Head of Linac; Andy Gorka, Head of Controls; Lloyd Lewis, Head of Computer Controls; Bob Lykken, Chief Mechanical Engineer and Everette Frisby, Chief Electrical Engineer. At the time of my appointment we were having a lot of trouble with the fabrication of the ring magnet yoke blocks at Baldwin-Lima-Hamilton. The quality control during fabrication was inadequate, resulting in out-of-tolerance laminations and lots of shorts between laminations. To make things worse, shortly after my appointment Foss resigned to go back to Carnegie Tech. So I was beset by problems on both fronts. I resolved the second problem by recruiting Ron Martin from TRG as Associate Director. For the first problem we instructed BLH as to the proper quality control during the fabrication including setting up a clean room, and, in addition, sent Herb Ross to live at BLH as an in-house quality inspector. Only then did we begin to receive useable magnet yoke blocks. BLH would never have been so cooperative and the work could have dragged on for years except for our promise that the cost would be renegotiated. They did claim excessive expenditure and the cost was finally settled long after the ZGS was completed.

The construction of an accelerator is like a continuous battle. There are problems arising every day, some technical, some fiscal, some organizational, some personnel, but all are critical. After a while one gets the feeling that the whole job is just solving one critical problem after another. But somehow there was progress. We got our first Linac beam in October 1962 and started the ring magnet power supply in June 1963. One August 1, 1963, when the construction was officially complete the big rectifier tubes of the power supply were still not fully conditioned and the ring magnet could be pulsed only to the level of 2-3 GeV. Nevertheless, we started injecting beam; by that time the Linac was already operating quite routinely. Although declared complete, the final touch-up construction work was still proceeding during the day and the commissioning could only go on at night. During the nights of the first couple of weeks, although the injected turns looked alright on the fluorescent screen, we could not accelerate the beam. On August 16, I had to leave for Dubna, USSR, to attend an International High Energy Accelerator Conference. While I was away the beam was accelerated to about 3 GeV under the efforts of John Martin, Ron Martin, Larry Ratner and company.

What happened was that we were so confident of the quality of the ring magnets that we did not bother to install either correction windings or correction dipoles. Luckily, toward the end of construction we decided to implement Martyn Foss' earlier suggestion of using the steel anchor plates of the main magnet coil as a one-turn correction winding. If we shorted across the anchor plates, the current induced in this one turn during pulsing was adequate to produce a large orbit shift at injection. So in each one of the eight octant magnets we had this one-turn coil shorted across a variable resistor to adjust

for the proper amount of correction. These anchor plate windings
turned out to be crucial in correcting for the orbit distortions and
obtaining an orbit which is closed inside the vacuum chamber. With
this orbit we were then able to circulate and accelerate the beam.

From then on it was just further conditioning of the rectifier
tubes to power the ring magnets to higher and higher levels. Finally
on September 18, 1963, the beam was accelerated to the top of the
full 12.5-GeV ramp on two pulses before the still-shaky rectifier
tubes arced down again. The oscilloscope trace of one of the two
pulses was caught on polaroid film by an alert technician with a
fast finger and we reached full design energy. The intensity was
miserable; only about 10^{10} protons reached the top. But there was
the usual celebration and party.

The next months were devoted to improving the intensity. Toward
the end of 1963 we were able to get something like $2-3\times10^{11}$ protons
per pulse. And it turned out that we were to be stuck at this intensity for a long, long time - being limited by the transverse
(vertical) instability. But this intensity was certainly enough for
a formal dedication of the machine. We worked on a crash schedule
to arrange for the dedication before the end of 1963 because, as we
reasoned at the time, then people would remember that the ZGS was
completed in 1963 instead of 1964. At the time, a year's difference
seemed a lot. Looking back now some 16 years later, it really would
not have made any difference whether it was 1963 or 1964.

The dedication was held on December 4, 1963. Both Glen Seaborg,
Chairman of the AEC, and George Beadle, President of the University
of Chicago, were present. The principal speaker at the evening
banquet was G. Bernardini of CERN. The whole event was a gala
affair. But since it came on the heels of the assassination of
President Kennedy, the atmosphere was unavoidably tinted with an
undertone of grief and guilt.

Disaster struck early in 1964. The outer vacuum system developed a leak and the inner vacuum chamber collapsed. Because of the
very wide aperture (32 inches) the relatively thin inner vacuum chamber was not designed to withstand the full atmospheric pressure,
hence we had to use the double vacuum system - a design with which
I never felt very comfortable. In case we lost the rough outer
vacuum, to avoid collapsing the inner vacuum chamber a set of
electrically operated safety valves were supposed to open up and
equalize the pressure in the inner and outer vacuum. The electrical
circuit actuating the safety valves failed to function and the full
atmospheric load on the inner vacuum chamber caused it to collapse.
In a series of mishaps over a period of about a month the vacuum
chambers in at least three octant magnets were collapsed. We tried
to push the vacuum chambers back up the best we could from the inside and operated for about six months with a vertical aperture of
$3\frac{1}{2}$ inches instead of the full $5\frac{1}{4}$ inches. Eventually the octant
magnets were taken apart, and the vacuum chambers fully repaired.
We also decided to glue the top and bottom of the chambers to the
magnet poles for support and put in simple puncture diaphragms between the inner and the outer vacuum - they rupture automatically
to equalize the pressure when the pressure differential between the

inner and outer vacuum becomes too large. On hindsight, these simple fail-safe rupture diaphrams were what we should have had in the first place, instead of the fancy electrically operated valves which depended on the proper operation of the whole electrical and mechanical systems to provide protection.

Coherent beam instabilities were something very new in 1964. Although the theory was simple and straightforward, I was never sure at the time that these phenomena could be "really happening." All through 1964 and 1965 a group of people from MURA led by Fred Mills came to help improve the intensity. Working with John Martin they approached the problem systematically and cautiously. Their study of the beam behavior finally convinced them (and me) that the beam intensity was limited by a vertical resistive wall instability. A series of vertical oscillation dampers were designed and built. Finally a damper with adequate bandwidth and power was installed in the middle of 1966 and we broke the barrier of $3-4 \times 10^{11}$. Almost immediately the intensity shot up to some 2×10^{12}.

Going back to organization, in 1958 Roger Hildebrand, Associate Laboratory Director for HEP, invented and helped to organize the ZGS Users Group, and Ned Goldwasser was elected its first chairman. This organization was very helpful in bringing to bear the talents and efforts of faculty members of the participating Midwestern universities. About a year before completion of construction we started accepting experimental proposals and formed two committees - the Program Committee (chaired by Roger) to approve or reject experiments, and the Scheduling Committee (chaired by me) to schedule approved experiments on the accelerator. The latter committee always seemed superfluous to me, because scheduling is certainly best done by one person, and cannot be done by a committee. This is indeed the way it developed. The leader of the Experiment Planning and Operations Group would work out the weekly and the monthly schedules, and brought only special problematic points up for discussion at the weekly meeting of the Scheduling Committee.

In addition to the ZGS, PAD operated two large bubble chambers - a 30-in. high precision hydrogen chamber built by MURA and the University of Wisconsin, and a 40-in. heavy liquid (freon or propane) chamber built by the Univesity of Michigan. The Division also designed, built and operated all beam line elements - quadrupole and bending magnets, spectrometer magnets, particle separators, liquid hydrogen targets, the neutrino horn, etc., etc. After the completion of the ZGS, PAD was reorganized toward operation. The Division was centered on four large operations groups: the ZGS Operations Group, headed by R. Daniels, operated and maintained the ZGS; the Experimental Planning and Operations Group, headed first by H. Blewett and then after 1965 by P. Livdahl, scheduled and helped to plan, install and run the approved experiments; the 30-in. Bubble Chamber Operations Group, headed by L. Voyvodic, and the 40-in. Bubble Chamber Operations Group, headed by V. Sevcik and L. Oswald, operated and maintained these bubble chambers. All others were discipline-oriented support groups, such as the Physics Group, the Electrical and Electronics Group, the Mechanical Engineering Group, etc. In 1965 Fitzpatrick was transferred to help in a construction project of the Solid State Science Division and I recruited Royce Jones as an assistant director. With about 500 people, PAD was the largest division at Argonne.

It was clear from the beginning that it would be difficult to carry out a trail-blazing HEP experimental program on the ZGS, what with an energy less than half of that of the Brookhaven AGS and the CERN PS and coming into operation more than two years behind. We tried to make up for the deficiency by providing better and fancier experimental equipment immediately at the first operation of the ZGS. The 30-in. bubble chamber had superb resolution and was equipped with a very versatile high energy separated beam. The 40-in. bubble chamber was the largest of its kind at the time of first operation. We built and operated some large and sophisticated beam elements: particle separators, Cherenkov counters, spectrometer magnets, and a high performance meson-focusing horn for the neutrino beam. All this made the transition from construction to operation of the ZGS Complex a difficult and trying task. In addition to commissioning and improving the operation of the ZGS we had to shake down and commission the bubble chambers and a multitude of sophisticated beam equipment, and to mount the approved experiments, all the while continuing with the design and construction of other equipment. With capable assistance from participating universities, PAD shouldered the titanic responsibility. I still remember how very proud I was of the performance of each and every member of the Division. Nevertheless, the birth pain during that period must have been obvious. Eventually everything was made to work, and by 1966 we were completing some 20 experiments during the year, several of which yielded significant results, such as the Michigan-Argonne pp scattering experiment led by Alan Krisch. But I believe that the ZGS really earned a firm place in the annals of HEP research with the work done on the 12-ft bubble chamber and the experimental program using the very unique polarized beam.

Through the continued efforts of a few dedicated people the intensity of the ZGS increased steadily. As early as 1964, in response to Al Crewe's request to compare the capability of the ZGS to that of the FFAG machine proposed by MURA, I claimed that the ZGS could eventually deliver some 10^{13} protons per pulse. The highest intensity ever obtained was 7.5×10^{12}, not far from the claim. All these values should be compared to the design intensity of 1×10^{12} protons/pulse given in the 1956 proposal.

In 1965 MURA's funding was terminated and PAD was supposed to absorb all MURA staff members who were willing to come to Argonne. This is sort of like a corporate take-over. Without precedents, I had to invent procedures as we went. But out of the total effort PAD got only some three or four junior staff members from MURA.

In 1964 Roger Hildebrand resigned and Bob Sachs took over as Associate Laboratory Director for HEP. Also starting from 1964 the interest in the 200-GeV accelerator mounted steadily. Many members of PAD participated in the early design work and eventually in the site selection throughout 1965-1966. The choice of Weston as the site was announced in December 1966. The operation of the ZGS

Complex had reached maturity in 1966, and I resigned in 1967 to go to the National Accelerator Laboratory, leaving the ZGS in the able hands of Ron Martin.

THE CONSTRUCTION OF THE ZGS

A. V. Crewe
The University of Chicago, Chicago IL   60637

Well, I guess I am the man in the middle here, Lee Teng left a
hole of about three years in his discussion, and I just happen to
be the guy who filled in that hole.

You can say that this is a kind of bittersweet occasion. It
brings back a lot of fond memories. It is rather sad that the whole
thing must end, but of course everybody knew that one day it would
have to end. It is probably not very necessary to point out to a
group like this that it is very difficult for some of us to look
backwards. We spend all our time being terribly interested in what
we're doing today and what we think we're going to do tomorrow, so
that looking backwards is generally not a very profitable occupation
for a scientist. As a result of that I have no records at all of
the ZGS. I have no file cabinet with stories of the ZGS; to my
knowledge I don't even have a single slide of that occasion. As I
say, the interest for most of us is looking ahead.

In being asked to give this talk, I had some decisions to make.
Should I try to recount facts (which I don't believe is actually
possible in the absence of these records), or should I be content
with giving you some impressions. Should I really tell it as it
was (again, if that's possible), or should I tell what happened
with the benefit of some very valuable hindsight. Are we just here
to have a kind of a happy occasion when we remember fond memories
or should I tell you some of the more serious incidents. I was not
able to really resolve all these issues in my mind but since I have
no records, whatever I say here is an impression of what I now
think really took place or possibly what should have taken place.
One's memory gets colored, of course, by subsequent events.

My first involvement with the construction activity was at a
time that Roger Hildebrand was asked by Norman Hilberry to become
Associate Director for the Laboratory for High Energy Physics.
Roger and I have been friends for many years now, and he asked me
to take over the construction of the ZGS because Jack Livingood was
going back to the Physics Division. Well, it was clear that Roger
was sincere in asking me to do it, and that Norman Hilberry had
great faith in Roger's judgment and went along with the idea. It
was a difficult decision for me. After all, at that time I was a
very young man - young and foolish, I suppose - and had just begun
a career as an Assistant Professor at the University of Chicago,
and building a machine, no matter how important a machine, is just
no way to ensure getting promoted in any university. Eventually,
however, the challenge of trying to build a machine was so intriguing that I decided to do as I was asked by Roger, and Roger decided
to do as he was asked by Norman Hilberry. To this day, however, I
do not understand how anybody came to the decision to ask me to do
it.

ISSN: 0094-243X/80/600018-07-$1.50 Copyright 1980 American Institute of Physics

It was a challenge. I was trying to remember the numbers and I
think that at that time - $29 million is the number that hangs in my
memory, Jack, of the budget at that time, but that could be wrong.
It is not possible to mention the names of everybody that was in the
Division at that time, and I hope that those of you whose names are
not mentioned will forgive me for not doing so, it is certainly not
because I don't remember you. But at that time there were just four
anchor points in the Division; just four people, really, that I had
to rely upon. Those were Martyn Foss and Lee Teng, the magnet designer and theoretician, the two who conceived the machine. There
was John Fitzpatrick, who was the Associate Director of the Division;
he was the man in charge of buildings and contracts. And David
Manson, the Assistant Director, who was the guy who simply knew how
to get anything done in the Laboratory. Well, obviously the first
task was to get the confidence of these four people. I don't know
whether I really succeeded, but they never expressed any overt concern about my presence, and I can only say that they gave me their
full and complete cooperation right from day one.

Well, as you've heard, a few decisions had been made, and my
memory again may not agree with the actual facts. The basic concept
was already there, the cost estimate was allegedly reasonably firm,
the architect-engineer (Sverdrup and Parcel) had been chosen, and
their scope had been determined. The site was reasonably firm. The
basic problem was to build the machine as quickly as possible. Well,
the task was perfectly clear. It was to review all the design, finding what areas were strong and what weak; begin the process of fixing
the parameters and fixing the necessary decisions; review the schedule and the cost estimate and change either of them, or determine
what had to be done to meet them. Look at the staff (which initially, as Jack has pointed out, was very small), and find areas of
strengths and add to the weaknesses; find out how other people had
built machines, or were building machines, and use whatever ideas
seemed good, and throw away whatever ideas seemed bad.

One of the most difficult problems at that time was trying to
understand the relationship between the AEC and the architect-
engineer and the Laboratory and the Division. It was a complex
relationship, and we tried to find out in particular what these
relationships had to be - had to exist - and which could be changed.
Well, of course, none of this review and design and fixing of parameters could proceed in any logical manner at all, because time was
much too important a factor. We had to go around and around this
cycle in what seemed like endless sequences in order to fix parameter after parameter. Because of this I can't give you any accurate history, everything depended upon everything else. The best
that can be done is to pick out various areas to give you some memories that will, of course, overlap in time.

I should, however, mention the political climate. You've already heard some of it. The AGS and the CERN machines were being
built, and our task, clearly given to us by Congress, was to build
a machine a little bit bigger than the Russians, and a little bit
faster and with more intensity. The general idea of Congress was

to move the United States far ahead of the Soviet Union. It's difficult now to understand why that was, and I was always absolutely convinced that Congress really did not understand the difference between high energy physics and nuclear weapons. And, of course, nobody bothered to explain the difference. I believe that to be a true statement, and I feel rather ashamed of participating in what I believe was a deliberate delusion.

The idea of the ZGS was to put a large accelerator in the Midwest, so that people wouldn't have to travel very far, a concept which does not seem a very reasonable one today. But nevertheless that was the general idea. Nobody has really mentioned this problem, and I thought very carefully as to whether I should, but the fact remains that the Midwest was politically divided. It was so badly divided that I believe that no one machine - any decision to build a machine anywhere - could have pulled the Midwest together. Essentially it was a division between those associated with MURA and those associated with Argonne. It was totally unknown to me at that time, and I tried never to get involved in it, but it did exist. One would like to gloss over this point, and I will cover it rather quickly, but the fact remains that it should be taken into account as a matter of history. It is sad to relate, for example, that at the time we began the construction of the ZGS in 1958, not one single person from MURA came to help. They did come much later on, when the machine became a reality. I must add, however, that none of this ill-will was personal. On personal matters we were all friendly, and sincerely friendly, and I was not aware of any personal animosity. The problem was essentially institutional, it appears, rather than personal. In fact, it is that problem which probably resulted in Roger and I coming to Argonne in the first place. The problem was to build a machine and to make it available to everyone in the Midwest, or indeed to anyone anywhere with a good proposal to do an experiment. The decision to put it at a big security operation like Argonne, which for many years had been a closed laboratory with a high fence around it, presented obvious difficulties. It was difficult to demonstrate to other universities, to the universities in the Midwest, that the machine in fact would be made available on a perfectly free basis to anyone. The only way to do that was to put some people from the universities in charge, and the only university that Argonne could really talk to on a sincere basis was the University of Chicago, and that's why we ended up here. Well, all that is another story, and I believe the Midwest was ultimately unified much later. In fact, I got involved in that as Laboratory Director, but only at considerable cost and a great deal of distress. But that is another story, and perhaps not relevant.

We started travelling around the world, looking at how other machines were built, and we got a great deal of help from everyone else. I might mention just for sake of the record, that there is one other ZGS machine, and that is the one which was built by Heyn, in Holland; built in the basement of the Technische Hochechult in Delft. A fascinating machine which he was using essentially to train students and provide them with thesis material and we profited by getting some of those students here. It was a very small machine, however.

The problems we had in common with other accelerators were many, not questions so much of magnet design, but questions of buildings, shielding, experimental facilities, etc. I should mention that the AEC had been very nervous about giving the facility to Argonne, at least apparently so, because they had determined the scope of the architect-engineer, who had been given the ultimate responsibility even for things like the magnet, the power supply, and the coils. That was obviously an extremely impractical thing to do. The design of the magnet was changing every day, the ideas on how to build the coil were changing, and therefore the power supply was changing all the time. There was no way in which we could convince the architect-engineer that we knew what we were doing. Their idea of building something is that the customer knows what he wants and they build what the customer wants. Well, this customer really didn't know what he wanted until the very last day when the decision had to be made. So ultimately we had to take that resonsibility away from the architect-engineer, but it wasn't without a struggle.

You might be interested also to know that the design criteria that we had to live by were established by the federal government, and they were exactly the design criteria used for building Post Offices. As a result, we were not allowed to have any kind of architectural input. We couldn't have any distinguishing architectural features in the building. It had to look as much like a Post Office as possible. You have no idea of how much time we spent in deciding how many stalls there should be in the toilets, and in particular how one should design the ladies' toilets. I can remember considerable discussions, for example, on whether the water cooling tower should have sprinklers for fire protection. You may laugh, but it does. At least it did at that time.

But of course it was the machine itself that occupied most of our time. Decisions were made just about every day, and, of course, each decision that we made and nailed down affected the outcome of any discussion we had the day after. Hopefully these discussions were logical, but I'm quite sure that a lot of anachronisms remain. In retrospect, of course, we were considerably ahead of commercial technology and that was one of the difficulties. We used mercury rectifiers, as you've heard; solid state rectifiers were just around the corner, but they hadn't come. Another example would be the control room itself, which was sized to take control systems having tube circuits. You will recall that transistors were just about moving into the business at that time. I'm sure the control room now could be about one quarter the size it is.

But by far the most difficult decisions were those involving the magnet, the coil, and the power supply. Let me give you a few incidents from history, at least my recollection of history. We needed a generator that was capable of being pulsed, and the problem was that every time you pulse a generator there are enormous forces on the windings and eventually these windings will come

loose and start rolling around inside the generator and you've lost
the generator. So the problem was how to wedge these windings down,
and Allis Chalmers had expressed great experience in this matter because they had enormous experience with pulsed generators. The
trouble was that when we added up the list of their total experience
it amounted to a few weeks' running time on the ZGS. They were a
little dismayed at that, but eventually they came out with a good
design, and I guess the generator has never fallen apart. The flywheel was another problem. Building a big flywheel to soak up the
energy was somewhat of a difficulty. The only people who had considerable experience with flywheels was the U.S. Navy, and they always made their flywheels out of bronze. It turns out that bronze is
perhaps the worst material you could possibly use because it crystal-
lizes under stress and several flywheels have exploded and sunk the
ships in which they were installed. This is clearly one example of
tradition that keeps going on regardless of actual every-day experience. We ended up specifying steel and apparently it has worked.

One of the most difficult problems was the coil. The coil itself had to be very strong, but the mechanical engineers needed very
little space to hold the coil together. And that's exactly what
Martyn Foss wanted - most of the space between the pole faces packed
with copper. The electrical engineers wanted an enormous amount of
space for the insulation to protect the coil against discharges to
ground. I remember a large number of discussions on that problem,
and the issue simply couldn't be resolved. I remember one day having
the electrical engineer's handbook waved under my nose; "there's the
design number." I looked and indeed he was absolutely correct; it
needed this enormous amount of insulation. But on further thought I
decided to look into where the guidelines came from! It turned out
that the large amount of insulation they needed was because occasionally power lines are struck by lightning, and the additional voltage
surge requires the additional insulation. It seemed unlikely to us
that lightning would ever strike the magnet inside the ring building,
and so we got away with a much smaller amount of insulation than the
handbook specified. I guess Underwriters Laboratory would never insure the ZGS.

You might be interested in other things, too. For example, the
shielding between the ring building and the experimental area. We
needed an enormous amount, several thousand tons of high-Z material
to fill that gap. I remember being offered several thousand tons of
depleted uranium for that task. Unfortunately, it was in the form
of uranium hexaflouride, and we didn't quite want to handle the
problem of converting it into metal. Finally, however, we located
a source of cheap metal. It was just at the time the Navy was phasing out their battleships, and there was a junkyard in Charleston,
West Virginia, where they kept spare parts for battleships. If a
battleship got damaged, either by being struck by the enemy or by
banging into one another (which was much more likely), they needed
a spare part and these spare parts were scattered over several
thousand acres of land in Charleston. It turned out that it was
feasible to go and slice off some straight pieces and ship them

back. I take my hat off to those who solved the logistics problem
and the jigsaw puzzle of putting all that stuff together to make a
good shield.

Another story that I distinctly remember was the preaccelerator,
the 750-kV power supply, which went out for bids. When the bids came
back the spread was about one order of magnitude in price. The cheap
one was the obvious one, but looking at the most expensive (which I
will tell you came from General Electric) showed that the problem
was to maintain the voltage at 750 kV during a pulse. As you extract
protons, the voltage tends to drop and the problem was to hold the
voltage up. General Electric solved the problem very neatly by having the world's largest stack of condensers. A stack of condensers
that would not fit into this room, for example. Of course the successful bidder just put a little regulator on the booster system,
which cost about $10,000, and the machine worked.

Later on, as the machine progressed, the decisions made, and the
contracts let, we had a little more time to look around at other
problems. The experimental areas, for example. We began working on
quadrupoles, bending magnets, and separators. And you may be interested that we started polarized proton investigations at that time.
I can remember asking Dave Cohen, who didn't have much to do one day,
to see if he could find out whether a polarized proton would survive -
whether the degree of polarization would survive - from the preinjector all the way through to full energy, and we were gratified to
find out that the polarization did survive. That was only the first
investigation, and I was very happy to hear later that this had been
pushed so well in later years.

The machine, of course, was built by people, and it is to their
credit that the machine has run so well and so long. The original
group was very small and we added really quite rapidly. I can't
mention them all by name. Some of them walked in the front door,
some of them we advertised for; they came in a whole variety of different ways. I can just give you a couple of examples. I can remember to this day Andy Gorka walking into my office one day saying
he was looking for a job. I said I needed a controls engineer, and
he said "I'm a good one," and I said "you're hired," and he was.
The conversation lasted about that long and it was one of the best
decisions I ever made. Another one I can remember was Finley Markley
(who I haven't seen here today). I can remember being terribly impressed by our need for glue. It appeared that we were overwhelmed
with sticking material A to material B, and the problem was how the
devil we could fasten these two things together, and the answer always came out "we'll glue it together." We didn't have anybody who
knew anything about glue, and I can remember asking Finley Markley
to go out and make himself a world's expert on glue. He did, and he
solved all our problems, or rather all those problems related to glue.
The vacuum chamber was glued together; indeed the whole ZGS is glued
together.

I think that any analysis of the ZGS would show that its primary characteristic is that of very sound engineering. Speed was
always uppermost in everybody's mind, but nobody sacrificed quality

of engineering for getting the job done with expediency. Lee has already mentioned the vacuum chamber. We did know at that time that this was a weak area, that it had to be watched very carefully. I must say that I don't think anybody thought that the control system would break down. We did think, however, that the vacuum chamber would disintegrate because of radiation of the glue in there.

Our greatest area of uncertainty was always the question of scheduling. We always made out schedules, and of course everybody puts an optimistic schedule down, and then his boss adds a little bit on, and then I used to add a little bit on, and we were never right, anyway. The PERT scheduling system was invented toward the latter years of building the ZGS, and we set up our own version, and the consequences were actually disastrous. For example, I remember one of the greatest features was that there simply wasn't time to drill and tap all the holes in the vacuum chamber, put the bolts in, and tighten them. It simply was not feasible to get all that done within the schedule we'd established. And so we came up with a much more reasonable schedule, and I think that it was more or less adhered to.

I must say partly because of that I was somewhat grateful when I was taken out of the ZGS and put into a slightly different area, and was able to turn the responsibility over to Lee Teng. That of course is not quite true! I was really sad to go because I'd spent three and a half years or thereabouts of my life totally dedicated to the ZGS, with one aim in mind, I wanted to sit in that control room and twiddle all those damn knobs and make that machine work. I was never able to do that, which is one of the greatest regrets in my life. But having turned the thing over to Lee Teng, I should just say that I don't know all the mistakes I made in building that machine, but I made one very sensible decision. That was never to ask Lee Teng what those mistakes were.

THE UNIVERSITIES' ROLE

E. L. Goldwasser
The University of Illinois, Champaign, IL   61820

It is a real treat to be here before a group of old friends and
to share some nostalgia. When Alan Krisch called me a few months
ago and asked whether I would participate in this happy occasion I
much too quickly, as you do on occasions like that, said yes, being
confident that I had a large file of all of the materials documenting the events that I had been connected with in the course of
working with Roger Hildebrand and with Albert Crewe in the organization of the Users Group. Most of you know that I left the University of Illinois in 1967 and went to Fermilab for eleven years,
leaving my files behind me. Recently when I said that I'd like to
look at those files I was informed that they had been put in dead
storage in a building behind the Physics Department and that the
building had burned up a few years ago. Well, that took an enormous weight off my shoulders, because I didn't have to be confined
to facts at all, I could just rely on my memory, which is note-
worthily faulty. Unfortunately, I have some people here on the
platform with me who may have facts, and they will undoubtedly correct my errors.

First I would like to set the stage, as it was back in the 50's,
because things were so different back in those days from what we are
accustomed to today that it's a little hard even for those of us who
lived through it to remember how it was. In the period 1945 to 1959
physics in general and high energy physics in particular were enjoying an era of unprecedented prestige and growth. The government was
urging young people to go into those fields, trying to expand our
capabilities. Sputnik had come in '57 and the push was to make
things happen more quickly, to make more people get into the act,
and of course universities had a very important role to play. Students were being encouraged to go into physics, and the universities
were trying to meet that demand and to compete for that demand. High
energy physicists were telling the university authorities that no
university could be whole or strong if it didn't have a strong physics department, and that no physics department could be strong if it
didn't have a very strong high energy physics group. There was nothing unusual about that, and it's still true today, but what was unusual in those days was that universities <u>believed</u> it. No university
could imagine itself expanding or even surviving in the future if it
didn't have a very strong group in high energy physics. As a result,
some of us found that we had a remarkable influence with the administrations of our universities. Thanks to that experience, as I now
sit as a member of a university administration, I know just how to
handle that kind of thing.

In any case, in that context we really had a very serious problem at Midwest universities. We had had strength in the Midwest -

ISSN: 0094-243X/80/600025-07$1.50 Copyright 1980 American Institute of Physics at Chicago, at Illinois - with accelerators in the early 1950's and late 1940's which had led the Nation and were leaders in the development of the particle physics of the day. As accelerators became more expensive and more demanding of resources and people, the West Coast and the East Coast enjoyed a very large share of the support that was available. The Midwest universities found themselves facing an era which seemed to be more and more difficult, with the best high energy physicists looking toward the east and west, not only to do their experiments, but also to find employment. It isn't the most desirable thing to commute to the East Coast and the West Coast. It is more convenient to find a job there, and since universities were looking for faculty people there was quite an exodus from the Midwest to the east and west.

I think about 30% of the Ph.D.'s in physics were produced in the large Midwest universities in that era. Many of us felt that the universities at which an important segment of our scientific population was getting educated were in jeopardy as a result of this exodus to the East and West Coasts. Those of us who had decided to remain here felt that we had to have a strong and large facility in the Midwest which could be used by the physicists who were remaining at these universities.

Roger Hildebrand described the ZGS approval, and alluded to the very puzzling interaction between the MURA activity and effort and the one here at Argonne. I thought one of the things I would try to do is to put into some kind of perspective the way that whole situation developed and how it was finally resolved, because one of my unique experiences was to be involved in all of the relevant meetings and deliberations. I was on sabbatical leave in Italy in 1957 and 1958, and I returned in the fall of '58. It was sometime in November that Roger Hildebrand came to me and asked if I would try to organize the university physicists to cooperate with and to work with the people at Argonne in developing the accelerator and the associated facilities with which all of us would be able to do high energy physics after the ZGS came on the air. My sabbatical in Italy had been spent at Rome, where the accelerator at Frascati was just being commissioned, and upon returning to the United States I found that the ZGS was in very much the same state. So it was quite a natural thing for me to become involved in preparations here and to try to organize the university people to become involved here.

At that time one of the transitions that was occuring was that the accelerators, which back in Lawrence's day had been built by a single person or a small group for the use of a single person or a small group, - those accelerators had become so large that no single individual could make full use of the facilities that he may have been responsible for. A variety of mechanisms were being tried out to involve a wider group of experimenters in the use of these very large facilities.

At Berkeley most of the evolution was in the form of increasing the size of the group at the Radiation Laboratory (the Lawrence Radiation Laboratory, now the LBL); they had a very large in-house group and they felt that to be the most effective way to do high energy physics. Brookhaven (BNL) was managed by AUI (Associated

Universities, Inc.), perhaps the first university consortium established to manage a new laboratory, and BNL soon began to turn toward
high energy physics. An accelerator was built which was shared initially by the member universities of AUI, about ten or so in number.
They were all on the East Coast. As things developed the Brookhaven
Laboratory began to entertain proposals from physicists from further
abroad. It really was the first example of an accelerator that developed a strong component of user participation. Still, even at
Brookhaven there wasn't a very well-organized system through which
the using physicist could really participate in the laboratory's
planning of the facilities.

When Roger Hildebrand and I first talked in 1958 we spent quite
a bit of time discussing the possibility that the potential users of
this facility could really take a major responsibility for some of
the jobs, some of the facilities that were going to be built for
eventual experimental use. The mechanism that we developed here -
the Users Group, which became formalized within the next year - had
served as a model for other groups, at Brookhaven, then on the West
Coast, and finally at Fermilab. I would like to make it clear that
it was Roger's imagination and initiative which started a user activity here at Argonne. We all owe a lot to him for his vision.

During the time of construction of the ZGS there had been established the "Ramsey Panel," with which most of you are familiar. That
panel was set up jointly by the General Advisory Committee of the AEC
and by the President's Science Advisory Committee. It was given the
task of considering the program in high energy physics in the United.
States for a period of the order of ten years. The committee was
established in 1962. It took a year for it to go through a set of
meetings and interviews with people from all over the country. The
ZGS program was presented to and discussed by that committee. The
possible directions for the future were discussed.

As you all know, the committee came out with a recommendation
that there were two frontiers for high energy physics, both of which
they felt were of some importance. One was in the direction of high
energy and the other in the direction of high intensity. The MURA
proposal was the single one before the AEC which aspired to go toward
high intensity. The Ramsey Panel report made it clear that their
choice between the frontiers of energy and intensity went strongly
to the high energy activity. They recommended the construction of a
200-BeV accelerator as a first step, to be followed some years later
by an 800- to 1000-BeV accelerator. The first one was envisioned to
be built at Berkeley and the second one - the next generation toward
high energy - was imagined to occur at Brookhaven. All that seemed
left to the Midwest was the ZGS which was coming on in the next year
or so, plus the possibility of establishing a new frontier in high
intensity with the MURA machine.

In November of 1963 President Kennedy was assassinated and within a month of that time President Johnson called a meeting in Washington of a group of people from the Midwest to discuss the decision
which was going to be made about the MURA accelerator. I had the
interesting experience of participating in that meeting. President

Johnson heard us out, at least in form, but then pulled a piece of
paper out of his pocket and read to us a decision which obviously
had been made prior to the meeting. The MURA project was terminated.
The senators and representatives who were present with the high
energy physicists at that meeting expressed their deep concern for
the health not only of high energy physics, but for universities in
the Midwest. They made a very strong plea that something should be
done to bring new initiatives into the Midwest. I thought it might
be interesting to quote for you from some of the correspondence of
that day. In a letter dated January 16, 1964, from President Johnson to Senator Humphrey, it was stated (in those days presidents
were really worried about high energy physics): "I would hope and
expect that the fine staff of MURA would be able to continue to
serve the Midwest through the universities and at Argonne, and I have
asked Glen Seaborg to use his good offices in that direction. I have
also asked him to take all possible steps to make possible an increase in the participation of the academic institutions of the Midwest in the work of the Argonne Laboratory. He has outlined for me
a concrete proposal to accomplish this. I share fully your strong
desire to support the development of centers of scientific strength
in the Midwest, and I feel certain that with the right cooperation
between government and the universities we can do a great deal to
build at Argonne the nucleus of one of the finest research centers
in the world." And then in a letter from Senator Humphrey to President Elvis Stahr of the University of Indiana it was stated, and this
is a quote from Senator Humphrey's letter (he seems to talk about
himself in the third person): "Senator Humphrey placed great emphasis,
we believe correctly, on the necessity of taking all practical steps
to get the Midwest universities behind the Argonne National Laboratory and to find ways to give them a greater voice in the program
for management at the Argonne National Laboratory." As you can imagine, those letters from the President and from the Senator from Minnesota were very influential.

In 1962 and 1963 Roger Hildebrand and I were working hand-in-
hand as a university representative and as an Argonne representative
to develop user participation at Argonne. We had some very active
groups from universities designing beams and other facilities for
use here. The group which had been deeply involved and supportive
of the MURA activity came from the same universities. The people at
MURA felt that somehow or other Argonne, in trying to maintain the
support that it would need for the ZGS, had undermined their efforts
and their activities toward getting the MURA machine built. There
had grown up an understandable antagonism between the groups at the
two laboratories. Fears were awakened in the minds of some of the
experimenters who were working at Argonne. Their MURA colleagues at
their own universities suggested that maybe they would not really be
able to use the ZGS, once it was built. A feeling of malaise developed in the university community. There was a great concern that
the Argonne machine might never come to fruition or that it would
not really be available to them in the way in which Albert Crewe and
Roger Hildebrand and I had all indicated we believed it would.

During this period one peculiarity was that Roger and I both served on the MURA Board of Trustees, so there were many interconnections between the activities that were going on here to support the Argonne program and those that were going on to support the MURA program.

The upshot of the Johnson and Humphrey letters was that a committee was set up. I call it the Williams Committee because it was chaired by John Williams of the University of Minnesota. It was a seven-member committee, and as I remember it there were three members from MURA, President Elvis Stahr of the University of Indiana; Bernie Waldman, who was then Director of MURA; and Vice-President Peterson from Wisconsin, who was an officer of MURA. From AMU, which was a university consortium that had to do with university cooperation with Argonne, there was John Williams and I; from the University of Chicago, there was Warren Johnson (Vice President of the University of Chicago), and from Argonne there was Albert Crewe.

That committee met over many months pondering two important questions: 1) how to increase the confidence of the university physicists in the Argonne program, and 2) how to construct the organization at Argonne so that it would have some chance of attracting MURA personnel to Argonne to participate in this program. Although construction of the ZGS was far along at that time, the kind of help that was available at MURA, where there were some of the most talented accelerator physicists in the country, could still have been used down here.

The Williams Committee came up with an organization of which I have never been terribly proud. It is the tripartite organization which now manages Argonne. The University of Chicago is still the prime university managing the day-to-day operation of Argonne; there is a new university consortium called AUA (Argonne Universities Association) with about 30 to 35 university members, which is supposed to participate in the establishment of policy at Argonne; and then there is, of course, the Department of Energy (previously the AEC).

I am not sure that the establishment of that organization was really of any help. The principal intent at the time was to try to increase the confidence, particularly of the MURA physicists, in the availability of the program at Argonne to university groups so that they would come down here and join the effort.

The Williams Committee report was finally issued in September of 1964. It was only a few months later, in June of 1965, that a site committee was appointed to find a site for a new 200-GeV machine. That then became a very important factor in the thinking of the MURA people. They had the opportunity of waiting to see what would happen in the site selection for the 200-GeV machine. They were interested in the possibility that the new laboratory would offer an opportunity that might be more attractive to accelerator designers than the ZGS, which was already far along toward completion and its design well established.

So it is now clear that the Williams Committee solution did not accomplish its principal purpose - namely, the bringing of a large number of the MURA people to Argonne to help in the final stages of construction of the ZGS. But it did accomplish something which may have been less tangible, but which was perhaps more important. It focused attention on user interest in Argonne, on the need of Midwest universities to participate actively in the program, and on the talent residing at those universities, available to be tapped for the new ZGS program.

I should now like to spend my closing minutes describing the early days of the ZGS Users Group. That Group was founded in November of 1958, and it actually became active in 1959. I shall read from the minutes of the second meeting of the Users Group a statement I made concerning the purposes of that group: "Prospective users of the Argonne machine are interested in learning about progress on the planning and construction of the ZGS and its associated facilities. At meetings of this type members of the Argonne staff will report on the progress of their work.

"The main motivation for building this particular machine in the Midwest is to bring a high energy research facility into a section of the country where in recent years that field has been falling behind. There is, then, a large number of physicists who plan to do research at Argonne but who are not yet active in the multi-BeV field. They have been working on the Chicago Cyclotron, the Illinois Bevatron, etc. Meetings of this type are designed to help bring these people up-to-date on the theoretical and technical problems which are faced by those who are most active in the field today. This will be accomplished both through scheduled talks and through informal contacts. In choosing speakers no effort will be made to draw equally from talent in the East, the West, and the Midwest. At the first meeting there turned out to be a predominance of local talent, at the second meeting there has been a large majority from East and West.

"The important criterion is that there is a certain body of information and experience upon which prospective users would like to draw. Whenever it can be found locally, that source will certainly be used; however, by the very nature of the high energy program in this country during recent years most frequently the East and West will serve as a source of supply. As active high energy research programs multiply and develop in the Midwest this situation will gradually change."

It is hard to imagine that it really wasn't paranoia that was creating language of this kind; it was the real situation in those days. Things have changed dramatically since then. I won't read any further from those minutes, but I do want to say in closing that it was a difficult period, from 1958 through 1964 when the machine actually came on the air, when those of us who had been designing beams and bubble chambers could really begin to get physics data. I know that I, in particular, and I am sure that Roger, on his side, found ourselves in a complicated interaction with the MURA activity, with the Williams Committee, etc. often pushing things in conflicting directions. In 1964, however, I wrote to Roger, in spite of all of the things I had been doing to make his life complicated and difficult, saying that: "Over the past five years you have done a tremendous job in organizing the high energy physics program at Argonne and you have always striven to ensure that it will be a live program, that it will enjoy the full participation of high energy physicists in the region, and, perhaps to a lesser extent, eventually in the Nation."

That was what I felt. It was a complicated time, but the productivity of this machine was well worth the effort and high energy physics in the Midwest is here to stay. Thank you.

EARLY ADVENTURES SURROUNDING THE ZGS

Roger H. Hildebrand
The University of Chicago, Chicago, IL   60637

It is some fifteen years since I last spoke to this group, or at least to its older members. My remarks today will be about the preceeding years. I will begin at August 25, 1958, and work backwards. It was then that Chancellor Kimpton asked me to serve as Associate Director of Argonne National Laboratory with overall responsibility for building an accelerator, for enlisting the cooperation of midwestern scientists and for initiating a high energy physics program. Three days later Director Hilberry formally appointed me to that position, and two weeks later at my initiative, he appointed Albert Crewe to serve as director of the Particle Accelerator Division. So there we were, in some old wooden building a few hundred yards to the west of this site, with responsibility for running the show. We were sustained by the courage of foolhardy youth and by the knowledge that we were inheriting a far better show than was appreciated by anyone outside the project. I will give a brief personal account of how we came to be there and what we did.

I had been sent from Berkeley to Chicago in 1951 on a job for which the first step was to extract a deuteron beam from the new cyclotron which Herbert Anderson had built. I found to my embarrassment that a small but very satisfactory beam somehow extracted itself - I still cannot say how - and that the apparatus I had so carefully inserted for the purpose worked best when I took it out and hid it in a trashcan behind the laboratory. My reputation as an expert in beam extraction began with that accomplishment.

Some months later when I joined the University of Chicago faculty I knew, because Herb Anderson told me, that I was expected to get busy and extract a really potent proton beam. I spent a little time studying the Tuck-Teng extraction scheme, but it didn't look easy. My own experiments used the neutron beam (which also extracted itself) and I succeeded in putting off proton extraction from month to month because Fermi and Anderson and their associates were using pions to made the hyperon resonances.

Then in 1954 while I was at a conference in Glasgow, I heard that someone at Liverpool had extracted a good proton beam from a synchrocyclotron. I went down to have a look and found that the essential component of their system was a Yorkshireman named Albert Crewe with whom I had some fruitful conversations. When I got back to Chicago and told Herb Anderson what I had learned he immediately picked up the phone, called Liverpool and hired Crewe. That is how I produced the world's finest external proton beam.

You must remember that for a very few golden years in the early 50's, the cyclotron at the University of Chicago and the betatron at the University of Illinois were, indeed, the finest accelerators in the world. But the demands of hyperon and kaon physics and the competition from East and West forced everyone in the Midwest to choose between leaving and building something new. Many chose to leave.

The scale of the cooperative effort that would be necessary for the whole enterprise of building, equipping, and using a machine of the next generation dawned very slowly on me as I found myself in the unique position of serving both as a member of the MURA Board of Directors and as a consultant to Argonne. In April of 1957, at Jack Livingood's invitation, I had agreed to form a consulting group to work on the design of beams and experimental areas. There followed about a dozen reports of which we can still find seven; most of them by Crewe, Hildebrand, Kruse, Warshaw, Wright, and York.

Everything we did made life more difficult for Jack Livingood, Martyn Foss and Lee Teng and the others in the machine design group. We wanted longer straight sections. We wanted holes through the magnets to extract small-angle beams. We wanted to extract protons in two straight sections and mesons in a third. We wanted each beam to be independently adjustable in energy and pulse length. And, of course, we wanted a lot of expensive magnets with expensive power supplies. In the end all of these things appeared in the design, in the budget, and in the hardware.

One consequence of our consulting work was a growing appreciation that the group Jack Livingood had assembled at Argonne was first rate and that their machine should be a powerful research tool. But our confidence was not generally shared in Washington or at other laboratories. The machine experts questioned our estimates of space charge limits and of cross sections and extraction efficiencies. They questioned the effectiveness of using external protons to produce secondary beams. They questioned the realization of the high and uniform fields required in the Foss magnets, and they questioned the reliability of the rather complicated thin-walled vacuum chamber. (They were right about that.)

The people at Brookhaven proposed that the ZGS design be discarded and that instead Argonne should assemble an alternating gradient machine using the Brookhaven design with half their number of magnets. They were probably right that half an AGS is cheaper than a ZGS and they were right that copying someone else's design is quicker than making your own. But there already _was_ a ZGS design and there was no one at Argonne willing to assemble half an AGS. To do so would insure that we could never do anything better than Brookhaven. We were right about the performance of the ZGS, but that took years to demonstrate, and our critics were successful builders whose warnings and counter-proposals were taken seriously. By the late summer of '57 the outlook for funding the ZGS was bleak.

An opportunity to turn the situation around came in September. Dr. Hillberry invited me to go with him and Jack Livingood, Martyn Foss, Lee Teng, and Ed Crosbie to a meeting in Washington (Sunday, September 29, 1957) of a subcommittee of the GAC, the General Advisory Committee to the Atomic Energy Commission. The task of the subcommittee was to consider the fate of several proposals including Argonne's. We faced an imposing array of commissioners, competitors, and critics.

There sat GAC chairman Keith Glennan, Admiral Bennett of ONR, Alan Waterman of NSF, and Tom Johnson and Charles Faulk of AEC. There sat Lee Hayworth, Ken Green, Ernest Courant, Ed Lofgren, Al Weinberg, Robert Livingston, Jesse Beams, and my former professors McMillan and Panofsky.

I was called upon to do much of the talking for Argonne. My job was to defend the ZGS as a machine for users. I was questioned on the extraction of beams, the layout of experimental areas, and on the physics we hoped to do at 12 GeV. I was well primed because of my consulting work and I saw, or thought I saw, that my professors were smiling at my efforts. I recall with great satisfaction that when we left the room Martyn Foss said he wanted to shake my hand. We knew it had been a critical meeting and we thought it had gone well.

Later we learned that on the following day the GAC drafted a recommendation "that the Commission proceed with urgency to authorize the Argonne National Laboratory to do all things necessary to design and build the 12.5 BeV machine they have proposed for construction" - just the words we wanted to hear. The Commission and Congress then proceeded to act with what in Washington must be considered an urgent pace and in something less than a year the project entered the new phase which Kimpton and Hilberry had asked me to head.

I said earlier that Crewe's appointment followed mine, and it did, but in fact I would not have accepted my job if I had not known that he would accept his. Our first day started well. Jack Livingood assembled the entire staff and with the strongest and most generous expressions of support turned the job over to us. With equal confidence I then turned the Accelerator Division over to Crewe.

From then on we lived in a continual state of crisis. The particular crises of concern to this panel were those which had to do with using the machine. On the entire site there was not a single person who had done an experiment above cyclotron energies. On that score the place was barren. We made some good appointments but it seemed a long time before we succeeded in recruiting people like Tom Fields and Gale Pewitt and Malcolm Derrick, who would take on major responsibility for experimental facilities. It was not feasible even to set up the HEP Division until June of 1959.

The task of engaging Midwest talent was made difficult by the history of strained and sometimes bitter rivalry between Argonne and MURA; a circumstance especially poignant to me because of my personal friendship with the MURA staff. We had somehow to convince potential users that they were welcome, desperately welcome, that the whole project was real, that we would be able to solve the management and budgetary problems of complicated joint ventures. (My own confidence on that point, and many others, came from promptly hiring Don Getz.) We had to convince them that experimental schedules would be determined in some intelligent way still to be developed.

We knew that even with intense cooperative effort there was barely enough time to get ready. To mobilize the university talent, from which Argonne was so completely isolated, it seemed necessary

that the mobilization itself should invoke respected outside responsibility, and for that I called upon Ned Goldwasser. I will let him
tell what he did and will only remark that all of us are deeply in-
debted to him. The MURA bubble chamber and the Michigan freon-
propane chamber were conspicuous examples not only of outside talent
but also of outside confidence in Ned's mobilization. Both projects
were initiated and far advanced before there was any formal assurance that there would be money to finish them. Both involved shared
Argonne/University responsibilities which brought many warnings to
me from experienced people about budgetary chaos.

I was also warned by experienced project directors about the
hazards of relinquishing any authority in matters of scheduling.
But those proved to be the hazards of mapping unfamiliar territory
and of discerning those matters in which the director's authority is
actually crucial. With Ned and then Keith Symon of MURA and then
Kent Terwilliger of Michigan we constructed ways of operating which
seemed at least good enough for a start and we supposed that our
successors would reshape them to work when the experiments really
began.

The start-up, as you know, came in late '63 with Lee Teng then
in charge of the Accelerator Division. With the machine working and
the first experiments ready or nearly ready, we entered a new phase
which seemed to me to require new talent and that brings me to my
last important contribution to the project: I made a trip to
Madison and somehow persuaded Bob Sachs to take my job.

On that proud note I shall end my remarks this morning.

STARTING THE ZGS SCIENTIFIC PROGRAM

Robert G. Sachs
The University of Chicago, Chicago, Ill. 60637

The start-up of the ZGS Scientific Program came at a decisive time in the history of U.S. high energy physics and it offered an opportunity to establish the participatory style of operation that has become a characteristic of the field throughout the world. The accelerator had been born in controversy* over the lack of a major high energy physics facility providing for the specific needs of the physics departments at the great Midwestern universities. If those universities were to maintain their traditional role as major centers of physics educating a large fraction of the nation's physicists, it was essential that they continue to be highly competitive in this rapidly developing field, but the facilities to do so were not available to them.

Some of the larger research departments in the Midwest had established research programs in high energy physics by making arrangements to carry out occasional experiments at the accelerators at Berkeley or Brookhaven. This involved a radical departure from traditional research methods since it required faculty members, research associates and students to spend very extended amounts of time off campus (transportation each way required a full day in those days). More troublesome was the fact that those machines were quickly saturated by their own constituencies and even within those constituencies there was difficulty in providing opportunities for new people to break into leadership roles.

But the most troublesome problem was the inability of even the most successful Midwestern groups to influence important scientific decisions concerning the nature of the facilities and programs at the accelerators. (An extreme example of such a decision would be the Brookhaven decision in favor of an AGS improvement program over the alternative of building intersecting storage rings.) This was certainly on the minds of the physicists who pushed so strongly for an accelerator to be built under MURA management. And in late 1963 when I was asked to come here from the University of Wisconsin to take responsibility for the ZGS program as Associate Laboratory Director for High Energy Physics, we were confident that a very high intensity proton accelerator would be built in Madison by MURA. Nevertheless, I accepted the position because the ZGS existed and offered a great opportunity to develop a research program fitted to the aims of our community during the seven or eight year period required to build the MURA machine.

By the time I arrived at Argonne, on February 1, 1964, President Johnson had eliminated the MURA machine from the AEC budget and it was clear that the ZGS was all we were going to have for a long time. Therefore the high energy physics community of the

ISSN:0094-243X/80/600 03604$1.50 Copyright 1980 American Institute of Physics

Midwest turned all of its considerable energies to trying to satisfy
its needs at the ZGS. In fact, they elicited an instruction to the
AEC from President Johnson that the AEC see to it that Argonne was
responsive to the large community of potential users represented by
the MURA interests. Responsiveness meant not only access to the
machine on the basis of scientific merit but also influence on scientific decisions that could be definitive to the research interests of
this growing group of high energy physicists.

Fortunately for me, my predecessor, Roger Hildebrand, had anticipated this need and, from the start of ZGS construction had set
about working with nearby high energy physicists to develop methods
for establishing with their cooperation as sound a scientific program as possible. The principal spokesman for what became known as
the "ZGS Users" was E. L. Goldwasser, and Roger and Ned developed a
form of governance for the ZGS program which has set the pattern
for other high energy physics programs. It was up to me to implement their plans in establishing a scientific program at the ZGS.

Certain aspects of the early program were predetermined for me
as a result of arrangements and agreements that Roger had already
worked out in accordance with these plans. In particular, responsibility for the design and construction of several major pieces of
equipment had been farmed out to high energy physics groups at the
universities: the 30-inch HBC (U. of Wisconsin, W. D. Walker), the
7° separated beam (U. of Illinois, Northwestern, Wisconsin, Argonne;
Goldwasser, R. Ammar, T. Fields, M. L. Good, M. Derrick), the 40-
inch heavy liquid BC (U. of Michigan, Dan Sinclair). As an aside,
I should say that these efforts were all successful in the end; the
30-inch chamber is presently an important facility at Fermilab, the
only one of its kind. However, these facilities were not brought
into successful use without pain, and in the process I learned about
the importance of engineering in the construction of reliable equipment. I also learned about the importance of proper management both
in getting them built and in providing for their continuous and
reliable operation. We stumbled many times before Lee Teng, Hildred
Blewett and Royce Jones put various fleas in my ear and we rooted
out some of our managerial weaknesses.

Although there were no high energy physicists at Argonne when
construction of an accelerator was initiated, there was a formidable engineering and technical staff which turned out to be invalu-
able in getting not only the ZGS, but also associated items of
equipment built and operating. We simply had to learn how to make
the best use of available talent.

The modus operandi governing the scientific program was built
on a foundation of a broadly based ZGS Users' organization and
consisted of a complex of advisory committees, a program committee
and a review committee, all later supplemented by various AUA board

representatives and committees. I remember being asked by Panofsky,
who was chairman of the ZGS review committee, to describe our system
for arriving at a decision on a facility or program. After I had
spent a half hour taking the committee through our labyrinth, Pief
threw up his hands in horror and said it would never work. Well it
did, because we worked at making it work.

As a theorist, I had the advantage of knowing so little about
matters to be decided as to be free of strong prejudice, and the
disadvantage of not knowing enough about experiments to trust my own
judgement. Of course, the pressures were enormous. The Jekyll and
Hyde character of many a good friend was brought out by the mere
submission of a proposal for an experiment. I am not so sure that
some of them did not try to take advantage of my ignorance. Fortunately I had the help of a program committee made up of first-
rate physicists who could advise me on the distinctions among proposals ranging from very good ones to pure fantasies. Although
the final decisions were mine, they were made only after listening
carefully to the committee. I don't think I ever made a decision
to which the majority was opposed, but I also never took a vote.

The Users Advisory Committee often gave me advice I could do
without, but played a key role in major decisions. Among them was
the decision to go ahead with what turned out to be the 12-foot
hydrogen bubble chamber. Many experienced people said that it
would be much too risky a venture for an organization lacking
experience in bubble chamber construction. At the time, even the
30-inch chamber had not yet proven itself and we were viewed as
rank amateurs in the field of bubble chamber design and construction.
Nevertheless the Users Advisory Committee insisted that we find a
way to design and build the first of an entirely new generation of
chambers.

Luckily I had the advice of Tom Fields and Roger Hildebrand
which led me to Gale Pewitt as the man to do the job. The result
probably can be described as the most successful bubble chamber
design, construction and operating experience on record since Luis
Alvarez's early work.

Altogether, I had an enormous amount of help from the community,
too many people to mention, and owe to them whatever success I had in
dealing with the problems and promise of experiments in high energy
physics. It required a year or so of fits and starts to bring the
accelerator and the facilities up to anything like their specified
levels of performance but, as you will hear, a truly significant
high energy physics program grew out of the effort. And we made a
point of providing opportunities for the younger members of our
constituency when they proposed promising scientific programs with
the result that the physics departments in this community were
greatly strengthened. When we were well on our way to a viable
research operation, it became clear that the responsibility for the

program should be in the hands of someone who could make technical judgements about experiments and facilities on the basis of his own experience and judgement, someone who would not be as dependent as I was on the help and advice of others. My successor, Bruce Cork, filled that role admirably and the successful physics program resulted from his good foresight and judgement.

The participatory process we went through to arrive at that point had been visualized by Hildebrand and Goldwasser and it was highly successful. Our experience came to fruition just at the time when the need for communal planning and action in high energy physics had grown to national proportions. The question of how to make plans for and decisions about the then-proposed 200-GeV national accelerator in a participatory fashion was very much in the air. At the time there was no national organization, nor even a complete list of those who would have a stake in the accelerator.

Our experience certainly influenced the handling of the national problem. For one thing, the ZGS experience provided a rather large microcosm of the system to be dealt with and, for another, we already had a list of some 200-300 high energy physicists with participatory inclinations. Since we had not limited ourselves to a local users community, but had taken the view that the ZGS was a national facility, the list was rather comprehensive. There was also a serendipitous situation in that, because of concern about the lagging influence of Midwestern physics, the Council of the American Physical Society had been persuaded by Don Kerst to establish a special Regional Secretariat of the Society, and I had accepted the position as Regional Secretary for the Central States in 1963, just before taking on the ZGS responsibilities.

This put me in a position to bring to bear the influence of the ZGS community on considerations of the APS Council about how to restructure the Society to provide for more participation by members. In particular we pressed successfully for amendment of the Constitution of the Society to strengthen the role of the Divisions. We immediately organized the Division of Particles and Fields (and the nuclear physicists immediately organized the Division of Nuclear Physics) to take advantage of this opportunity. I doubt that many members of the Division know that it was organized as part of the activity surrounding the ZGS, but it is one of the important contributions to the nation's high energy physics program resulting from our community's efforts to find the answers to some important questions.

* For a more detailed discussion see Leonard Greenbaum, A Special Interest (Ann Arbor, Mich.: University of Michigan Press, 1971), and my review of it, "The National Laboratories," Bulletin of the Atomic Scientists, Vol. XXVIII, No. 6 (June 1972), pp. 51-53.

Session III

The Peak Years

Bruce Cork
Lawrence Berkeley Laboratory, Berkeley, CA   94720

INTRODUCTION

For historians, history is composed of events that are well
documented. My discussion today will not be supported by direct
references, even though the high energy physics program at Argonne
National Laboratory is well documented by some two hundred or more
physics proposals, monthly "as run" summaries, seminars, and physics
results as published in the various international journals and conference reports. The contributions by the members of the High Energy
Physics Division, Facilities Division, Accelerator Division and
University groups were so great and so interwoven that it is difficult to single out the role of each individual. Also, the roles of
the Argonne University Association, the University of Chicago, the
Argonne National Lab and the sponsoring Atomic Energy Commission were
vital.

This symposium is valuable because it will reiterate the role of
the universities in inspiring and participating in large projects
that are of major interest to the national research programs. A
pattern has been set, goals established, and the results have generally been published. The joint enterprise of universities and
national laboratories with major support by an agency of the national
government has been a winning combination. However, it is a tenuous
collaboration and will require continued diligence if the collaboration is to continue to be successful.

I will arbitrarily assume for the "peak years" the period
beginning in 1968. By 1968 the Zero Gradient Synchrotron was producing approximately 10^{12} 12 GeV protons per pulse with a beam duty
cycle of greater than 500 m.s. every four seconds. Sufficient beam
lines had been designed and components were on hand, or on order, to
establish four secondary beam lines. The thirty-inch diameter
liquid hydrogen bubble chamber was a well-engineered operating
chamber; the forty-inch diameter heavy liquid chamber with a forty
kilogauss field had been used for some experiments. Also, the
twelve-foot diameter liquid hydrogen bubble chamber was well along in
construction, and plans were made for a second extracted proton beam
to include this new facility.

However, the high energy physics program at the various midwestern universities had been seriously limited because of inadequate
experimental facilities at the various East and West Coast laboratories. Also, the logistics problems associated with doing an experiment at a distant site were tolerable but time consuming. This trend
was reversed when facilities came into operation at the ZGS for
twelve simultaneous experiments. Of course, the 400-GeV accelerator
program at NAL, which was designed in 1968, had a major effect on
the ZGS program. Physicists wishing to prepare for the experimental
program at NAL were able to initiate a program at the ZGS, often in

ISSN:0094-243X/80/600040-13$1.50 Copyright 1980 American Institute of Physics

collaboration with other groups. This was sufficient incentive for the physicists and engineers to do a significant experiment and do it well. The result was a healthy demand for experimental time at the ZGS and maximum use of the facilities, including data analysis.

On July 1, 1968 I was committed to guide the high energy physics program at Argonne. We had completed the instrumentation for a high energy proton-proton experiment using cosmic ray protons at Mt. Evans. Because of the limited flux of protons, my move on to the East seemed natural.

I had been a member of the Argonne High Energy Advisory Committee for the years 1967-68. I regarded the role of the Advisory Committee as more than just a responsibility to approve or reject proposed experiments, but to assure that major effort and resources would be dedicated to complete the approved experiments. Fortunately, in spite of decreasing budgets, it was possible to complete over 100 major experiments in the five-year period 1968-73.

The choice of experiments in the succeeding six years was comparable and approximately 100 major experiments were completed in the last six years of the program.

This was possible because of the dedication of many individuals. It was also possible because of the beam characteristics of a proton synchrotron. The 12-GeV protons have an energy sufficiently above pion, kaon, and anti-nucleon threshold that the secondary beam yields in the forward direction are very favorable. The spectroscopy of these particles has been a rich source of information of particle physics. In fact, the information content has been so great that a large part of our talent has been devoted to detecting the multiplicity of events and analyzing the immense amount of data. The guidance of theoreticians has been valuable during this period.

THE ZGS ACCELERATOR

It has been possible at ANL to compete in the international high energy physics program because of many circumstances including the machine, the facilities, the choice of experiments and, of course, the devotion of the participants. It was a period of "small" experiments, each with typically five physics Ph.D.'s, two or three graduate students, three or four skilled technicians, and engineering support. Everyone knew everyone in his own group, and the weakness and strength of competing groups. The incentives for publishing the experimental results were high. A typical work week was 80 hours.

The characteristics of the ZGS were such that as many as twelve or more experiments could be operated simultaneously. Two simultaneous extracted proton beams plus one or more internal secondary beams allowed a very favorable distribution of secondary beams over a wide area (Figs. 1-4). The beam duty cycle and intensity, a few x 10^{12} protons per pulse, were adequate for this program; however, even more intensity would have been valuable, and a continued program of improvements was a part of the program.

One major activity was the installation of titanium vacuum tanks in each octant of the ZGS. This was a major engineering task that involved diffusion welding of the large panels of the titanium

42

Fig. 1. The Zero Gradient Synchrotron (ZGS). The research building is in foreground; accelerator is under circular mound of earth shielding in background. Experimental areas are peripheral.

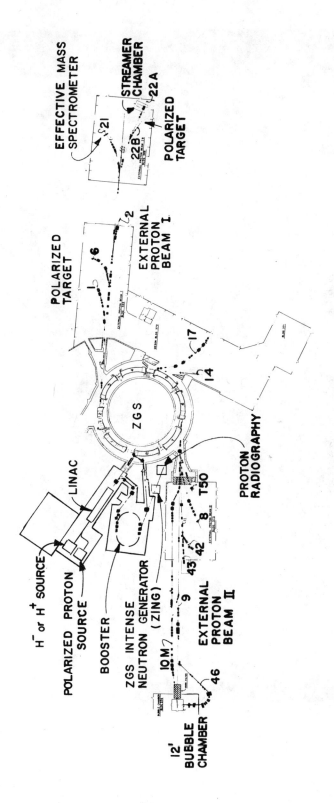

Fig. 2. Layout of experimental charged particle beam lines.

Fig. 3. Superconducting beam line to Effective Mass Spectrometer.

Fig. 4. Meson beam area.

vessel. Also, provision was made for adequate pole face windings so that detailed corrections to the ZGS magnetic field could be made. These tanks were made by a major air frame manufacturer and shipped, one to a railroad car, from the West Coast. The installation, completed in 1972, was very successful and very important to the improvements in beam intensity, control and reliability of the machine. Later improvements included more precise control of extracted proton beams, negative ion injection, a second pre-accelerator, and acceleration of polarized protons and deuterons.

The decision to accelerate polarized protons was made in 1971. It seemed a reasonable gamble that the spin-spin force would be significant at multi GeV energies. If true, an entire new field of particle physics would be possible. The technical problems associated with accelerating polarized protons were believed to have a solution. Since the protons made only one pass through the non-uniform magnetic quadrupoles of the 50-MeV Linac, depolarization in the Linac was not expected to be a problem. The uniform magnetic field of the ZGS was an asset and there was room in the straight sections for correcting magnets so that loss of polarization due to resonances could be avoided. The feasibility of the program became apparent when we visited the ANAC Laboratory at Auckland, New Zealand, where they had produced an engineered atomic beam source, and were prepared to manufacture a source for the ZGS. Polarized protons were accelerated to 6.0 GeV by the fall of 1973.

Major repairs included rebuilding of two large octant coils and construction of a new spare coil. Also, the beginning of a fatigue flaw in the rotor of the large motor generator main ring power supply was detected early and repaired. Each of these repairs required expert engineering, considerable ingenuity and technology.

The great multiplicity of beams and experiments required new instrumentation so that computer control of magnet currents and other functions was necessary and possible.

THE ZGS FACILITY

The design, installation and operation of the multitude of beams and instruments that are a part of these beams has been a responsibility of the Facilities Division. Generally, high intensity, separated beams have been a requirement for secondary beams. This required development of beams with good optical properties, good electromagnetic separators, and shielding.

With these beams it was possible to utilize many new instruments. These included the MURA 30" hydrogen bubble chamber, the Michigan 40" heavy liquid chamber, the ANL twelve-foot hydrogen bubble chamber, the ANL-Illinois streamer chamber, and various polarized proton targets. Some of the features of these instruments, and some of the experiments will be described at this symposium. Various university groups had a major role in the development of many of these instruments.

At the same time, the selected event spark chamber became a very effective instrument because it could be used "on line" to a small computer for data analysis. This was followed at a later time by proportional chambers and drift chambers. These chambers were built especially for a particular experiment with appropriate triggering, event selection, and data analysis. Also, intense neutral beams, both neutron and kaon, have been built and used in good statistics experiments.

A major and very successful development has been the "Polly" system of automatic scanning of bubble chamber photographs. This system, with "Pilot assist" in scanning and measuring will be valuable for many applications.

During this period cryogenic engineering has been a pioneering activity. With the success of the ten-inch diameter liquid helium bubble chamber, it was decided that a very large hydrogen chamber should be built. The twelve-foot chamber was built "on time"--comleted October 1969--and nearly "on budget"--only a few hundred thousand dollars of "extra" reactor funds were transferred to complete the chamber, buildings and beam lines. The large fourteen-foot diameter, eighteen-kilogauss superconducting magnet was the largest magnet of this type (Fig. 5 and 6) and was followed by the design and construction at ANL of the NAL fifteen-foot diameter, 30-kilogauss bubble chamber magnet. Both of these magnets continue to be very successful.

Smaller cryogenic magnets are now standard engineering practice. However, the design of rapidly pulsed high field superconducting magnets requires the maximum ingenuity of scientists and technicians.

THE HIGH ENERGY PHYSICS

In 1968 some of the compelling physics of particles and fields was meson spectroscopy, baryon spectroscopy, nucleon-nucleon scattering, diffraction scattering, and interference phenomena. In meson spectroscopy some thirty meson isomultiplets have been observed. Today we regard these as levels of a quark and an antiquark $(q\bar{q})$. Likewise, some fifty baryon isomultiplets are regarded as levels of a three quark (qqq) system. With more and better experiments "the plot has thickened". More quarks and more colors of quarks are required to explain the observed results. Surviving theories will have to be consistent with the experimental results. These results include some 200 major experiments completed since 1968 at the ZGS accelerator (Fig. 7).

These experiments were typically done with very good statistics, a requirement at this period of high energy particle physics. Bubble chamber experiments were able to compete because effective methods of automatic scanning and tracking were instrumented. Spark chamber experiments could select specific interactions, with good spatial resolution; scintillation and Cherenkov counter experiments could select events and operate at high instantaneous rates. These could also be used to provide a trigger for the streamer chamber, or the thirty-inch diameter hybrid bubble chamber, one of the first "hybrid" experiments.

An example of the complementarity of the experiments is the discovery of the $J=0^+$ states in the mesonic spectrum using the effective mass spectrometer; the same state was observed in the

Fig. 5. Twelve foot diameter liquid hydrogen bubble chamber. Superconducting magnet with liquid helium storage vessel in foreground.

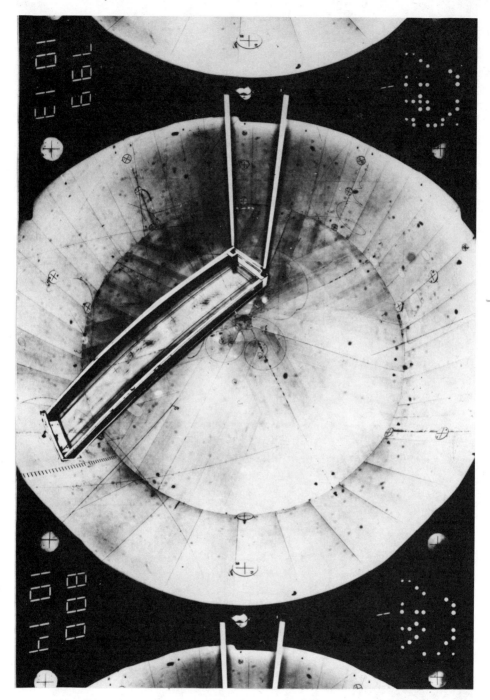

Fig. 6.   Tracks in twelve-foot diameter bubble chamber with
liquid hydrogen track-sensitive target in beam.

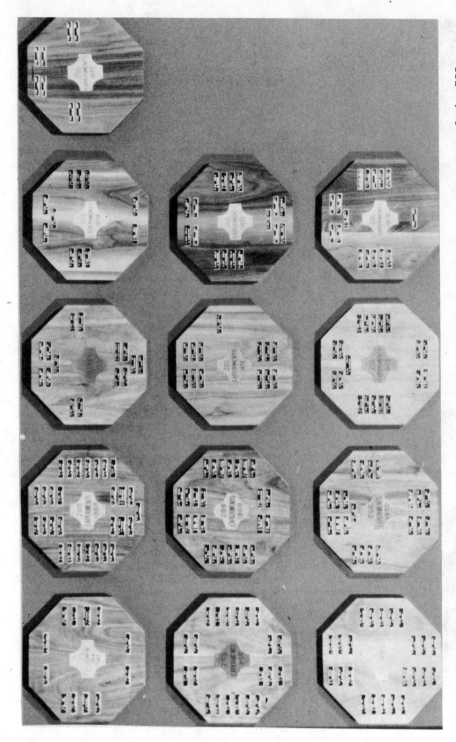

Fig. 7. Plaques showing title of experiments completed each year of operation of the ZGS.

the streamer chamber. An example of the selective trigger of the
effective mass spectrometer is the measurement of the ρ-ω interfer-
ance, done with very good statistics. As a result of this and other
experiments, detailed studies of the spectroscopy of hadronic states
were possible.

The twelve-foot diameter hydrogen bubble chamber with a fiducial
volume of 16 m^3 and a magnetic field of eighteen kilogauss was an
ideal detector of neutral and charged current neutrino events. This
was an important supplement to the low energy neutrino events from
other large chambers.

Other experiments, 6.5 GeV K^- and low energy $\bar{p}$ were done with
good statistics in the twelve foot chamber. Also, a joint Rutherford
Lab/CERN/ANL chamber experiment gave good tracks in a track-sensitive
target located in the twelve foot chamber.

A series of nuclear chemistry experiments has been done in the
high intensity circulating beam of the ZGS to observe the fission-
fragmentation process for heavy nuclei. The results are consistent
with the interpretation that there is no cascading of the secondary
hadrons in the nucleus, in agreement with higher energy experiments.

Also, a series of nuclear physics experiments has been completed
using the separated kaon beam and the low momentum pion beam. Kaon
x-rays from Al, Si, Ni, and Cu; and anti-protonic x-rays from ^6Li and
^4He have been measured.

The energies and relative intensities of nuclear gamma rays
associated with stopped kaons in Ni, Cu, Si, and Al have been measured. The spectra from Ni and Cu indicate that removal of one, two
or three alpha particles is the favored mode. This is interpreted as
some new aspect of nuclear structure. To further explore this
problem experiments were done with the interaction of high energy
pions with Ni, S, Ar, Ca, V and As. Results consistent with those of
stopping kaons showed that single and multiple alpha removal is
present a significant fraction of the time.

A critical weak interaction experiment was measurement of the
rate of decay of $K_L \rightarrow \mu^+ + \mu^-$. This experiment was done by producing
an intense beam of K_L in the extracted proton beam. The two muons
were detected in a double arm spectrometer. The production of
$(K^0_L \rightarrow \mu^+ + \mu^-)/(K^0_L \rightarrow all)$, was measured to be $(8.4^{+3.8}_{-1.8}) \times 10^{-9}$. This
is consistent with the unitarity limit.

The description of the nucleon-nucleon interaction has generally
been simplified by ignoring the role of the spin-spin interaction.
Recent experiments at the ZGS have demonstrated that many details of
the n-n reaction can be revealed by measuring the spin-spin part of
the reaction. The ZGS has been an excellent facility for these
measurements because the beam is approximately 70% polarized, the
targets can be approximately 80% polarized, and the beam-target
luminosity is high. The energy range 1 to 12 GeV has been explored.

The strong interactions are probed most deeply by measuring the
high transverse momentum events, over a range of 10^{13} in cross
section. This is believed to be a measure of the structure of the
nucleon. In contrast, the low transverse momentum events are a
measure of the size of the nucleon, approximately one fermi.

The spin-spin reaction is measured by comparing the scattering cross sections for nucleon spins in the same direction, in contrast to spins opposed. For high transverse momentum $p_\perp = 5$ GeV/c , the spin-spin correlation factor A_{nn} approaches 60%. This is a very large dependence on nucleon spin and is interpreted as a deep elastic, or inelastic, dependence on the spin of the valence quarks of the nucleon.

Polarized beams and targets thus allow nucleon spin to be a variable quantum number for inelastic as well as elastic scattering. Thus, the entire field of meson spectroscopy and baryon spectroscopy can be explored in more detail. For example, a diproton resonance of mass near 2.2 GeV is observed in proton-proton scattering. This resonance requires a polarized beam to separate the large number of other resonances.

In summary, the high energy physics program at the Argonne National Laboratory has been favored by an abundance of significant and feasible high energy particle physics experiments. The program has provided a vital laboratory for a generation of physicists, engineers, and technicians. The ZGS has been the right machine at the right place at the right time. Most of these results have been published in the international journals, at major conferences, workshops or symposia. Although there are gaps in our knowledge in the range of 1 to 12 GeV particle phyiscs, these gaps will have to be filled in some other manner.

Someone once said to Rutherford, "Lucky fellow, Rutherford, always on the crest of the wave". Rutherford's response: "Well, I made the wave, didn't I?".

We at least had the pleasure of riding along, sometimes a bit behind, but near the crest of the wave, supplying sufficient knowledge to maintain a wave.

THE MURA-ANL-FERMILAB 30" BUBBLE CHAMBER

W. D. Walker
Duke University, Durham, NC   27706

The project to construct the 30" chamber really began with the succession of heavy liquid bubble chambers that were constructed at Wisconsin beginning in 1955. These chambers ranged in size from a couple of inches diameter up to a 15" diameter propane-freon bubble chamber. In 1958, Joe Ballam and I were very interested in constructing a 30" heavy liquid chamber which could be used at Brookhaven with the AGS. Unfortunately, a proposal was made by Jack Steinberger to build such a chamber, and he subsequently did construct a chamber which eventually grew into the 30" hydrogen bubble chamber which was used for several years at Brookhaven. In 1959-1960, we were working closely with Bob Adair, who had constructed a very successful cylindrically shaped 15" bubble chamber with a liquid expansion. This was the chamber that was used in the experiments in which we discovered the ρ meson at the Cosmotron.

In 1959, a hydrogen bubble chamber project was started at Argonne under the leadership of R. W. Thompson. It happened that I knew Thompson rather well in that both of us had worked Cosmic Rays in the late 40's at the Echo Lake, Colorado, laboratory. Although I had the utmost regard for Thompson, I was dubious of the project. As a result of this fact, and the fact that Ballam and I were still interested in constructing a modest size chamber, we began to talk in the fall of 1959 about making a proposal to build a chamber which would be in the 20 to 40" size range. We heard that George Tautfest was proposing to build a 15" hydrogen chamber to be used at Argonne with the planned ZGS. As a result, a meeting was called involving the groups from Wisconsin, Michigan State, and Purdue which took place in mid-December of 1959. The people who attended the meeting were: from Purdue, George Tautfest, Pete Palfrey, Frank Loeffler, and Harry Fechter; from Wisconsin, Ugo Camerini, Jack Fry, Bud Good, Albert Erwin, and myself. Ballam, unfortunately, could not attend the meeting because of poor weather conditions in East Lansing. The result of the meeting was the proposal from MURA to the AEC which is appended to this paper.

The general concept of the chamber and also the scope of the chamber were essentially my decision; the idea being that one should build a chamber as large as one could, and still have a man-size chamber so that a person could pick up the various parts of the chamber. This didn't turn out to be true, but it set the scale of what was proposed. The chamber was to be a precision instrument which means that we would use the best optics available, which was a system in which light was shown through the chamber and brought up to a focus close to the camera lenses. Again using the idea of precision, the size of the magnet and the strength of the magnetic field were to be such that quite a high magnetic field would be available. This turned out to be something like 36 kilogauss for 5 MW of power.

ISSN:0094-243X/80/600053-17 $1.50 Copyright 1980 American Institute of Physics

The expansion system of the chamber was to be very similar to that used in the 15" chamber of Adair. In our case, there were 3 pistons operating simultaneously which produced the expansion of the chamber.

In the spring of 1960, we were able to begin to assemble a staff to design and build the bubble chamber. Probably the single most important person to be brought in was our chief engineer, John Mark. John Mark was contacted at Berkeley by Good, and was subsequently hired by MURA. He was very important for us because no one else on the project had previous experience in cryogenic engineering. Mark had just finished work on the construction of the 72" chamber, which really broke a lot of new technical ground. It should be remembered that large-scale cryogenics had its start in the late 40's and early 50's with the project involved in the construction of the H-bomb that was detonated at Eniwetok. The compressors and high pressure hydrogen system that we used on the 30" chamber were actually obtained surplus from Los Alamos where it had been part of that program. Probably the weakest part of the project was the lack of continuity between the people who built the chamber and those who eventually operated it. Various people who designed and built the chamber so far as cryogenics were concerned were John Mark; mechanical design was done by Mark and Jack Froehlich, who came from Argonne Laboratory; the magnet design was done by George Tautfest and Bob March; the electronics to control the operation of the chamber was built by Harry Fechter with a great deal of technical assistance from Bill Williams, who is currently at Fermilab. We decided that the chamber controls should be as nearly as possible totally solid state. This was quite a departure from the usual practice at that time in that previous chambers had large parts of the controls done by vacuum tubes rather than transistors. Fechter borrowed as much of the design as he could from the control systems used on the 72" chamber. The optics for the system were designed by W. F. Fry and Ugo Camerini. The system used is to illuminate each of the cameras with a ringed source which is focused about the camera lens. This has proved to be quite a nice illumination system, although we had some difficulty with reflections and had to have a more expensive coating system put on than we had originally proposed.

The design parameters of the chamber were pretty well settled on by the spring of 1961, and we began to place orders for the various parts at that time. The magnet iron was purchased from Allis Chalmers in Milwaukee, and the coils were fabricated in New Jersey by Westinghouse; the glass was purchased from Schott in Germany, and the refrigerator from the Cryogenic Engineering Company in Denver. The assembly of the magnet and chamber took place in Madison in a building on the campus of the university that was constructed for this purpose with funds from the National Science Foundation.

The construction and assembly was completed in the spring of 1963, and the chamber was moved to the ZGS that summer. Lou Voyvodic became associated with the project very soon after the move. Useful pictures were obtained with the chamber in the summer of 1964. I have suppressed probably many of the difficulties encountered in the process of getting the chamber into operation at Argonne. There were many difficulties associated with the cleanliness of the vacuum system. Another difficulty which seemed fairly major had to do with excessive friction in the expansion system, which made it difficult to keep the chamber sensitive for reasonable periods of time. This difficulty was overcome with the help of Royce Jones, who effectively helped in re-engineering some aspects of the chamber. The expansion system caused some difficulty, but because of the fact that we had relatively small and light pistons, it has been possible to multipulse the chamber both at Argonne and at Fermilab. This was a feature that was built into the chamber from the very beginning.

In retrospect, I believe that one of the things that I learned in the process of doing this chamber was that the construction of the device, and the training of the crew or the assembling of a group of people capable of running such a device, seemed to be almost comparable efforts so far as the length of time required. There were only one or two people who went with the chamber from MURA to Argonne. Once at Argonne the personnel seemed to flow in and out of the project with a relatively short time constant.

7° BEAM

I believe that it is fair to say that the prime mover in the construction of the 7° beam was M. L. Good from Wisconsin. Good had had a considerable amount of experience in the use and design of beams at Berkeley. The idea of designing such a beam came as a result of a conversation between Camerini and Good. The Illinois group had similar ideas about the necessity of constructing a beam for bubble chambers which resulted in a fusion of the Wisconsin and Illinois efforts. The fusion of these two efforts occurred at a summer study at Argonne. The basic concept of the beam was Bud Good's; however, the final detailed design of the beam was done by the Illinois group. The group at Illinois consisted of Ned Goldwasser, Uhle Kruse, Gulio Ascoli, and Bob Sard; Ray Ammar and Tom Fields from Northwestern were also heavily involved in the construction fo the beam. Bud Good and several students from Wisconsin, among them Don Reeder and Dick Koffler, were involved in assembling and testing the separators at Argonne. The beam turned out to be one of the nicest separated beams ever built, in that it was designed from the beginning without reference to where posts and building supports were located so that there was considerable freedom in the design of the geometric layout of the beam, and consequently, optically it was a very precise beam.

The motivation for the beam had to do with using high energy K^-. This was certainly Bud Good's motivation. The reason that one was interested in doing such a thing had to do with the fact that Gell-Mann had at the High Energy Conference held at CERN in 1962 predicted the existence of the Ω^-. This he did on the basis of the quark model and the masses of the particles of the decuplet which were known at that time. It was clear that it was very important to have a high energy negative K^- beam with which one might have some chance of producing such an object. As we all know now, the Ω^- was discovered by the BNL bubble chamber group in early 1964. The 30" chamber and the 7° beam were just in the process of getting started and subsequently quite a few Ω^- were found in the exposure of the 30" chamber to the 4-GeV K^- beam which was produced.

If we look at the whole effort of the bubble chamber plus 7° beam in retrospect, it seems to have been a very successful scientific enterprise. In contrast, the chamber has had a fitful life at Fermilab. When it was operated in the 7° beam, it was operated very efficiently; it must have taken over 10 million pictures. The person in charge of the operation of the chamber at Argonne was Lou Voyvodic from practically the time the chamber was moved down to Argonne. In the years between 1964 and the end of running at Argonne I counted that there were 64 Physics Review Letters published and 114 Physical Review and Nuclear Physics articles published from experiments done with the 30" chamber. These experiments involved 16 institutions. A large amount of exploratory work in resonance production was done. Detailed data on the characteristics of the f^o, g, and w' were obtained. We also obtained much data on the Regge character of the production properties of various resonances. All in all it is a satisfying memory to me at least.

Fig. 1. The 30" chamber during its assembly at Argonne. At this time a field map was being made by the University of Illinois Group.

PROPOSAL

FOR

AN INTERMEDIATE SIZE

HYDROGEN BUBBLE CHAMBER

MIDWESTERN UNIVERSITIES RESEARCH ASSOCIATION

January 12, 1960

STATEMENT OF PROPOSAL

The history of high energy physics indicates that visual techniques have been overwhelmingly important in the development of the subject. With the exception of the neutron, the neutral pi-meson, and possibly the anti-proton, all of the elementary particles discovered since 1932 have been discovered by visual techniques. In the case of the anti-proton, the important confirmatory experiments were done by visual techniques.

Until 1955, the instruments used were the cloud chamber, the diffusion chamber, and nuclear emulsions. Since that time, the development of the bubble chamber resulting from the work of D. A. Glaser with superheated liquids has significantly changed the quality of high energy physics. Today the bubble chamber technique has largely displaced the other visual techniques in high energy physics. The rate at which data is gathered is phenomenal compared with what could be done five years ago. This means that more and more difficult experiments are feasible. This can be illustrated by the process $\pi^- + p \rightarrow K^0 + \Lambda^0$ which has a cross section of $\sim 10^{-27}$ cm^2. In the period 1953 - 1955 about ten examples of this reaction were found using hydrogen diffusion chambers. Today experiments can be planned in which 10,000 such events are seen in a few weeks running time at a large accelerator. We can expect to probe reactions with cross sections of $\sim 10^{-31}$ cm^2 in the next few years. The amount of "bubble chamber time" expended at accelerators will probably continue to grow as rapidly as time is available. Beam sharing techniques will probably double or triple the available time on existing accelerators in the next few years.

At the present time there is under construction at the Argonne National Laboratory the Z.G.S. proton synchrotron with a design energy of 12.5 Bev. This accelerator represents a large investment of money and effort for the country as a whole and the midwestern community of physicists in particular. We feel that it is very important tha experimental apparatus be available to utilize the facilities which will become available in the near future. The facilities at the Argonne accelerator will eventually have four external proton beams as well as a number of exterior beams of secondary particles arising from interactions with targets placed in the internal beam. This number of external beams provides great flexibility in the utilization of machine time in that experimental apparatus may be set up while experiments are being run. From a logistic point of view, the number of bubble chambers that can be usefully utilized at the Argonne accelerator is relatively large. Since the experiments attempted will become more difficult as pointed out previously, the length of a run will become longer and it is imperative that a number of bubble chambers be available to experimenters. An illustration of the importance of a relatively large number of bubble chambers at an accelerator is the fact that even though there are three operating hydrogen bubble chambers at the Cosmotron, Shutt's chamber at Brookhaven is currently scheduled 18 months into the future. At the present time there is one moderately large hydrogen chamber ($\sim 40''$) and one small hydrogen chamber ($\sim 15''$) in the design or construction stage in the midwest and which are planned to be used at the Argonne accelerator.

We propose to design and build a hydrogen bubble chamber of intermediate size ($\sim 25"$) to be used with a fairly large magnetic field (~ 25 kilogauss). The usefulness of a chamber grows as either the square or the cube of the linear dimension of the chamber. The complexity of chamber construction and the associated magnet grows in a similar fashion. The dimensions (25" in diameter, 15" deep) of the proposed chamber represents a compromise between size and the capabilities of the group.

With the proposed chamber and magnetic field one should be able to make fairly precise ($\sim 10\%$) measurements on reaction products from primary particles of 4 or 5 Bev/c momentum. This is well above the energy range being investigated at the Cosmotron or the Bevatron at the present time.

The philosophy of the group is to do as little development work as possible on the components of the chamber and magnet. We intend to base the design on existing chambers with a view to make all operations of and on the chamber as simple as possible. Reliability and simplicity are the design goals. A secondary goal is to have an operating bubble chamber capable of doing significant physics in the shortest possible time (~ 2-$1/2$ to 3 years). We would then be well prepared when the A.N.L. accelerator becomes operative. Work on the project can begin immediately after authorization is given.

We would propose that MURA would retain prime interest in the chamber. This means that although it would initially be used at the Argonne it might be used eventually with a MURA accelerator. The chamber is to be viewed as midwestern community property. The Argonne National Laboratory has indicated interest and offered full cooperation in this project. (See exhibit A).

PROPOSED PROJECT

Design and construction of the bubble chamber and its associated magnet will be carried out at the MURA Laboratory in Madison, Wisconsin by physicists from MURA, Purdue University, the University of Wisconsin, and Michigan State University. The construction and testing period required is estimated to be 2-1/2 to 3 years. The cost of the project in addition to money available and budgeted through 1962 for the MURA Laboratory is estimated to be $370,000 which includes a 10% contingency on the chamber and magnet items.

I. Description of Chamber

The chamber body will consist of a stainless steel cylinder of inside dimensions of 27"or 28" in diameter and 15" deep with steel flanges for the support of the windows. The working liquid will be illuminated and photographed through two 4" thick tempered glass windows 28" in diameter clamped to the flanges of the chamber body. The 4" thickness of glass provides a safety factor of 10 for a pressure of 150 p.s.i.

Expansion of the chamber will probably be effected by a piston working directly on the liquid. The piston will be driven by a helium or nitrogen gas pressure system which will be closed, allowing recirculation of the gas and avoiding wastage.

The chamber temperature will be controlled by the pressure at which hydrogen is allowed to boil off from a reservoir in thermal contact with the chamber. The chamber will be surrounded by a copper radiation shield at liquid nitrogen temperatures and may have an additonal shield. The chamber and the radiation shields will be surrounded by a stainless steel vacuum tank designed to withstand the pressure rise which would result from a window failure.

In order to avoid the dangerous transfer of liquid hydrogen from Dewars to fill the chamber and to reduce the operating costs of the bubble chamber by a large amount, a hydrogen liquifier of 300 watts capacity will be built into the chamber.

II. Description of Magnet

The magnet consists of a pair of Helmholtz coils surrounded by iron to provide a return path for the flux. The volume over which the field must be provided is a cylinder 15" in radius and 31" long. To provide a magnetic field of 25 kilogauss over this volume requires approximately 25 tons of hollow copper conductor surrounded by 60 tons of iron. The power requirement is 5M watts. The estimated cost of the magnet including the magnet support is $150,000. We propose this rather expensive magnet in order to achieve simplicity in chamber illumination. The cost of our illumination system is thereby decreased and we believe that higher quality pictures will result. There is an additional safety feature of having a symmetric field which will stress the chamber less in case of power failure.

Personnel and Facilities

The duration of the construction and testing period of the bubble chamber is estimated to be 3 years. During this period the following people will devote a major portion of their effort to the project:

> J. Ballam, Prof. of Physics, Michigan State University
> G. W. Tautfest, Assoc. Prof. of Physics, Purdue University
> W. D. Walker, Assoc. Prof. of Physics, University of Wisconsin.

The following people will devote effort as needed and will be available for consultation:

> H. Blosser, Assoc. Prof. of Physics, Michigan State University
> U. Camerini, Assist. Prof. of Physics, University of Wisconsin
> A. R. Erwin, Assist. Prof. of Physics, University of Wisconsin
> H. R. Fechter, Assist. Prof. of Physics, Purdue University
> W. F. Fry, Prof. of Physics, University of Wisconsin
> Myron Good, Assoc. Prof. of Physics, University of Wisconsin
> F. J. Loeffler, Assist. Prof. of Physics, Purdue University
> T. R. Palfrey, Assoc. Prof. of Physics, Purdue University

R. R. Crittenden will be a Research Associate on the project with the expectation that additions will be made in this category when it appears desirable.

The supporting staff of engineers, draftsmen, machinists, and technicians required for the project will be made up from the MURA technical staff.

The detailed design of the chamber and magnet as well as the construction of the chamber electronics and optics will be carried on at the MURA Laboratory. It is anticipated that at the end of two years from the date of initiation of the project, assembly and testing of the chamber and magnet will be started at the University of Wisconsin High Energy Physics building which has been requested from the NSF with this use in view. If such a building is not available, the space would have to be rented. It is expected that final testing of the chamber at full magnet excitation will take place at A.N.L.

Estimated Material Budget *

Chamber

Chamber Body		$ 10K
Window Glass		10K
Vacuum Tank		15K
Expansion System		10K
Compressor (closed expansion He system)		5K
Tank (closed expansion He system)		3K
Expansion Control Plumbing		10K
Vacuum Pumps		7K
Hydrogen Liquifier		45K
	Total Chamber --------	$115K

Optics

Cameras		$ 20K
Lenses		2K
Lights		1K
Condensing Lens		10K
	Total Optics ----------	$ 33K

Magnet

Cu Coils 25 tons		$100K
Iron Yoke 60 tons		30K
Magnet Support		20K
	Total Magnet----------	$150K

Misc.

Control Electronics		$ 30K
Dewars		7K
	Total Misc. -----------	$ 37K

		$335K
	10% Contingency	35K
		$370K

Plumbing
2

Estimated Labor Included in MURA Budget

4 Physicist man years		$ 48K
5 Engineer 4 — 2 Engineer man years		20K
30 2 Draftsman man years		12K
10 Machinist man years		60K
8 Technician man years		48K
		$188K

1 Mech. Engineer
½ power Engineer   ⅓ Total Electric
1 Electronics

*All material and manpower figures were prepared with the assistance of Mr. Paul Hernandez, Chief Engineer of the 72" Bubble Chamber at the Lawrence Radiation Laboratory, and are based on a review of costs of existing chambers on which he worked and on which he graciously furnished cost and manpower data.

ARGONNE NATIONAL LABORATORY
Box 299 Lemont, Illinois

December 16, 1959

Professor William D. Walker
Physics Department
Sterling Hall
University of Wisconsin
Madison, Wisconsin

Dear Bill:

I was delighted to learn from our conversation this morning that your group is planning to build a medium sized bubble chamber for use at Argonne. This is exactly the sort of project we have wanted to stimulate by constructing the ZGS.

I think I was overly cautious in discussing the possibility of having an Argonne crew to operate the chamber. We will be glad to provide such an operating crew and to do anything else we can to help your project. It would probably be a good idea for an Argonne engineer to participate in the design of the chamber so that we will have someone thoroughly familiar with the apparatus when it comes time for us to do the operation. If you want to construct all or part of the chamber at Argonne we will be happy to provide our shop facilities and our cryogenic facilities.

I was pleased to note that you would invite others to use the chamber when it is operating provided that your group will have priority. I hope very much that you and other members of your group can visit us sometime between Christmas and New Year to discuss in detail the ways in which we can assist you. Please let me know when you can fix a definite date for a visit.

Best regards,

/s/   Roger

Roger H. Hildebrand

RHH:afk

Midwestern Universities Research Association
2203 University Avenue
Madison, Wisconsin

January 11, 1960

Dr. Paul McDaniel, Director of Research
U. S. Atomic Energy Commission
Germantown, Maryland

Dear Dr. McDaniel:

Submitted herewith is a proposal for the design, construction and test of an intermediate size bubble chamber. It is proposed that this chamber be used in conjunction with experiments on the Brookhaven AGS accelerator and the Argonne ZGS synchrotron as well as eventually being used with the MURA accelerator. The building of this chamber will complement the work in high energy physics being carried on in the mid-western community and will enhance the experience and skills of the MURA Technical Staff.

This project will go far toward establishing the close liaison between the MURA Laboratory and the high energy physicists which is so necessary for intelligent design and development of accelerators.

Very truly yours,

/s/ R. O. Rollefson

R. O. Rollefson
Director

ROR:rw

Enc.

70

EARLY STRONG INTERACTION COUNTER EXPERIMENTS

Kent M. Terwilliger
University of Michigan, Ann Arbor, MI   48109

ABSTRACT

The 17° beam and some π-p two body scattering experiments run in the beginning years of the ZGS are discussed.

INTRODUCTION

I would like in this talk to give you a bit of the flavor of the early counter experiments and their physics results. I will briefly describe the ZGS 17° beam line where the initial counter type scattering experiments took place, looking at the status of the line at machine turn-on, the equipment available, and the design of the beam for the counter program. This paper will then discuss a few of the early strong interaction counter experiments which turned out to be the start of major research areas at the ZGS. The experiments discussed will be limited to π-p two body scattering - most of the initial strong interaction experiments were of this type, and the kaon and proton work will be considered in other papers.

Two π-p elastic scattering experiments will be described. One, a systematic study of $\pi^\pm p$ angular distributions as a function of energy, and the second, $\pi^- p$ backward scattering, right at 180°, with a very detailed energy dependence. The elastic scattering field has proven one of the strongest ZGS areas. An early large magnet spectrometer experiment presented here, on associated production, $\pi^+ p \to K^+ \Sigma^+$, was the start of a major spectrometer program, and was a predecessor of the highly successful Effective Mass Spectrometer. Finally, two experiments will be described that began ZGS polarization physics: measurements of the asymmetry parameter in $\pi^- p$ elastic and charge exchange scattering using a polarized proton target.

17° BEAM

Experiments at the ZGS were first done using an internal target. Three main beams were planned for at turn-on: the 7° separated beam for the 30" MURA bubble chamber, the 17° beam for unseparated pion counter experiments, and the 30° beam for low momentum kaon physics. A layout for the initial stages of the three beams, which obviously coupled closely with the accelerator and shield wall, was worked out quite early, and gave the experimenters a well defined starting point for the detailed beam design required for their set ups. A schematic of the internal target region and planned initial stages of the three

ISSN: 0094-243X/80/600 070-12$1.50 Copyright 1980 American Institute of Physics

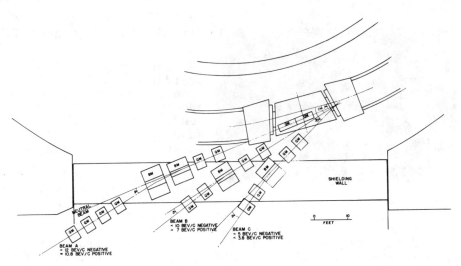

Fig. 1. Plans for internal target area and first stages of beam lines.

beams is shown in Fig. 1. Looking back over the ZGS user group meeting reports, I noted considerable discussion of the expected flexibility of the shield wall and beams - in fact some support posts originally in the design were removed to keep the area open for beam line changes. Once in position, however, the initial beam components became rather permanent - the basic three charged beam line arrangement remained for the life of the internal target, and only the 30° shield wall section had a major reworking.

The possible particle momenta for each beam, determined by the machine field and target position, were calculated by Larry Ratner and were readily available in the ZGS Users Handbook. Fig. 2 is a page from the Handbook showing the

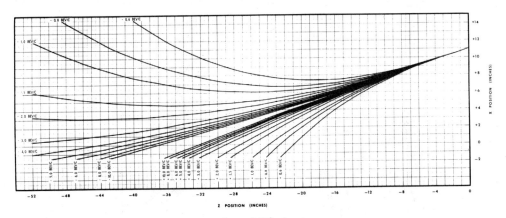

Fig. 2. Allowed momenta for 17° beam.

72

Fig. 3. Bending magnet BM105.

17° beam range. Also
available for initial
experiments was an
adequate supply of
bending magnets and
quadrupoles. Fig. 3
shows a standard ZGS
bending magnet, BM105,
being measured and
Fig. 4 its properties.
The magnets and quadruples were well measured and were
standard enough that experimenters could generally
take them as given units for a beam, and quite reliably
design the beam and predict its properties. From the
experimenters' sketches the ZGS staff would make layout

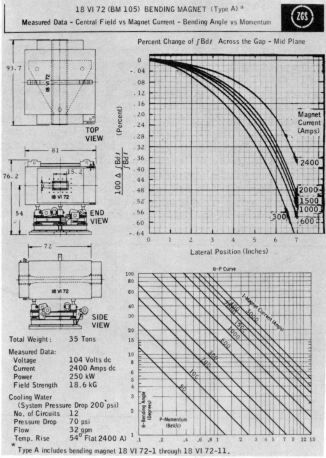

Fig. 4. BM105 properties.

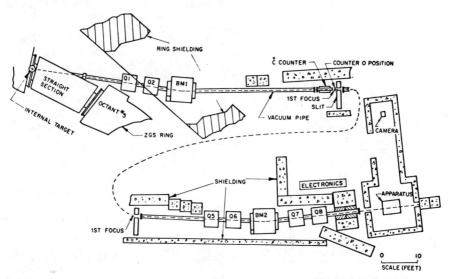

Fig. 5. Layout of 17° beam and set up of optical spark chamber elastic scattering experiment.

drawings and position the magnets and shielding. An experimenter could rely on top quality professional help right at the start of the ZGS. For the 17° beam Wayne Nestander's layout drawings with small specified positioning errors materialized on the floor within specs. The 17° beam we worked up was quite simple - a two stage beam (Fig. 5) with ∿ 1% momentum resolution, about 2→6 GeV/c, delivering up to ∿ 3 x 10⁵ π⁻ per 5 x 10¹¹ accelerated protons; the π⁺ intensity, with non-0° production, was much lower.

EXPERIMENTS

ELASTIC πp SCATTERING

$\pi^{\pm}p$ Elastic Scattering 2.3→6 GeV/c (E-3)

This Michigan experiment[1] used optical spark chambers and was designed to be as simple as possible for the initial turn on - measuring elastic scattering in the horizontal plane. The general layout is shown in Fig. 5. A top view of the spark chamber and a photo of spark chamber tracks of an event are presented in Fig. 6. The experiment was just a bit too simple it turned out - four gap chambers are pretty minimal. We also had just two views of each chamber at 90° which made it hard to handle more than one track, limiting the data rate. Also some anticoincidence counters could have suppressed

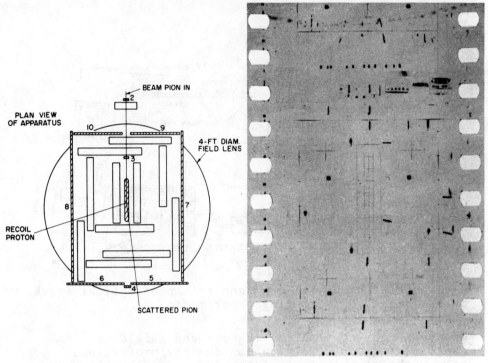

Fig. 6. Left, plan view of spark chambers, right, photo of an event, elastic $\pi^{\pm}p$ scattering experiment.

many of the inelastics. So, we learned. But since the physics was simple, too, we got reasonable data which rather surprisingly had not yet been obtained elsewhere. Our results, in Fig. 7 and 8, show that the structure previously observed at lower energy in π^-p elastic scattering was in fact quite a systematic phenomenon, having similarities to the structure observed in charge exchange scattering.

π^-p 180° Elastic Scattering, 1.6 to 5.3 GeV/c. (E-30)

Another elastic scattering experiment, done the following year in the same 17° beam line by Alan Krisch and his group,[2] is probably one of the most often cited experiments from the ZGS - because the group made a set of precision measurements in a region where the theorists found their models could do a lot of predicting: exactly backwards π^-p elastic scattering. This experiment, like the subsequent experiments of Krisch and his group, is conceptually very simple - designed with what looks like more than enough constraints-

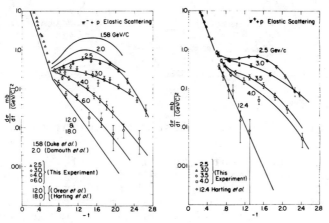

Fig. 7.   Energy dependence of angular distributions.

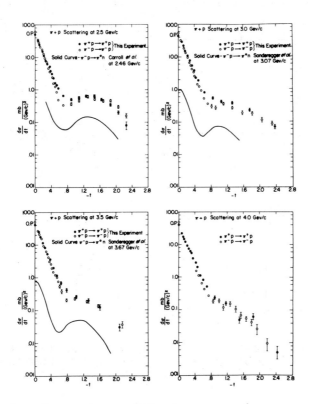

Fig. 8.   Comparison with charge exchange scattering.

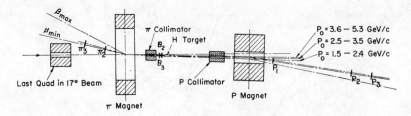

Fig. 9. Layout for backward elastic scattering experiment

an overdesign approach that generally pays off, since
there are usually problems one hasn't counted on. The
group's set up is shown in Fig. 9. With the magnets left
stationary they were able to do measurements over a wide
range of momenta by just changing fields and moving counters-
a style of operation the group has continued to use. I
remember looking at the experiment for the first time -
I was somewhat startled by the number of lines on the floor-
there was one line for each point, and there were clearly
a lot of points being planned. They got their points
all right, as seen in Fig. 10. The very rich structure
showed strong resonance effects-particularly at the

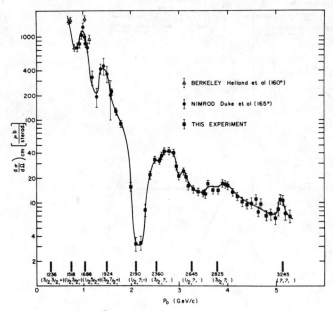

Fig. 10. 180° π⁻p elastic scattering cross section as a
function of π⁻ laboratory momentum. Resonance positions
are indicated.

center of mass energy of 2190 MeV. The destructive or constructive interferences with background enabled many quite clear parity assignments. The couple of standard deviation peak at 3245 MeV hasn't been seen by others - though nobody else has looked at exactly 180° with better statistics.

These two experiments were the first in a series of very productive elastic scattering experiments involving a variety of groups - it has been a strong program.

SPECTROMETER EXPERIMENT

Very early, well before ZGS turn on, Don Meyer was pushing for the construction of a general purpose spectrometer magnet for spark chamber experiments.[3] The final resulting magnet, SCM105, has turned out to be extremely useful (as in the previous described experiment). A schematic of the magnet, from the ZGS Users Handbook, is shown in Fig. 11. This magnet was used by Meyer and his collaborators for a very productive series of spark chamber experiments. I consider that the Effective Mass Spectrometer was at least in part an extension of

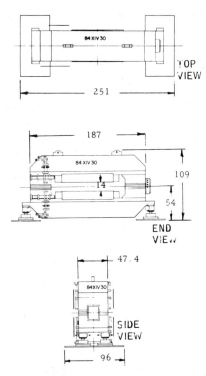

Fig. 11. Dimensions of the SCM105, a general purpose spectrometer magnet.

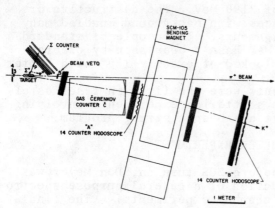

Fig. 12. Plan view of the wire spark chamber experiment $\pi^+ p \to K^+ \Sigma^+$.

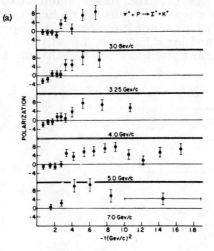

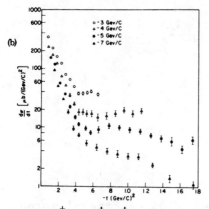

Fig. 13. $\pi^+ + p \to K^+ + \Sigma^+$ experiment.
(a) Polarization of the Σ^+.
(b) Differential cross section.

their experimental approach.

$\pi^+ + p \to K^+ + \Sigma^+$, 3 to 7 GeV/c (E147)

Their initial experiment[4] was in the External Proton Beam. Their layout is shown in Fig. 12. The forward K^+ momentum and angle determined the missing mass; the events in the Σ^+ peak were further selected by looking at the proton from the Σ^+ decay. The up-down asymmetry of the decay protons enabled the determination of the Σ^+ polarization. Their results are shown in Fig. 13. They saw a dramatic change in the slope of the differential cross section going from the diffraction region to $-t \geq .4$ (GeV/c)^2, and an apparently associated increase in Σ^+ polarization. This cross section and polarization behavior is not unlike that seen in pp scattering; a particularly interesting result for those looking at spin effects.

POLARIZED TARGET EXPERIMENTS

The study of polarization physics at the ZGS was started by Aki Yokosawa and his collaborators.

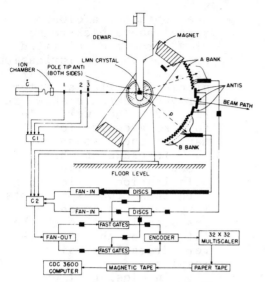

Fig. 14. Schematic of apparatus for π⁻p elastic scattering polarization experiment.

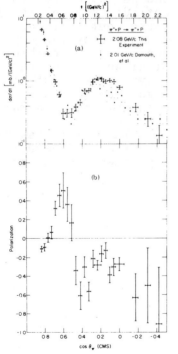

Fig. 15. π⁻p elastic scattering 2.07 GeV/c. a) Differential cross section. b) Asymmetry parameter.

Their first two polarized target experiments, carried out in the 17° beam, are discussed here.

Polarization in π⁻p Elastic Scattering, 1.7 to 2.5 GeV/c (E17)

The setup for their π⁻p elastic experiment[5] is shown in Fig. 14. In these first experiments they had to cope with a large background from the LMN polarized target (97% non-hydrogen)- a 50% background subtraction was required in this experiment. Their target polarization was ∿ 50%. Even with these limitations they obtained quite good data, as seen in Fig. 15. Data of this type over a range of energies enabled the group to do a phase shift analysis, with results shown in Fig. 16. As indicated, they were able to establish the spin and parity of the resonance at 2.07 GeV/c (2190 MeV) to be 7/2⁻. Clearly polarization was a powerful tool.

Polarization in π⁻p→π°n, 2→5 GeV/c (E 111)

Their second polarized target experiment, polarization in charge exchange scattering,[6] required a considerable increase in complexity, as evident in their layout in Fig. 17. With neutron time of flight information

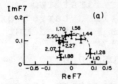

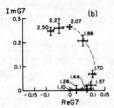

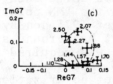

FIG. 16. The partial-wave amplitudes (a) A_{3+} and (b) A_{4-} obtained by starting from lower energy solutions and using a continuity condition. (c) The partial-wave amplitude A_{4-} obtained by starting from an optical mode.

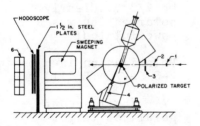

FIG. 17. Side elevation of the experimental arrangement. (1)-(4) are plastic scintillation counters with respective dimensions of 1 in. × 1 in., 1 in. × 1 in., 10 in. × 10 in. with a 1-in. × 1-in. hole, and 3.5 in. × 8 in. The lead-Lucite Cherenkov counter (6) consists of ten modules, each 8 in. × 16 in. × 4 in. Antico-incidence counters (7) protect the 52 recoil-neutron counters (8).

they obtained a clear signal above background of π° from the polarized proton. Their polarization data and fits to $J=(\ell\pm1/2)$ predictions from an interference model are shown in Fig. 18. They were able to make a choice for many of the resonances.

Yokosawa and his group initiated the ZGS polarization physics work with these two experiments. They have continued this research line through the lifetime of the ZGS with very dramatic physics results.

In closing, I want to say that Roger Hildebrand's vision and determination to have the ZGS wide open to outside users obviously worked out well, as indicated by the experiments I've described: Michigan, one of the universities most heavily involved with MURA, has been able to be a major ZGS user.

This work has been supported by the U.S. Department of Energy.

REFERENCES

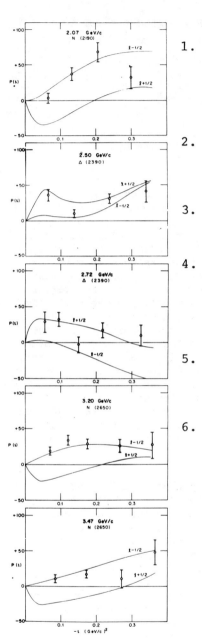

1. C.T. Coffin, N. Dikmen, L. Ettlinger, D. Meyer, A. Saulys, K. Terwilliger, and D. Williams, Phys. Rev. Lett. 15, 838 (1965); Ibid 17, 458 (1966); Phys. Rev. 159, 1169 (1967).

2. S.W. Kormanyos, A.D. Krisch, J.R. O'Fallon, K. Ruddick, and L.G. Ratner, Phys. Rev. Lett. 16, 709 (1966); Phys. Rev. 164, 1661 (1967).

3. D.I. Meyer, Argonne Accelerator Users Group Meeting, Argonne National Laboratory, Nov. 1962, p. 17 (unpublished).

4. S.M. Pruss, C.W. Akerlof, D.I. Meyer, S.P. Ying, J. Lales, R.A. Lundy, D.R. Rust, C.E.W. Ward, and D.D. Yovanovitch, Phys. Rev. Lett. 23, 189 (1969); Phys. Lett. 30B, 289, (1969).

5. S. Suwa, A. Yokosawa, N. Booth, R. Esterling, R. Hill, Phys. Rev. Lett. 15, 560 (1965); Phys. Rev. Lett. 16, 714 (1966); Phys. Rev. D1, 729 (1970).

6. D.D. Drobnis, J. Lales, R.C. Lamb, A. Moretti, R.C. Niemann, T.B. Novey, J. Simanton, A. Yokosawa, and D.D. Yovanovitch, Phys. Rev. Lett. 20, 274 (1968); Phys. Rev. Lett. 20, 353 (1968).

Fig. 18. Polarization in $\pi^- p \to \pi^0 n$ and fits to $J = (\ell \pm 1/2)$ predictions.

WHAT IS HIGH ENERGY PHYSICS ALL ABOUT?

Kameshwar C. Wali

Institut des Hautes Etudes Scientifiques,
Bures-sur-Yvette, France

and

Physics Department, Syracuse University
Syracuse, New York

I have the uneasy feeling that after all the exciting episodes
we have heard so far dealing with the history of the ZGS, my discussion will be somewhat of an anticlimax. I had no role of any
significance in the conception of the ZGS or its turning on; nor in
formulating its scientific programmes and policies. Hence I have no
nostalgic happenings to tell you, except those associated with the
"Happy Hour" every Thursday - those better wait for the happy hour
to come later this evening. For the sake of the record however, I
should note that the ZGS was responsible not only for the beginning
of   excellent high energy experimental research in the midwest,
but also for the establishment of a high energy theory group whose
size and stature has grown steadily. With the starting of the ZGS
and the growth of the experimental program, Argonne became a central
place for many theorists. Such activities as the summer visitors
program, one day topical symposia, lecture series etc., were instituted and they created an exciting and stimulating atmosphere.
I was indeed fortunate to have had the opportunity of spending seven
crucial years of my career, here at Argonne, during that period.

I

Now the question "What is high energy physics all about?", re-
ceives generally the answer that it is the physics of the fundamental
building blocks of all matter and their interactions. As such, not
only is it an integral part of all physics, but also its most exciting and dominating frontier field. The subtle interplay between
theory and experiment has led to many dramatic discoveries in both
theory and experiment throughout its history. Antiparticles,
neutrinos, pi-mesons are but a few examples of theoretical predictions, whereas neutrons, mu-mesons, strange particles, multitude
of resonances are likewise a few examples of experimental discoveries that belong to the recent past. More recently we have the
examples of quantum numbers like charm and beauty and particles like
J/ψ and T (upsilon) which live a thousand or so times longer than expected from the known properties of strong interactions.

As the frontier field in search of the fundamental building
ISSN:0094-243X/80/600082-08$1.50 Copyright 1980 American Institute of Physics

blocks and the laws that govern their binding, high energy physics has evolved through many "energy" phases. From the turn of the century to the thirties, it was the physics of the atom dealing with phenomena of a few electron volts (ev); from the thirties to the fifties, it was the physics of the nucleus in the million electron volt (Mev) region. Then onwards it has been the physics of the sub-nuclear particles and phenomena in billions of electron volts (Gev) range. Today we dare in high energy physics to contemplate theories and experiments in the 10^{15} Bev range! The pace at which the field progresses, the complexity and the ingenuity of the experimental techniques, the bigness (both in money and manpower) of the exper-imental technology is baffling and will continue to baffle not only the "outsider" but sometimes even the "insider". Think of the complexity of our present day experiments and the laboratory set ups that host them. Compared to these, the classic experiments of Thomson, Rutherford, and Millikan look like kindergarten experiments, which indeed they were, since they taught us the alphabets of the atomic and the nuclear world.

But we have come a long way from the days of the alphabets in our explorations of the atomic, nuclear and subnuclear worlds. Using higher and higher energy beams, we explore more and more of the interior of the subnuclear particles. Using different probes, we examine their different interactions. One may ask whether colliding one particle with another as violently as possible is a sensible way to know how they are made. ("To know how a swiss watch works, bang two of them and see what comes out", Feynman is supposed to have said about high energy physics). But there is no doubt we have made a great deal of progress this way. How else could we have found out, for example, that the subnuclear particles like the protons, neutrons, pi-mesons, etc., are not fundamental or elementary, but composites of simpler constituents? In what follows, I would like to describe qualitatively this simpler sub-subnuclear structure that we have currently come to believe.

II

The majority of high energy physicists believe today that we have come to the final stages in our search for the ultimate con-stituents. The ultimate constituents are quarks and leptons which behave as pointlike objects with no internal substructure. The quarks are the constituents of the ordinary, everyday hadrons like the pi- and K-mesons, protons and neutrons, and various mesonic and baryonic resonances. They come in three _colors_ and at least five different _flavors_. A sixth flavor is expected to be discovered any day and announced in a fashion that has become customary - by calling a press conference. I should remind you, when we talk about color and flavor of these objects, we do not mean color and flavor in their every day sense of the terms. They are just convenient substitutes for a more formal mathematical description of the degrees of free-dom we call internal, which we need to postulate in order to explain

some of their properties. For example, the color degree of freedom obeys the exact laws of an SU(3) symmetry group; the observed
hadrons are colorless combinations (SU(3) singlets) of the quarks.
We do not yet know what, if any, symmetry laws govern the flavor
degree of freedom. The leptons, like the quarks, come in a variety
of flavors; but unlike them, exhibit no color. The presently known
flavors of quarks are commonly referred to as up, down, strange,
charm, and bottom (or beauty). The leptonic flavors are the familiar
electrons, mu-mesons along with their distinct neutrinos and the
very recently discovered heavier lepton - the τ-meson which is most
probably accompanied by its own distinct neutrino like the others.
There appears to be a close correspondence between the quark and
the leptonic flavors:

QUARKS: u   d   s   c   b   t(?)
         (up)   (down)   (strange)(charm)(bottom or (top or
                                            beauty)   truth)

LEPTONS: ν_e e^- μ^- ν_μ τ^- ν_τ
         (electron (electron) (mu-   (mu-   (tau-   (tau-
         neutrino)   meson) neutrino) lepton) neutrino)

The quarks participate in all the known elementary particle interactions whereas the leptons take part only in the weak and electromagnetic interactions being immune to the strong interactions. The
quanta that mediate the electromagnetic and weak interactions are
the familiar photon and a set of particles ($W^\pm, Z^0$) known as the intermediate vector bosons yet to be discovered. The quanta of strong
interactions are the so-called gluons. Finally to complete the list
of the fundamental entities, I should also mention the necessity of
the existence of at least one scalar particle called the Higgs meson.

III

Until recently one classified the elementary particle interactions into four main categories:

1) Gravitational interactions, which are all pervading, but of importance only in the domain of the large. They are responsible for
the large scale organization of matter into planets, stars, galaxies,
and so on. We demonstrate the irrelevance of these interactions in
the microscopic domain, by usually calculating and comparing the
electric and gravitational forces between, say an electron and a
proton at some distance apart, and convincing our elementary physics
students that the gravitational interactions are indeed negligibly
small and proceed to dismiss their further consideration in elementary
particle physics. We also point out to our students that this is very

ironical - the gravitational interactions have such a beautiful and elegant theory due to Einstein, and that this unique theory should stand apart and be of no relevance to our understanding of the structure of matter is such a great pity.

2) Electromagnetic interactions, which are responsible for the binding in the atoms, binding of atoms in the molecules, binding molecules and molecules to form complex inorganic and organic compounds and hence responsible for life itself as we know. They govern the domain of every day experience - "the middle ground" as Dyson likes to characterize it, which includes "atoms and electricity; light and sound: gases, liquids and solids: chairs, tables, and people". The theory that describes these interactions at the basic level is the quantum-electrodynamics. It is not as elegant as Einstein's theory of gravitation, but it is a remarkably successful theory in providing unambiguous answers with any desired degree of accuracy. Unfortunately all its successes are based on a perturbative approach which works because the fine structure constant which characterizes the strength of the electromagnetic interaction is indeed small. Also infinities occur in the calculation of physical quantities, infinities which have to be carefully identified and removed by what is called a renormalization procedure. Within the framework of such a theory, fundamental quantities like the masses and charges are to be taken as given empirical parameters with no hope of ever predicting them. So, although one would be quite happy with a theory like the quantum-electrodynamics for other interactions, one hopes eventually for a better theory.

3) Weak interactions, which were discovered in the radioactive β-decay of nuclei. They are responsible for the instability of the neutron and similar instabilities of other elementary particles. They play an important role in nuclear reactions and in the generation of energy by burning nuclear fuel in the interior of the stars. It was the weak interactions that led Pauli to postulate the neutrino - the most perplexing and the most elusive of all the elementary particles. It is in the weak interactions that we had to give up some of our cherished conservation laws, such as parity and charge conjugation combined with parity. A phenomenological model due to Fermi based on Pauli's neutrino was quite successful in explaining the low energy phenomena associated with these interactions. Although the model was based on the analogy with electromagnetic interactions, a theory like the quantum-electrodynamics could not be constructed for the weak interactions. The reason for this is that the relevant quanta (intermediate vector bosons) have to be electrically charged and massive unlike the photon which is electrically neutral and has zero mass. Any attempt to calculate higher order effects ran into grave troubles and the situation appeared hopeless till recently.

4) Strong interactions, which are responsible for the forces that

lead to stable nuclei. They have nothing to do with the electrical or weak interaction forces. It is in the domain of these interactions that one needed new concepts; the usual conservation laws such as energy—momentum and angular momentum stemming from the relativistic invariance of the interactions were not sufficient. Isotopic spin, strangeness, charm and so on had to be introduced.

One of the main characteristics of the strong interactions was the proliferation of the so-called elementary particles, namely the baryonic and mesonic resonances (which incidentally led to the expansion of Rosenfeld's wallet card to wallet book industry). This "population" explosion of the resonances made the concept of elementary particle very difficult. Which is elementary? which is composite? For a while one toyed with the idea of nuclear democracy - all are equally elementary or all are equally composite. Then higher and higher symmetries, Regge trajectories, dual models and so on. In the last two decades, the strong interaction regime has seen the birth and death (or fading away) of many theoretical breakthroughs,each claiming that a fundamental understanding was just around the corner. It did not turn out that way. Each such attempt provided only a fragmentary and only a partially satisfactory account of some observed aspects, and usually it provided only a new jargon and some new calculational tools. With so many particles or resonances around, the situation was not very different from the one around the turn of the century. The idea that the diverse observable properties of matter can be explained in terms of a few ultimate building blocks, an idea at least two thousand years old, had led to the atomic hypothesis. The invisible atoms explained successfully many laws of chemistry; they were responsible for the development of the kinetic theory of gases and were the unseen agents behind the Brownian motion. Yet the atomic model was too complicated - a closer study revealed too many different types of atoms with different chemical and physical properties, different emission and absorption spectra and so on. The idea that the atoms of different elements cannot all be different, and that they should have a common substructure became indispensable. That indeed it was so was established by Thomson by splitting the indivisible atoms and showing that they were all made from two constituents, the electrons and the nuclei.

IV

The quark model, like the atomic model, arose out of the multiplicity of the elementary particles, the resonances of the fifties and the sixties. Quarks were supposed to be the basic constituents of all the hadronic states. A fundamental triplet of quarks and an anti-triplet of anti-quarks that comes naturally with it were sufficient to explain many features of hadron spectroscopy based on Gell-Mann's eight fold way (SU(3) symmetry). The baryons and their excited states were made from three quarks, the mesons from quark-anti-quark pairs. Simple non-relativistic considerations involving these combinations in different angular momentum states went a long way in classifying a large number of states, in predicting new states and correlating their decay rates. However, such studies faced a basic difficulty. If the quarks are fermions, three identical quarks cannot exist in the lowest angular momentum state, at the same time belonging to certain SU(3) representations (e.g. 8, 10) without violating the Pauli's exclusion principle. But such states did certainly exist. To overcome this difficulty, the idea of color was introduced. Instead of one basic triplet, three basic triplets each with a different color degree of freedom (known popularly as red, white and blue) were introduced. The requirement that the observed hadrons be "colorless" combinations removed the difficulty. Then came the 1974 November revolution — the discovery of J/ψ particle followed by the discovery of "upsilon" (T) a few years later. These discoveries required still newer degrees of freedom. "Charm" and "bottom" (or beauty) are the currently used terms to describe these new degrees of freedom. While "charm" was anticipated to explain the absence of some decays of K-mesons, neither the "beauty" (or bottom) nor the additional lepton known as the τ-meson had compelling reasons for their existence. As I said before, the "flavor" space is the least understood, beginning from the simple question: "Is the number of such flavors of quarks finite of infinite?"

The idea that quarks were the constituents of hadrons received support from another type of experiments — deep inelastic scattering of electrons from protons. In these experiments, we probe the deep interior of the protons by electrons which interact only with the electrically charged constituents. It soon became evident that to explain the surprising experimental results, one had to imagine the proton to be made up of point-like constituents which were initially called partons. But evidence started building up which enabled the partons of the deep inelastic scattering to be identified with the quarks of the hadron spectroscopy. Today the terms partons and quarks can be used interchangeably.

V

If subnuclear particles have a substructure, our ideas about their interactions have to change. The observed interactions of the hadrons have to be derived from the interactions between quarks and quarks, quarks and leptons and so on. In the last ten years, a new theory of electromagnetic and weak interactions based on a unified description of these interactions has emerged. This new theory is mainly due to Salam and Weinberg, although many other physicists have contributed in an important way to the various ideas that have gone into synthesizing this theory. The unification provided by this theory is comparable in many respects to the synthesis of light, electricity and magnetism due to Maxwell. For example, Maxwell added the "displacement" current in empty space to the physical currents and charges in order to construct a consistent field theory

that embodied Faraday's and Ampère's experimental observations.
This modification introduced a conserved current which implied
charge conservation. It led to the prediction of the electromagnetic waves in empty space, the velocity of propagation of
which was exactly equal to the velocity of light waves. In the
same way, the most important element of the new theory is the
addition of a neutral current related in a specific way to the
charged currents of the weak interactions. The theory is a gauge
theory like the ordinary electromagnetic theory, but the combined
weak and electromagnetic charges and currents are described by a
non-Abelian Gauge theory based on the group $SU(2) \otimes U(1)$. The
helicity states of quarks and leptons are assigned to definite representations of $SU(2)$; they, therefore, form multiplets (weak
isospin multiplets) and carry a hypercharge quantum number due to
$U(1)$ (weak hypercharge). The quanta of the unified field are four
massless vector bosons, three of which acquire mass by a spontaneous symmetry breaking mechanism while the fourth remains massless
to be identified with the photon. The spontaneous symmetry
breaking known as the Higgs mechanism is accomplished by a scalar
doublet which is described by four real fields. Three of the scalar
fields combine with three of the transverse vector fields to make
three massive vector bosons. One scalar field remains to represent
a true Higgs meson. This intricate way of making the vector bosons
massive preserves the renormalizability of the starting zero mass
theory. The difficulties of constructing a weak interaction theory
along the lines of quantum electrodynamics are resolved and we
might say, we have a quantum electro/weak dynamics. Physically it
means that the electromagnetic and weak interactions are delicately
linked in cooperating with each other to cancel the unwanted infinities. At certain high energies, the distinction between them
would disappear and we will witness new phenomena. At low energies,
the theory so far has withstood all the tests and there does not
appear to be any doubt concerning its validity.

Similarly the picture of strong interactions has to be modified.
If quarks are the fundamental constituents, what we used to call
strong interactions cannot be the primary interactions. They should
be looked upon as secondary or residual interactions derivable from
more basic primitive interactions between quarks, much the same
way the Van der Waal's forces between neutral atoms are derived from
the basic Coulomb forces. Indeed a very good candidate for such
a theory has emerged recently; it is the quantum chromodynamics,
which is a gauge theory like the quantum electrodynamics, but more
complicated in that it is a non-Abelian Yang-Mills-type gauge
theory. The quanta are eight photon-like colored "gluons" which
are responsible for the strong binding of quarks into "colorless"
hadrons. The non-Abelian nature of the theory predicts that at
high energies, the quarks behave like free point-like constituents,
consistent with the parton picture in deep inelastic scattering.
At the same time, at low energies, the theory is expected to provide strong binding so that the quarks are eternally confined in-

side the hadrons. These two seemingly opposite pictures are
referred to as "asymptotic freedom" at short distances and "infra-
red slavery" at large distances.

VI

Thus we have a remarkably successful gauge theory to describe
electromagnetic and weak interactions in a unified way. We have a
very promising candidate for strong interactions in quantum chromodynamics which is also a gauge theory. Ways of unifying electro/
weak interactions and strong interactions are being seriously considered. Such a grand unified theory will be described by one
gauge group characterized by a single coupling constant. Quarks
and leptons will be even on a more equal footing. As a consequence,
protons may not be absolutely stable! Looking even beyond, one
might look for a grand, grand unified theory which includes gravitational interactions as well.

VII

This has been indeed a very brief and sketchy survey of the
present day high energy physics. There are many aspects of the
field that I have not touched upon. However, let me conclude by
posing the question "are we close to a final theory of structure
of matter?", and making a few comments on it. The answer to the
question is probably "No". First of all there are too many basic
fields; at least five, possibly six flavors of quarks, each flavor
with three different colors; six leptons. The internal degrees of
freedom like color and flavor have neither a space-time nor a
dynamical origin. They have to be assumed at the outset, which
suggests a possible further substructure for the quarks. Already
several such substructures (Prequarks, Maons, Primons, Rishons,
Quips) have been suggested in the literature. Besides, in the
present theories, quark and lepton masses, various coupling constants, mixing angles and phases are arbitrary parameters. A good
deal of experimental input and theoretical approximations and guess-
works go into their determinations, which makes it difficult to
distinguish the input assumptions from the predictions of the theory.
To put it another way, more often than not, there are no precise
quantitative tests by means of which one can unambiguously establish
a given theory. Our present theories of structure of matter may be
grandiose, abstract, complicated, and crazy but "not crazy enough",
as Pauli is alleged to have said about Heisenberg's unified theory.
They lack the type of basic simplicity which characterizes Einstein's
Theory of Gravitation. Somehow looking at our theories, one does
not have the feeling expressed by Herman Weyl about General Relativity: "It is as if a wall which separated us from the truth has
collapsed. Wider expanses and greater depths are now exposed to the
searching eye of knowledge, regions of which we had not even a
presentiment. It has brought us much nearer to grasping the plan
that underlies all physical happening".

THE EXTERNAL PROTON BEAMS AND THE
PROTON-PROTON EXPERIMENTS

Lazarus G. Ratner

Argonne National Laboratory, Argonne, Ill. 60439

The ZGS was one of the first accelerators in which extracted primary proton beams and experimental halls were part of the initial design. Special features incorporated into the ring made it possible to permanently position extraction magnets, and there was no need to plunge them as in other machines. Figure 1 shows schematically the displacement of the injected orbits, which is produced by the use of tuning magnets in each of the four straight sections. The high-energy orbits are radially tilted compared to the more strongly bent low-energy orbits. This allowed the placement of the permanently positioned extraction magnet to bend the protons out of the machine without using up injection aperture.

From the first thoughts on experimental facilities, the use of sequential foci along the beam lines, as well as beam splitting, was envisioned. The original design called for three, and possibly four, external areas. Financial considerations dictated that only one of these would be built initially. This was the External Proton Beam I (EPB I). Later, with the funding of the 12-foot bubble chamber, EPB II was built. Figure 2 shows EPB I and EPB II. One can see also the channel that was left for a branch beam in EPB I, and there also was the possibility of a branch beam being built in EPB II. For some time during the heyday of High Energy Physics (HEP) funding, plans were made for the use of the EPB I branch to produce a beam for the MIT 500-ℓ bubble chamber. Although considerable effort was expended on this project, it disappeared, along with the big budgets for HEP.

There were several extraction processes used for the EPBs, and the first one turned out to be unexpected. When targeting first started for the secondary beams in the meson area, we discovered that we could extract a "parasitic" beam to the EPB from the production target without too much loss in intensity for the meson area. Early tuning for E-1 was done with this beam. In April 1964, we got the first beam out of the machine using the originally-designed energy loss (Piccioni) target. A fast spill was later obtained by use of the beam bumper magnet and target scheme devised by J. H. Martin, T. Novey, S. Suwa, and A. Yokosowa. Later on, we did resonant extraction, initially on $\nu_x=1$, and then on $\nu_x=2/3$. Typically, targeted extraction efficiencies were about 25% to 30%, and resonant and fast bump extraction about a factor of two more. There was always a nagging uncertainty in these numbers, and when loss monitors in the ring were used as a criterion, it appeared that these efficiencies were low by about a factor of 1.3. We were never absolutely certain about this, but it may be that we did achieve extraction efficiencies as high as 80% in some modes.

Another important feature of operation in the EPBs was the ability of the ZGS to extract at any energy on a front porch for a

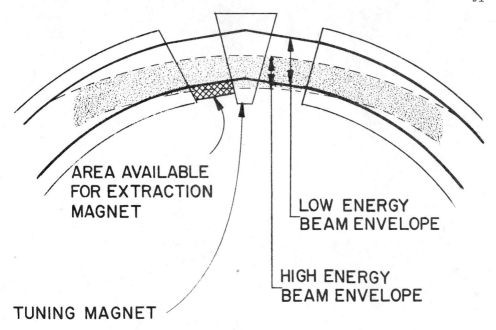

Fig. 1.   Injected Orbit Displacement

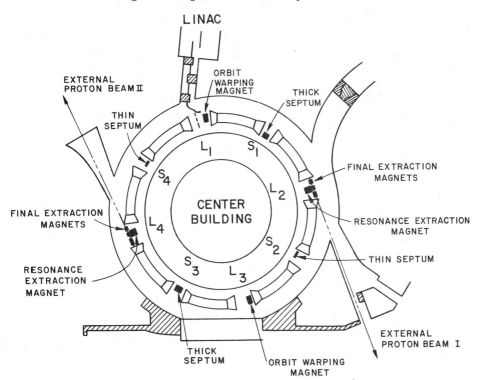

Fig. 2.   ZGS and Extracted Beams

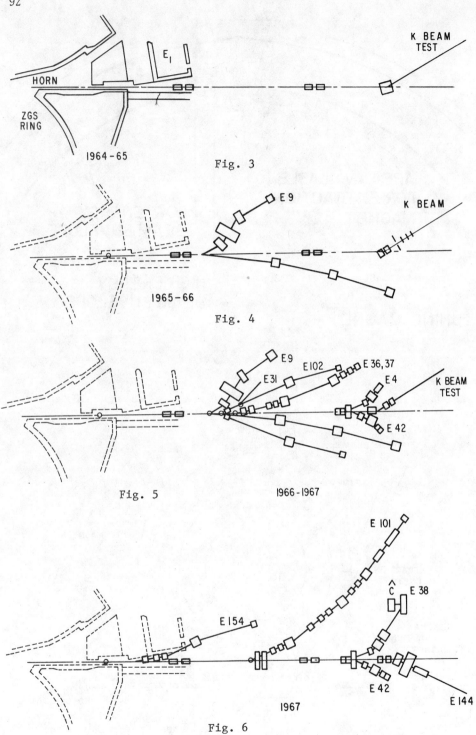

Figs. 3, 4, 5, 6. Changes in Beam Lines in EPB I from 1964 to 1967

given set of experiments and to carry the rest of the beam to full
energy for use in other areas. With the coming of the polarized
beam, the extraction system was made even more versatile by putting
in programmable magnets in the transport lines. This enabled spills
at different energies to be done on each ZGS pulse. During its
lifetime, the ZGS extracted beams at over 60 different momenta from
800 MeV/c to 13.4 GeV/c. Also, for a two month period in the twi-
light of its life, the ZGS produced polarized deuteron beams, which
were extracted at several momenta from about 2 GeV/c to 12 GeV/c.

So, in the beginning, there was beam and the first configuration of EPB I as it was in 1964, as shown in Fig. 3. The first
physics activity here was by the neutrino group, who measured the
production of $K^{\pm}$ and $\pi^{\pm}$ (E-23) in order to determine the neutrino
spectrum. Within the next year, there was the addition of experiment E-9 and a stopping K beam, as shown in Fig. 4. Figure 5 shows
the changes occurring in 1966 when E-102, E-36-37, E-42, and E-21
were installed. In Fig. 6, we see E-154 replacing E-9 and E-101
replacing E-102 and E-36-37. Also shown are E-144 and E-38, which
replaced E-21. Figure 7 shows 1968's layout with E-147 and E-125
replacing E-154, and the addition of E-182 and a test beam. Also,
a septum magnet was installed as a beam splitter. As you can see,
EPB I had good usage, and there was a long list of experiments waiting their turn.

On February 12, 1969, the installation of EPB II was completed
and the experimental program started there in March. EPB II also
had its share of experiments and again added to the complexity of
tuning and running many simultaneous experiments in two separate
areas. A development of prime importance in this regard, was the
addition of a 48-channel integrator and electronic scanner for the
Segmented Wire Ion Chamber (SWICS). This allowed quantitative, normalized beam profiles at many stations to be seen on each ZGS pulse.
This was connected to the CDC-924-A computer in the Main Control
Room (MCR) for on-line monitoring of beam spot size and position.
With this information and with the use of a program for the 924-A,
one could calculate emittances and do beam transport calculations to
aid in tuning up the beam lines. The computer also generated a set
of data giving the position and FWHM size of the beam in both planes.
This digital information was transmitted to the experimenters, as
well as to the beam tuners in the MCR. The SWIC capability allowed
for efficient tuning of the multi-foci beam lines, as well as giving
the experimenters necessary information on spot size and position.

In 1971, an addition was added to EPB I -- the Annex -- and by
October the area looked like Fig. 8. We still had all the experiments in EPB I but, in addition, we had several more operating in the
Annex. Perhaps we can get a better overall view of the area by considering the following moving picture. We follow the proton beam as it
starts coming out of the machine in the L-2 area, passes through the
channel in the shield wall, and enters the multi-beam area of EPB I.
It then passes through the open area and into the Annex and finally
ends as the last protons end up in a beam stop. Needless to say,
EPB II looks similar and had the 12-foot bubble chamber as its final
component. Figure 9 shows a layout of the ZGS areas and the complex

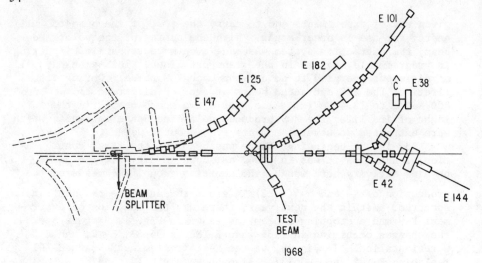

Fig. 7. Beam Lines

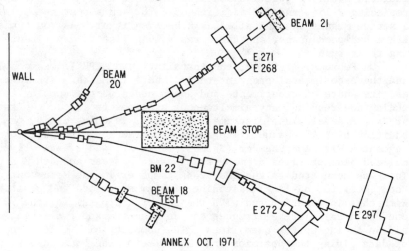

Fig. 8. The EPB I Annex Lines

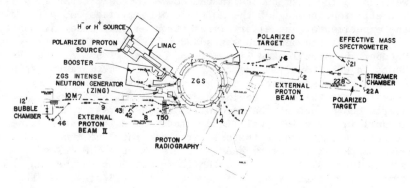

Fig. 9. Overview of ZGS Showing EPB I, EPB II Beam Lines
and the 12' Bubble Chamber

of beams can be seen in EPB II with beams 9 and 10 feeding the 12-foot bubble chamber. Well, these are the EPBs, and we'll now take a short look at what went on there.

There were many experiments and many well-known experimenters. Experiments completed in the first few years were done by such people as T. Novey, R. Lundy, D. Jovanovic, A. Krisch, A. Wattenberg, R. Schluter, T. Romanowski, A. Roberts, N. Booth, R. Heinz, V. Telegdi, D. Meyer, E. Beier, E. Steinberg, H. Neal, L. Pondrom, to name only spokesmen. The proton areas accounted for about 150 experiments during the lifetime of the ZGS. Since there are talks on the Bubble Chamber, Effective Mass Spectrometer, Streamer Chamber, and polarization experiments, we'll try to discuss the other proton-proton experiments that have been published to date. This can't, in any sense, be a review talk, but we would at least like to present the authors' conclusions on their experiments. This subset of experiments is listed in Fig. 10.

Both E-9 and E-154 by the Anderson group were Missing Mass Spectrometer experiments whose layout was seen in previous figures. The experimenters concluded that bosons appeared to be produced directly and not as decay products of nucleon isobars. Their data confirmed the existence of a prominent peak in $P+P \to d+\pi^+$ in the forward cross section at $E_{C.M.} \approx 2.9$ GeV/c and showed a new shoulder at $E_{C.M.} \approx 3.8$ GeV/c. They found no evidence for higher mass bosons and no evidence of a δ^+ as seen in some other experiments. These experiments provided new data to test and refine the one pion exchange (OPE) and one nucleon exchange (ONE) models.

E-25 was a search for dibaryon resonances in the mass region of 1.9 to 3.9 GeV and was done by measuring the charged-pion production at 180° in P+P collisions at 12.5 GeV/c. Two body reactions of the form $P+P \to \pi^\pm + x^{+,+++}$, where x as a dibaryon resonance would result in a peak in the pion-momentum spectrum. The experiment observed a known X^+, the deuteron, and measured the cross section at 0°. No other statistically significant enhancement was observed. An upper limit at 0° of 1.5 μb/sr was placed on dibaryon production for a width of a possible resonance of ≤ 200 MeV. There was also no evidence for a previously reported 2520 MeV X^{+++}.

E-102 measured the differential cross section for proton-proton elastic scattering at 90° in the center-of-mass system from 5.0 to 13.4 GeV/c with very small momentum intervals, 100-200 MeV/c. There were several conclusions from this experiment. (1) Since it measured the energy dependence of the differential cross section at fixed angle, it was sensitive to resonances and, as in the previous experiment, it found no dibaryon resonances from 5.0 GeV/c to 13.4 GeV/c. (2) It tested the statistical model in two ways. First, it saw no evidence for Ericson fluctuations which would imply that there were no overlapping intermediate states and thus makes it hard to justify a statistical model. Second, it tested a prediction of Hagedorn on the behavior of the 90° cross section and found that the data disagreed by 10 or more standard deviations. (3) It found a sharp break in the 90° cross section. By parameterizing the data against the

E-9   P-P Isobar Production - H. L. Anderson, S. Fukui,
D. Kessler, K. A. Klare, M. V. Sherbrook, H. J. Evans,
R. L. Martin, E. P. Hincks, N. K. Sherman,
P. I. P. Kalmus

E-25   Baryon No. 2 Resonances - R. C. Lamb, R. A. Lundy,
T. B. Novey, D. D. Yovanovic, R. Lander

E-102   Proton-Proton Elastic Scattering at 90° and Structure
within the Proton - C. W. Akerloff, R. H. Hieber,
A. D. Krisch, K. W. Edwards, L. G. Ratner, K. Ruddick

E-145   Production of Pions, Kaons, and Antiprotons in the
Center-of-Mass System in High Energy Proton-Proton
Collisions - L. G. Ratner, K. W. Edwards, C. W. Akerloff,
D. G. Crabb, J. L. Day, A. D. Krisch, M. T. Lin

E-154   Study of I=1 Bosons with a Deuteron Missing Mass Spec-
trometer - H. L. Anderson, M. Dixit, H. J. Evans,
K. A. Klare, D. A. Larson, M. V. Sherbrook, R. L. Martin,
K. W. Edwards, D. Kessler, D. E. Nagle, H. A. Thiessen,
C. K. Hargrove, E. P. Hincks, S. Fukui

E-182 &   Inelastic High Energy Proton-Proton Collisions -
E-216   C. W. Akerloff, D. G. Crabb, J. L. Day, N. P. Johnson,
P. Kalbaci, A. D. Krisch, M. T. Lin, M. L. Marshak,
J. L. Randolph, P. Schmueser, A. L. Read, K. W. Edwards,
J. G. Asbury, G. J. Marmer, L. G. Ratner

E-199   Differential Cross Section in Large Angle Elastic
Proton-Proton Scattering from 1.5 to 5.0 GeV/c -
B. B. Brabson, R. R. Crittenden, R. M. Heinz,
R. C. Kammerud, H. A. Neal, H. W. Paik, R. A. Sidwell,
K. F. Suen

E-300   Some Features of the Reaction $P+P \rightarrow \Delta^{++}$ (1236)n at
6 GeV/c - J. D. Mountz, Gerald A. Smith, A. J. Lennox,
J. A. Poirer, J. P. Prukop, C. A. Rey, O. R. Sander,
P. Kirk, R. D. Klem, Irwin Spirn

E-365   Elastic Scattering Near 90° Center-of-Mass -
K. A. Jenkins, L. E. Price, R. Klem, R. J. Miller,
P. Schreiner, H. Courant, Y. I. Makdisi, M. L. Marshak,
E. A. Peterson, K. Ruddick

Fig. 10

variable $\beta^2 P_\perp^2$, the authors were able to obtain a good fit for the 90° data, as well as the data at all other angles. This showed three distinct regions in the cross section that could be characterized by regions having radii of 0.9 F, 0.52 F, and 0.34 F. However, the fit especially near 90° was not perfect and, during the next couple years after the experiment, it was thought that spin effects could have a large influence. This led directly to the initiation of the polarized beam which has produced a great deal of new and unexpected results and indeed shows that the cross section in pure spin states are different. Even a very simple ansatz about the spin effects at 90° gave a striking fit to the data over 12 orders of magnitude, as shown in Fig. 11.

E-145, E-182, E-216. This group of experiments received its impetus from the previous results. One wished to test if the three regions might be diffraction scattering associated with three different types of inelastic processes. In particular, could π, K, and $\bar{p}$ production each occur from a different one of the three regions? Although the answer to the latter question turned out to be no, this set of experiments initiated the $P+P \to K^\pm_{\substack{\pi^\pm \\ p^\pm}}$ + anything reactions (inclusive interactions) which occupied the time of a good deal of experimentalists and theorists over the next few years. Although production spectra had been measured previously, these were "Beam Survey" types of experiments which measured production in the Laboratory frame to provide information for the design of secondary beams. There had been few attempts to relate this to the theory of strong interactions. Inelastic scattering dominates the total cross section and, since it might possibly drive the elastic scattering, one was looking for some simple relationship between the elastic and inelastic cross sections. As with the elastic cross sections, the inelastic cross sections showed a gaussian dependence with, however, only one break and with different slopes than the elastic. The slopes seemed to be about the same for all types of particles and, in particular, a very sharp forward peak was observed in the pion spectra. The pion production cross section was well represented by the sum of two gaussians with the slope of the large-momentum-transfer production equal to $3(GeV/c)^{-2}$, which was also about the same for kaons. The inelastic scattering cross section for protons showed a less prominent forward peak and again an $e^{-3P_\perp^2}$ dependence. The proton cross section seemed to be independent of inelasticity over a large range.

E-199 measured large-angle proton-proton elastic scattering at momenta from 1.5 to 5.0 GeV/c and thus complemented the measurements of E-102 which had taken data from 5.0 to 13.4 GeV/c. Their data indicated a break at small-t close to $-t=0.8(GeV/c)^2$ and another less pronounced slope change near $-t=3(GeV/c)^2$. These breaks have shown up even more clearly in the polarized beam measurements of pure spin cross sections. The authors tried to interpret the data in terms of

98

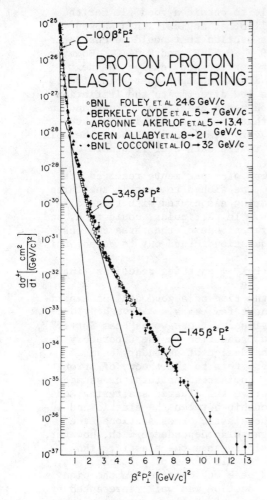

Fig. 11.   Fit to P-P Elastic
Scattering Data

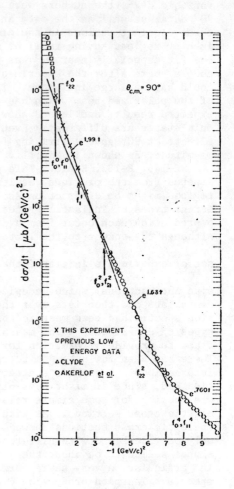

Fig. 12.   Plot Showing Partial Wave
Amplitudes Near Breaks in P-P
Cross-Section

several models: optical, statistical, Regge-pole, and diffractive. None of the fits were very convincing. The optical model fits with various absorbing discs and, spin effects neglected, required the disc radii to vary in order to fit all energies. The introduction of a spin flip term did not help, since the angular distribution required a different value than for polarization data. The authors felt that the physical interpretation of this type of fit was rather nebulous due to the complex spin structure of the P-P system. Another fit was to the Krisch model, as shown in E-102. As we pointed out before, this was not a totally satisfactory fit and did not fit the low energy data of E-199. Above P_{Lab}=4.0 GeV/c the

agreement improves considerably. The authors concluded that their deviation from the Krisch model was perhaps a manifestation of the nondiffractive nature of the data at low energy.

We have mentioned dibaryons previously, and there had been a model that predicted breaks corresponding to dibaryon masses of 2.2, 2.6, 3.4, 3.9 GeV for breaks at $-t \approx 0.8$, 1.7, 4.1, 6.5. The authors did a Regge calculation to predict the energies at which dibaryon resonances might occur. Several assumptions were made, and Fig. 12 shows their data with the partial-wave amplitudes indicated by arrows. Near $-t \approx 6.5$ $(GeV/c)^2$, there are maxima in three amplitudes. Maxima are approximately equally spaced about the break near $-t$=0.8 $(GeV/c)^2$ and also about $-t$=3 $(GeV/c)^2$. Of course, this is not very conclusive, since there are many assumptions made in the calculations and one would need a good deal more evidence for the dibaryon hypothesis.

E-300 essentially found that in the $P+P \rightarrow \Delta^{++}$ (1236)+n reaction, the invariant mass and momentum-squared distributions were in agreement with the predictions of the Chew-Low one-pion-exchange model. Since the Chew-Low model only considers the kinematics of OPE, models to give form factor and t distribution were also used. The data were extrapolated from the physical region to the pion pole, and it was found that the Dürr-Pilkuhn and Benecke-Dürr models, in conjunction with quadratic extrapolation in t, reproduced the known on-mass-shell

dependence of the cross section for $\pi^+ P$ scattering.

E-365 was a test of the dimensional counting rule. Since hadrons are believed to be composite particles, a model for large-angle hadron-hadron scattering could involve scattering of the constituents (for example, quarks) of the projectile and target particles. A dimensional counting rule predicts the asymptotic energy dependence of the differential cross section at large fixed angles in this con-

text. The rule predicts that at constant C.M. angle, $d\sigma/dt \propto S^{-n}$; n=m-2 and m is the number of quarks in the initial and final states of the interaction, S is the square of the C.M. energy, and t is the square of the four-momentum transfer. This experiment found that the PP data are inconsistent with one value of n over the entire energy range. Previous experiments had reported breaks in the slope of the $90°$ cross sections plotted against a variety of parameters, and this disagreement with the expected n = 10 is reflected in the data of this experiment. The energies at which these breaks occurred, however, might not be high enough for the dimensional counting rule to apply.

The authors did fit the 90° data for S > 12 GeV^2 but several other forms could fit the data equally well. These PP results gave a verdict of not proven for the dimensional counting rule.

To sum up, the EPBs with 150 completed experiments were busy places with many simultaneous experiments and, indeed, sometimes with many simultaneous arguments on who got how much beam, and who was interfering with whose experiment. All in all, however, a great body of data was accumulated, and it will do its share in elucidating the nature of elementary particle interactions.

THE STREAMER CHAMBER AND K PHYSICS

A. Abashian

National Science Foundation, Washington, D.C. 20550

ABSTRACT

A view of the ZGS experimental program during the 1960's is presented, with particular emphasis upon the study of weak interactions as exhibited in neutral K meson decays and upon the study of strong interactions using the Illinois-Argonne streamer chamber facility.

INTRODUCTION

Earlier in this symposium, you heard presentations made by physicists involved in the planning of the ZGS and the initiation of the experimental program with the facility. One of the hopes of those planners and administrators was that of a strong university user involvement in the research program. Having been an active user of the ZGS, I would like to reflect upon and give a perspective to the ZGS from the user community viewpoint. In what I shall say in my talk, I will attempt to give a flavor of what was happening in the research program. I shall not attempt to be comprehensive or encyclopedic in discussing all that ensued in the areas of K physics and the Streamer Chamber facility. I apologize to those who may feel slighted as a result, if their work is not included.

The talk consists of two separate parts: K Physics and Streamer Chamber Physics. The section on K physics will not include all experiments dealing with the K meson but will, instead, be confined to those dealing with the weak decay processes and almost exclusively, those performed using electronic techniques. The section dealing with the Streamer Chamber will be hardware oriented with a brief discussion of some of the accomplishments of the experiments performed using the facility.

HISTORICAL BACKGROUND

Upon passing my preliminary exam in 1955, I went to my thesis advisor Leon Madansky, professor of physics at Johns Hopkins University to discuss what might be a suitable topic for my thesis. He asked what I would like to do, and I responded by saying that I wanted to work in high energy physics. His response to that was Hopkins did not conduct research at that time in that area but that perhaps he could arrange to send me to the Brookhaven Cosmotron. He had spent the previous year there working with Rod Cool and Oreste Piccioni and felt I might work with them.

$\theta^+ - \tau^+$ PUZZLE

In a short time, I found myself at Brookhaven, impressed by all the activity and excitement going on. After about a week, I went to a seminar in which the then labled θ-τ puzzle was being discussed. In those days, physicists had seen two strange particles produced in strong interactions called the θ and the τ. These particles had a number of striking similarities as shown in Table I.

Table I Properties of θ^+ and τ^+ Mesons

	θ^+	τ^+
Strangeness No.	+1	+1
Masses	equal	
Lifetimes	equal	
Spin	0	0
Decay Mode	$\pi^+\pi^0$	$\pi^+\pi^-\pi^-$
Intrinsic parity	even	odd

In almost every way the θ^+ looked like the τ^+ except that the parities inferred from the decay states appeared to be opposite. How were the particles related if they were not the same? By 1956, the answer arrived from an experiment on nuclear beta decay by C. S. Wu of Columbia and her collaborators, motivated by theoretical speculations of T. D. Lee and C. N. Yang. The answer was two-fold; the θ^+ and τ^+ are the same particle (K^+) and more importantly, the parity operation is <u>not</u> conserved in the weak decays. Thus the situation remained for another eight years.

CP AND THE K SYSTEMS

There is a general theorem, called the CPT theorem which states that in any process, the combined operation of charge conjugation (C), parity (P) and time reversal (T) is an invariant. Since parity (P) was found to be non-invariant in weak processes, at least one other quantity needed to be non-invariant. Shortly after the discovery of parity violation, charge conjugation (C) was also found to be violated. In reformulating the theory of symmetries, it was then hypothesized that the combined operation CP was an invariant as well as T.

In the strong and electromagnetic production of K mesons, K^0 mesons or their anti particles, $\overline{K}^0$ mesons are produced. If one performs a CP transformation upon these states the following occur:

$$CP \mid K^0 > = - \mid \overline{K}^0 > \tag{1}$$

$$CP \mid \overline{K}^0 > = - \mid K^0 > \tag{2}$$

The K^0 and $\overline{K}^0$ states are obviously not invariant under the CP operation and cannot, therefore, be the states which decay weakly.

A superposition of the K^0 and $\overline{K}^0$ states can be constructed which are CP invariant as follows:

$$K_S^O = \sqrt{\tfrac{1}{2}} \{ \ | \ K^O > + \ | \ \overline{K}^O > \} \text{ and} \qquad (3)$$

$$K_L^O = \sqrt{\tfrac{1}{2}} \{ \ | \ K^O > - \ | \ \overline{K}^O > \}. \qquad (4)$$

The first, K_S^O , is allowed to decay into the $\pi^+\pi^-$ state, even under the CP operation while the K_L^O is odd under CP violation and is therefore prohibited from decaying into the $\pi^+\pi^-$ state, but is allowed to decay into three body final states such as $\pi\ell\nu$.

Since the ZGS was not yet operative in 1961, Al Wattenberg and I decided to set up a K_L^O beam at the Brookhaven AGS and to study the decays of those particles. Al had detected evidence for K_L^O mesons the summer before at Brookhaven and had concluded there were sufficient numbers of K_L^O mesons there to study the decay properties of these particles, with particular emphasis upon the form factors. In late 1962, we set up a magnetic spectrometer with spark chambers in the first K_L^O beam at the AGS. By 1964, most of our reconstruction programs were working to where we could study the three body Dalitz plots of $K_{\mu 3}^O$ and K_{e3}^O decays.

In the same beam we had set up, Fitch and Cronin set up an apparatus in 1963 to investigate the "Adair effect" in which anomalously large numbers of K_S^O mesons were reported to be regenerated by K_L^O interactions in a hydrogen bubble chamber. By the summer of 1964, both the Princeton group [1] and independently our group at Illinois [2] had evidence that the anomalous effect was instead due to the fact that K_L^O mesons did decay into the $\pi^+\pi^-$ state with a branching ratio of the order of 2×10^{-3} and that CP was in fact not an invariant quantity in weak decays. This discovery lead to the focus of the ZGS program in K physics for the next several years.

ZGS PROGRAM IN K PHYSICS

The discovery of CP violation quickly led to the following list of questions:

QUESTIONS

1. Why is there a CP violation; what is it due to?

2. Does time reversal hold; to what level?

3. Are there other manifestations of CP violation? ($K^O \to \pi\ell\nu$)

4. Can the study of K^O decays shed light on the existence of weak neutral currents?

$$K_S^O \to \mu^+\mu^- \qquad (5)$$

$$K_L^O \to \mu^+\mu^- \qquad (6)$$

5. Do weak currents always obey the $\Delta S = \Delta Q$ rule; can there be $\Delta S = -\Delta Q$ currents?

6. What are the detailed properties of the $K_L^0 - K_S^0$ system?

What are ΔM, decay amplitudes, relative phases, form factors, rare Ξ decay modes of the K_L^0, K_S^0 complex?

Many answers were provided (although not exclusively by ZGS experiments). The following conclusions were reached:

1. CP violation is consistent with the concept of a superweak force (Wolfenstein) which was not invariant under CP. We still do not know what the violation is due to (Illinois, Chicago); however, since experiments have been insensitive to the predictions of a super-weak theory.

2. No evidence has yet been obtained for time reversal violation. It must be small. (Yale, Illinois)

3. No evidence exists for weak neutral currents to about the 10^{-5} level compared to charged currents. (Illinois)

4. No evidence exists for $\Delta S = -\Delta Q$ currents in K decay. (Illinois)

5. Real and imaginary parts of the form factors of K mesons were measured (ξ parameter). (Yale, Wisconsin, Argonne, Illinois)

6. CPT theorem was tested and found to be valid. (Illinois)

Later, a lack of $K_L^0 \to \mu^+\mu^-$ at the 10^{-8} level was reported by a group at LBL. The concept of charm was introduced to explain the absence. The experiment was repeated at the ZGS.

7. $K_L^0 \to \mu^+\mu^-$ at $\sim 10^{-8}$ level was found, negating the LBL measurement. (Chicago)

Figures 1 and 2 show examples of the publications which emerged from work performed at the ZGS, showing the nature of the on-going research.

In conclusion of this part of my talk, I would like to state that the ZGS program competitively contributed to a better understanding of the nature of CP violation, neutral currents, symmetry principles, validity of selection rules and to the structure of K mesons. Almost all of such work was carried out by an active university user community with strong and enthusiastic support of the ZGS and its management.

STREAMER CHAMBER PROGRAM

The second portion of my talk will be devoted to a brief discussion of the streamer chamber facility with which I was involved.

In about 1968, upon the completion of our last K decay experiment, Rich Orr and I decided that perhaps we might extend our

VOLUME 17, NUMBER 12 **PHYSICAL REVIEW LETTERS** 19 SEPTEMBER 1966

EXPERIMENTAL INVESTIGATION OF CP VIOLATION IN K_{e3}^0 DECAYS*

L. J. Verhey, B. M. K. Nefkens, A. Abashian, R. J. Abrams, D. W. Carpenter,
R. E. Mischke, J. H. Smith, R. C. Thatcher, and A. Wattenberg
Department of Physics, University of Illinois, Urbana, Illinois
(Received 20 July 1966)

A spark-chamber experiment has been performed which measures the ratio $R = (K_L^0 \to \pi^+ + e^- + \bar{\nu})/(K_L^0 \to \pi^- + e^+ + \nu)$. On the basis of 1539 identified K_{e3}^0 decays, no variation of R is found over the Dalitz plot. The corrected integrated value is $R = 0.93 \pm 0.05$.

VOLUME 30, NUMBER 20 **PHYSICAL REVIEW LETTERS** 14 MAY 1973

Muon Polarization in the Decay $K_L^0 \to \pi^- \mu^+ \nu_\mu$, an Experimental Test
of Time-Reversal Invariance*

J. Sandweiss, J. Sunderland, W. Turner, and W. Willis
Yale University, New Haven, Connecticut 06520

and

L. Keller†
Argonne National Laboratory, Argonne, Illinois 60439
(Received 8 January 1973)

We report the results of a high-statistics scintillation-counter electronics experiment measuring the muon polarization in the decay $K_L^0 \to \pi^- \mu^+ \nu_\mu$ by the method of spin precession. We present results for the real and the imaginary parts of the form-factor ratio.

VOLUME 18, NUMBER 4 **PHYSICAL REVIEW LETTERS** 23 JANUARY 1967

DETERMINATION OF THE PHASE OF THE CP-NONCONSERVATION
PARAMETER η_{+-} IN NEUTRAL K DECAY*

R. E. Mischke,† A. Abashian, R. J. Abrams,‡ D. W. Carpenter,§ B. M. K. Nefkens, ‖
J. H. Smith, R. C. Thatcher, L. J. Verhey, and A. Wattenberg
Department of Physics, University of Illinois, Urbana, Illinois
(Received 8 December 1966)

VOLUME 25, NUMBER 15 **PHYSICAL REVIEW LETTERS** 12 OCTOBER 1970

PRECISE DETERMINATION OF THE K_L-K_S MASS DIFFERENCE BY THE GAP METHOD
(UNIVERSITY OF CHICAGO-UNIVERSITY OF ILLINOIS CHICAGO CIRCLE COLLABORATION)*

S. H. Aronson, R. D. Ehrlich,† H. Hofer,‡ D. A. Jensen, R. A. Swanson,§ and V. L. Telegdi
The Enrico Fermi Institute and Department of Physics, The University of Chicago, Chicago, Illinois 60637

and

H. Goldberg and J. Solomon
Department of Physics, University of Illinois at Chicago Circle, Chicago, Illinois 60680

and

D. Fryberger
Stanford Linear Accelerator Center, Stanford, California 94305
(Received 13 August 1970)

Figure 1.

Volume 19, Number 2   PHYSICAL REVIEW LETTERS   10 July 1967

MEASUREMENT OF THE RELATIVE PARTIAL DECAY RATES FOR $K^{\pm} \to \pi^{\pm} + \pi^{+} + \pi^{-}$†

C. R. Fletcher, E. W. Beier, R. T. Edwards, W. V. Hassenzahl, D. Herzo,
L. J. Koester, Jr., C. Mencuccini,* and A. Wattenberg
Department of Physics, University of Illinois, Urbana, Illinois
(Received 19 May 1967)

Volume 23, Number 11   PHYSICAL REVIEW LETTERS   15 September 1969

MEASUREMENT OF THE RELATIVE PHASE OF THE $K_L \to \pi^+\pi^-$
AND $K_S \to \pi^+\pi^-$ DECAY AMPLITUDES BY "VACUUM REGENERATION"*†

D. A. Jensen,‡ S. H. Aronson, R. D. Ehrlich, D. Fryberger,§ C. Nissim-Sabat,‖ and V. L. Telegdi
The Enrico Fermi Institute and Department of Physics, The University of Chicago, Chicago, Illinois 60637

and

H. Goldberg and J. Solomon
Department of Physics, University of Illinois at Chicago Circle, Chicago, Illinois 60680
(Received 26 June 1969)

The nonexponential $\pi^+\pi^-$ decay of neutral K's, due to the interference of K_S and K_L in this common channel, has been quantitatively studied using a beam which is preponderantly K^0 at $t = 0$. The value $(40 \pm 5)°$ is obtained for $\arg \eta_{+-}$ ($\eta_{+-} = \langle \pi^+\pi^- |T| K_L \rangle / \langle \pi^+\pi^- |T \times |K_S \rangle$), using $\Delta m/h = (M_L - M_S)/h = 0.538 \times 10^{10}$ sec^{-1}; allowing for the current uncertainty in Δm, the error in $\arg \eta_{+-}$ becomes $\pm 12.5°$. This result agrees with the predictions of that class of theories which give $\arg \eta_{+-} \simeq \tan^{-1}(2\Delta m/h\Gamma_S)$, in particular the "superweak" theory, for which this prediction is exact.

Volume 28, Number 22   PHYSICAL REVIEW LETTERS   29 May 1972

Measurement of Decay Correlations in $K_{\mu 3}^+$ Decays*

C. Ankenbrandt,† R. Larsen, L. Leipuner, and L. Smith
Brookhaven National Laboratory, Upton, New York 11973

and

F. Shively
Los Alamos Meson Physics Facility, Los Alamos Scientific Laboratory, Los Alamos, New Mexico 87544

and

R. Stefanski
National Accelerator Laboratory, Batavia, Illinois 60510

and

R. Adair, H. Kasha, S. Merlan, R. Turner, and P. Wanderer‡
Yale University, New Haven, Connecticut 06520
(Received 20 January 1972)

Volume 17, Number 11   PHYSICAL REVIEW LETTERS   12 September 1966

TEST OF TIME-REVERSAL INVARIANCE IN $K_L^0 \to \pi^- + \mu^+ + \nu$†

R. J. Abrams, A. Abashian, R. E. Mischke, B. M. K. Nefkens, J. H. Smith,*
R. C. Thatcher, L. J. Verhey, and A. Wattenberg
University of Illinois, Urbana, Illinois
(Received 29 July 1966)

We have measured the polarization of the muon in the decay $K_L^0 \to \pi^- + \mu^+ + \nu$ using spark chambers. On the basis of 1458 events, the component of polarization transverse to the decay plane was found to be -0.05 ± 0.18, which is consistent with zero, in agreement with time-reversal invariance.

Figure 2.

studies of weak interactions and study Ξ decays. Because of low production rates and short lifetimes, new techniques needed to be developed for their study. We felt the detector should do the following:

1. Be capable of handling high rates.

2. Have a large solid angle for acceptance.

3. Have high measurement accuracy.

4. Provide a relatively easy task for pattern recognition.

From these considerations was born the idea of building a streamer chamber at Illinois for use at the ZGS.

Recognizing the enormity of the task in building a large enough device and placing it into a magnetic field, we approached Bruce Cork with the idea of making the streamer chamber a facility at Argonne and of obtaining Argonne assistance and cooperation in its construction and operation. Bruce encouraged us and provided us with technical assistance in the form of Sepp Prunster who built the Blumlein and Jerry Watson who took over its operation upon its completion, and a number of technical people. Bob Sachs, through his acceptance of the 40" bubble chamber from the CEA had, by that action, provided for the magnet which later would house the streamer chamber.

At Illinois, we assumed responsibility for the construction of the Marx generator, the transmission line, and the chamber body itself. Construction of the liquid hydrogen target was a joint effort of formidable magnitude because of electrical breakdown which would result in the rupture of the target vessel. The problem was finally solved when we recognized that either a very good vacuum or a very poor vacuum around the target flask would prevent a breakdown. We opted for the latter by encasing the target vessel in a PVC foam box.

About six strong interaction experiments were conducted with the facility during the period of 1972-1977. Groups from Illinois, Chicago, Notre Dame, Lehigh, Strassbourg, Michigan State and Argonne used the facility. Some of the more prominent results were the discovery of a new scalar meson, backward production of vector mesons and a study of $\pi-\pi$ scattering.

Figure 3 shows some of the publications which emerged from the results of that program.

CONCLUSIONS

I would like to conclude my remarks with a few general comments and observations on working at the ZGS as a user. I found the following to be true:

The machine was a reliable one, well operated and maintained. There was outstanding cooperation between the university user community and the laboratory. The staff was highly competent at

NUCLEAR INSTRUMENTS AND METHODS 115 (1974) 445–456; © NORTH-HOLLAND PUBLISHING CO.

DESIGN AND DEVELOPMENT OF A 1.5 METER STREAMER CHAMBER SYSTEM*

A. ABASHIAN, N. BEAMER, B. EISENSTEIN, J. D. HANSEN†, W. MOLLET, G. R. MORRIS,
T. O'HALLORAN, J. R. ORR+, D. RHINES, P. SCHULTZ§ and P. SOKOLSKY

University of Illinois at Urbana-Champaign, Urbana, Illinois, U.S.A.

and

S. J. PRUNSTER** and J. WATSON

Argonne National Laboratory, Argonne, Illinois, U.S.A.

and

M. BUTTRAM

Iowa State University, Ames, Iowa, U.S.A.

Received 12 June 1973

A 1.5 m × 1.0 m × 0.6 m Ne–He streamer chamber and high voltage pulsing system capable of 1.0 MV pulses with widths continuously variable from 10 to 20 ns has been developed by a University of Illinois – ANL group. The design and construction of the dc fast charge regulator, Marx generator, Blumlein pulse shaping network, transmission line, and chamber are described. Studies on the effects of increasing field emission and injection of free electrons into the pulse shaping network gap on pulse amplitude stability are reported. Under normal operating conditions, pulse amplitude stability of 2–3% rms has been achieved. Photographic and pulse monitoring systems are described and operating experience over two experimental runs is summarized.

PHYSICAL REVIEW D   VOLUME 13, NUMBER 1   1 JANUARY 1976

Backward production in $\pi^- p \to p\pi^+ \pi^- \pi^-$ at 8 GeV/c*

A. Abashian, B. Eisenstein, J. D. Hansen,† W. Mollet, G. R. Morris,‡ B. Nelson, T. O'Halloran, J. R. Orr,§
D. Rhines, P. Schultz,‖ P. Sokolsky,¶ and R. G. Wagner
University of Illinois at Urbana-Champaign, Urbana, Illinois 61801

J. Watson
Argonne National Laboratory, Argonne, Illinois 60439

N. M. Gelfand
*University of Chicago, Chicago, Illinois 60637
and University of Illinois at Urbana-Champaign, Urbana, Illinois 61801*

M. Buttram
Iowa State University, Ames, Iowa 50010
(Received 21 July 1975)

PHYSICAL REVIEW LETTERS

VOLUME 41   31 JULY 1978   NUMBER 5

Measurement of the $\pi^+\pi^- \to K_s^0 K_s^0$ Scattering Cross Section

N. M. Cason, A. E. Baumbaugh, J. M. Bishop, N. N. Biswas, V. P. Kenney, V. A. Polychronakos,[a]
R. C. Ruchti, W. D. Shephard, and J. M. Watson
University of Notre Dame, Notre Dame, Indiana 46556, and Argonne National Laboratory, Argonne, Illinois 60439
(Received 29 March 1978)

VOLUME 36, NUMBER 25   PHYSICAL REVIEW LETTERS   21 JUNE 1976

Observation of a New Scalar Meson

N. M. Cason, V. A. Polychronakos, J. M. Bishop, N. N. Biswas, V. P. Kenney,
D. S. Rhines, and W. D. Shephard*
University of Notre Dame,† South Bend, Indiana 46556

and

J. M. Watson
Argonne National Laboratory,‡ Argonne, Illinois 60439
(Received 29 March 1976)

Figure 3.

assisting users, and included the plastics shops, machine shops, surveyors, floor technicians, and secretaries. The ZGS contained a pleasant, congenial and co-operative atmosphere in which to work. I am sure that I speak for all users in conveying these feelings and wish to express our thanks to all who have contributed so importantly to the success of the ZGS research program.

REFERENCES

1. J. Christenson et al, Phys. Rev. Letters 13, 138 (1964).
2. A. Abashian et al, Phys. Rev. Letters 13, 243 (1964).

THE POLARIZED TARGETS*

A. Yokosawa

Argonne National Laboratory, Argonne, Illinois   60439

INTERESTS IN SPIN PHYSICS AROUND THE TIME OF ZGS GROUNDBREAKING

Around the time of the ZGS groundbreaking, active discussions had been held to propose unique experimental facilities to be constructed for counter physics. The importance of spin physics in the strong interaction has been emphasized by Albert Crewe. In 1960 accelerator physicists had already been convinced that the ZGS could be unique in accelerating a polarized beam; polarized beams were being accelerated through linear accelerators elsewhere at that time.

However, there was much concern about going ahead with the construction of a polarized beam because i) the source intensity was not high enough to accelerate in the ZGS, ii) the use of the ZGS would be limited to only polarized-beam physics, that is, proton-proton interaction, and iii) p-p scattering was not the most popular topic in high-energy physics.

In fact, within spin physics, π-nucleon physics looked attractive, since the determination of spin and parity of possible πp resonances attracted much attention. To proceed we needed more data beside total cross sections and differential cross sections; measurements of polarization and other parameters were urgently needed.

Polarization measurements had traditionally been performed by analyzing the spin of recoil protons. The drawbacks of this technique are: i) it involves double scattering, resulting in poor accuracy of the data, and ii) a carbon analyzer can only be used for a limited region of energy.

In 1961 we heard that solid-state physicists had succeeded in polarizing protons in a crystal: Professor R. V. Pound et al. at Harvard University and Professor C. D. Jeffries et al. at Berkeley. (Later we learned of similar work independently done at Saclay by Professor A. Abragam et al.) We also learned that high-energy physicists at Berkeley (O. Chamberlain et al.), at Rutherford (J. Thresher et al.), and at CERN (C. Rubbia et al.) were planning to make use of a polarized proton target.

HOW DOES THE TARGET WORK?

In the water of hydration in the crystal $La_2Mg_3(NO_3)_{12} \cdot 24 \, H_2O$ (in which 1% La^{3+} ions are replaced by paramagnetic Nd^{3+}), dynamic polarization of the protons occurs when one saturates a microwave "forbidden" transition that flips a proton and a Nd^{3+} ion.

This was done by placing the crystal in a strong magnetic field, cooling it to a temperature of 1.0 K, and irradiating it with microwave energy (see Fig. 1). Proton polarization can be achieved up to $\sim$ 90%.

---

*Work supported by the U.S. Department of Energy.

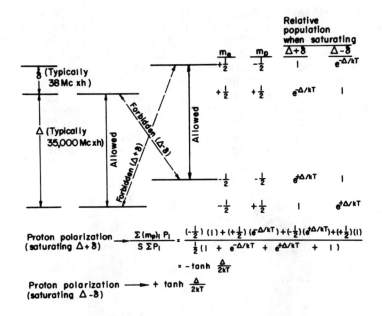

Fig. 1   Energy level diagram of an electron spin 1/2 with a proton-
spin 1/2 in a magnetic field. The relative populations are
given in column at right.

GETTING READY TO BUILD A POLARIZED TARGET

The main advantage of the use of a polarized target is that one
can measure polarization by single scattering. But the only material
available in 1962 was a crystal lanthanam magnesium nitrate, La_2Mg_3
$(NO_3) \cdot 24 H_2O$. We had to be convinced that we could carry out πp.
elastic scattering measurements using such a crystal at π incident
momentum as high as 2 GeV/c; counting on the Fermi-momenta spread
due to bound protons, we could select events that occurred only with
free protons that are polarized. We demonstrated the feasibility
through the use of Monte Carlo calculations.

Some theorists thought it would be wasteful to measure an asymmetry effect at the high energy of 2 GeV/c, and they predicted that
the asymmetry would be zero. But it seemed worthwhile experimentally
to explore this type of measurement.

The polarized-target device, the product of scientists in many
disciplines, seemed a unique facility, but was regarded as a terribly
unusual and difficult project. Fortunately, the laboratory management favored a proposal that A. Moretti, T. Khoe, S. Suwa and
A. Yokosawa build a polarized target.

CONSTRUCTION OF THE FIRST POLARIZED TARGET AT ARGONNE

Finding that there were competing experiments at Berkeley, CERN,
and Rutherford, we were anxious to proceed with this project as

quickly as possible. Fortunately, Professor C. D. Jeffries agreed to serve as a consultant to this project. One of the first things we had to settle was selecting a magnet with extremely fine uniformity. Although we wanted to have a horizontal scattering plane, we quickly realized no such magnet was commercially available. Finally, Varian Associates agreed to build a custom-made magnet with a vertical scattering plane, that is, with a horizontal magnetic field.

By working closely with Professor Jeffries, we made excellent progress on the project. High-quality technical assistance was extended, and crystal growing was accomplished by a devoted student. Target polarization was achieved up to 90%. A picture of the crystal immersed in the helium-4 cryostat is shown in Fig. 2.

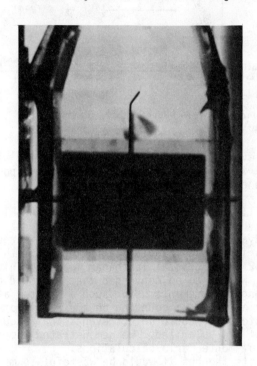

Fig. 2   Picture of the LMN crystal in the cryostat.

EXPERIMENTS WITH THE LMN TARGET

Our first proposed experiment with the polarized target was to clarify the spin and parity of the possible N*(2190) resonances, revealed in $\sigma^{Tot}(\pi^-p)$ as shown in Fig. 3, with the collaboration of a University of Chicago group (Professor N. Booth et al.). Because the experiment used a "dirty target" which was unfamiliar to many high-energy physicists, a low-priority status was initially assigned to the measurements. Fortunately, we found a huge spin

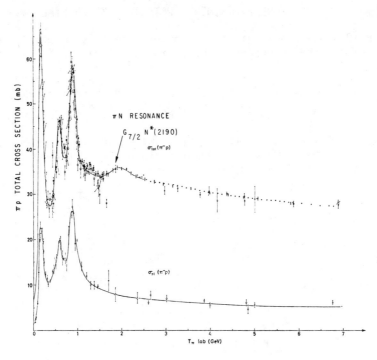

Fig. 3 π⁻p total cross section.

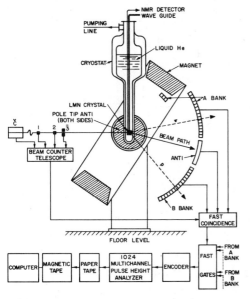

Fig. 4 The first experiment with a polarized target at ANL.

effect in π⁻p plastic scattering from 1.6 to 2.5 GeV/c. The results promptly led to our determining that the spin and parity of N*(2190) is $G_{7/2}$. The experimental setup is shown in Fig. 4.

EXPERIMENT WITH A POLARIZED BEAM AND A POLARIZED TARGET IN 1965

In 1965, while we were preparing for the next series of experiments, we took the target to the University of Chicago cyclotron, where we had made measurements of the spin-spin correlation parameter, C_{NN}, using polarized beams at several energies.

SOME MORE EXPERIMENTS WITH THE LMN CRYSTAL

In 1966, with new detectors (shower detectors, neutron counters) and a new on-line computer, we began to investigate reactions such as

i) $\pi^\pm p^\uparrow \rightarrow \pi^\pm p$ at up to 5.15 GeV/c. We found mirror symmetry in the $\pi^+ p$ and $\pi^- p$ world (see Fig. 5).

ii) $\pi^- p^\uparrow \rightarrow \pi^\circ n$ at up to 5 GeV/c. We destroyed a "naive" Regge pole model which was popular at the time. (The experimental setup is shown in Fig. 6.)

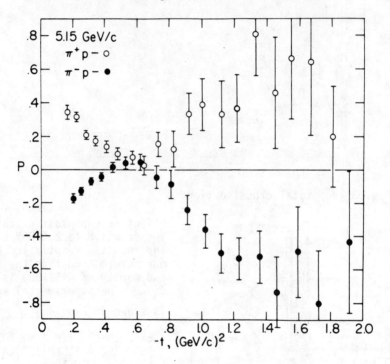

Fig. 5 Polarization in $\pi^\pm$ elastic scattering.

POLARIZED-TARGET DEVELOPMENT AND CONSTRUCTION OF THE 2ND TARGET

As we performed our experiments, we felt the urgent need for improvements on the target in two areas:

i) to look for a material with higher hydrogen content; this would require intensive investigation on hydrocarbon materials.

ii) to construct a new magnet and a new cryostat so that we would have a horizontal scattering plane.

Fig. 6 Experimental setup for πp charge-exchange measurements;
        52-element neutron counters located 186" above the polarized
        target, lead Lucite Cherenkov counters to catch the forward-
        going γ's from π°, and a flag in the center of the picture
        which reads "divine wind" written by D. Javonovich.

The search for new material was rather successful. From among
several candidates available, we decided to use ethylene glycol (a
hydrocarbon material originally developed at Saclay). Polarization
was achieved up to 40%.

The design of the horizontal cryostat was similar to the one by
P. Roubeau (Saclay), and a new magnet (vertical field allows the
horizontal scattering plane) was built by Varian Associates (see
Fig. 7).

A new target was used to search for exotic resonances in the
k^+p system by measuring polarization in the reaction as

$k^+p\uparrow \rightarrow k^+p$. We found an exotic resonance Z* (1890).

The Varian Associates magnet was then used by two ZGS users:

i) Michigan State - Ohio State (Abolins et al.) for
    $np\uparrow \rightarrow pn$ up to 6 GeV/c

ii) EPB-1 Experiment (A. Krisch et al.) for $p^\uparrow p^\uparrow \to pp$ from 2 to 13 GeV/c

Fig. 7   Conventional magnet for the second polarized target (vertical field allowing the horizontal scattering plane).

INVENTION OF NEW TARGET AT HELIUM-3 TEMPERATURE AND CONSTRUCTION OF THE THIRD TARGET

Although ethylene glycol is a cleaner material than the LMN crystal, we obtained polarization of only 20 to 40% during the experiments. Then in 1968, physicists (D. Hill et al.) at Argonne in the fields of high-energy physics and solid-state physics jointly discovered that by cooling the sample to 0.5 K, 80% of polarization could be obtained. The investigation was done in a ^{3}He cooled refrigerator. (A similar effect was discovered about the same time at Saclay, France.) Note that the running time per measurement in order to obtain the required experimental accuracy is proportional to the square of the target polarization.

This accomplishment prompted the decision to build a new helium-3 cryostat.

At this time we also made additional improvements on the target and detection system: i) construction of a new superconducting magnet (by H. Desportes et al.) which allowed us to have a longer target (5 → 10 cm) and a larger gap (3.8 → 5 cm) so that more room would be available for pole anticounters; ii) replacement of the scintillation hodoscope with a multiwire proportional chamber (MWPC) system.

The third target is shown in Fig. 8 At first we had used the target for a πp backward-scattering experiment and then nucleon-nucleon scattering using a polarized beam. When both the beam and target were polarized in the vertical direction, we determined the parameter C_{NN} = (N,N;0,0). We carried out measurements up to 6 GeV/c.

Fig. 8 Superconducting magnet for the third polarized target with proportional wire chambers.

Together with a superconducting spin-flip solenoid, it was possible to measure three spin measurements:

$$p^\uparrow(S) \; p^\uparrow(N) \to pp^\uparrow(S), \text{ parameter } H_{SNS} = (S,N;0,S),$$

and simultaneously

$$p^\uparrow(S) \; p \to pp^\uparrow(S), \text{ parameter } K_{SS} = (S,0;0,S),$$

and

$$pp^\uparrow(N) \to pp^\uparrow(N), \text{ parameter } D_{NN} = (0,N;0,N) \; .$$

Spin directions N, S and L are defined in Fig. 9. After these measurements, the superconducting magnet was sent to the Rice group (J. Roberts et al.).

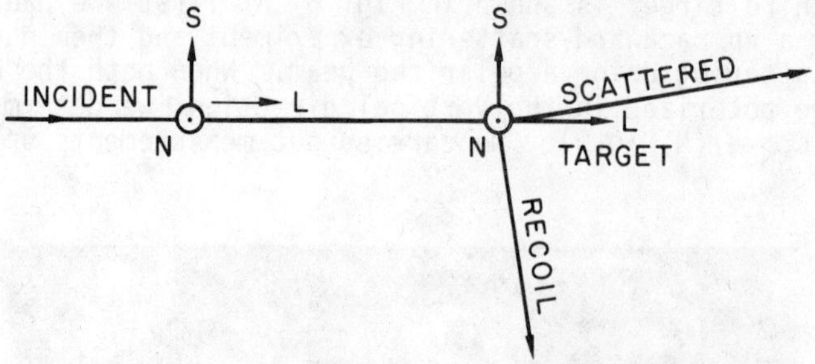

N: NORMAL TO THE SCATTERING PLANE
L: LONGITUDINAL DIRECTION
S = N x L IN THE SCATTERING PLANE

Fig. 9   Spin directions N, L and S.

POLARIZED TARGET AT FERMILAB

A target identical to the third polarized target was built for use at Fermilab. Polarization measurements were carried out in $\pi^{\pm}p$, pp elastic scattering up to 300 GeV/c.

CONSTRUCTION OF THE 4TH POLARIZED TARGET ("R AND A" TYPE)

It had been obvious for some time that in order to determine scattering amplitude in the pp system, we needed to measure several observables (a minimum of nine for five amplitudes). For this, we needed $\vec{S}$ and $\vec{L}$ type polarized targets. The magnet constructed (by H. Desportes, R. Wang et al.) is shown with the fourth target in Fig. 10.

Our original intention was to carry out complete measurements from 6 to 12 GeV/c in the pp system. However, we covered low energies for two reasons: i) the economy of a summer run and ii) interesting theoretical predictions at lower energies (E. Berger et al.).

BIGGEST SURPRISE IN SPIN PHYSICS

We carried out total cross-section measurements with a longi-tudinally polarized beam and a longitudinally polarized target:

$$\Delta\sigma_L = \sigma^{Tot}(\underset{\leftarrow}{\rightarrow}) - \sigma^{Tot}(\overset{\rightarrow}{\rightarrow}).$$

The results are shown in Fig. 11. The structures are interpreted as diprotons and dinucleons. Since then, we observed the evidence for the existence of diprotons in many other channels.

Fig. 10

The fourth
polarized target
($\vec{S}$ and $\vec{L}$ type).

Fig. 11 Nucleon-nucleon
total cross-section
measurements for
pure longitudinal
initial spin states
(I=0 and I=1)

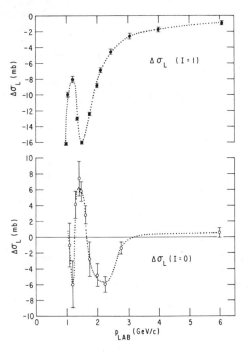

Here we remark briefly on some theoretical background of dibaryon resonances. For example, the MIT bag model predicts a dibaryon resonance, which is made of 6 quarks inside a cavity, depicted in Fig. 12.

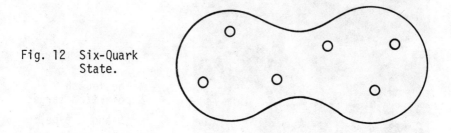

Fig. 12 Six-Quark State.

POLARIZED NEUTRON TARGET WITH A DILUTION REFRIGERATOR

If in the hydrocarbon target, one replaces protons by deuterons then we have a polarized deuteron target in which neutrons are polarized.

It was found at CERN that when a deuterated carbon target is cooled down to 0.03 K by a dilution refrigerator to a helium-3 temperature, one increases polarization by 25 to 40%. Recently we constructed and operated polarized deuteron targets with a dilution refrigerator and used these targets for the measurement of $\Delta\sigma_L(pd)$. From $\Delta\sigma_L(pd)$ data we deduced the value of $\Delta\sigma_L$ in the I=0 system.

Surprisingly, we observed a structure in $\Delta\sigma_L(I=0)$ at 1.5 GeV/c, where we found $\Delta\sigma_L(I=1)$ as shown in Fig. 11.

SUMMARY OF THE POLARIZED TARGETS

We summarize a layman's view of the target and associated experiments in Fig. 13. We also show the appreciation of polarized targets comically expressed by some theorists in Fig. 14.

ACKNOWLEDGMENT

The design, construction, and operation of the polarized targets called on the skills and cooperation of a large number of Argonne's scientists (D. Hill et al.), engineers, designers, technicians, and machinists. The results of their combined efforts led us to discover interesting phenomena in physics. Finally, we acknowledge the enthusiastic support we have received from the laboratory management for many years.

121

Polarized Targets are the greatest
invention since ·······

High Energy Phenomenology Conference
Pasadena California (1971)

Fig. 13 Appreciation of the polarized target.

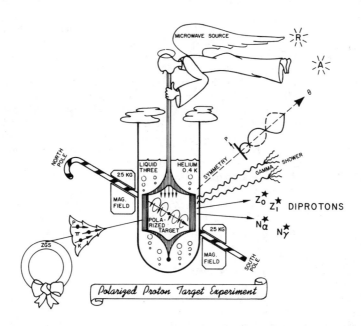

Fig. 14 Layman's view of the polarized target.

THE EFFECTIVE MASS SPECTROMETER

D. Ayres

Argonne National Laboratory, Argonne, IL 60439

ABSTRACT

The history and major accomplishments of the Effective Mass Spectrometer (EMS) are described. In the eight years since the EMS turned on, 21 experiments have been completed by groups from nine institutions in 32 months of operation. Over 400 million triggers have been recorded on magnetic tape, resulting in 29 journal publications to date. A list of experimental proposals for the EMS and a sampling of results are presented.

THE EARLY DAYS

As I was struggling to gain some sort of historical perspective on the Effective Mass Spectrometer for this talk, it occurred to me that I would be speaking to you on a personal anniversary of some significance. Just ten years ago tomorrow I was arriving at Argonne to begin work with the new Spectrometer Group, which then consisted of Bob Diebold, Art Greene, Barry Wicklund and myself. Looking back through my notes of that time, I found that the Effective Mass Spectrometer was no more than a collection of back-of-the-envelope scribblings, Fermilab was only a muddy hole in the ground, Meson Spectroscopy was in its heyday, the A_2 was split, and gasoline was 35 cents a gallon!

Meson spectroscopy, and the split A_2 in particular, had a profound effect on the design of the EMS, and determined many of the properties which have been responsible for the success of the spectrometer. Thus, one of our design goals was to be able to study the $K^*(1420)$, in case other tensor mesons might also be split. To this end we labored to obtain simultaneously good effective-mass resolution and large acceptance at high masses, and to design a short beamline with lots of K mesons, good particle identification, and good momentum resolution. Although the $K^*(1420)$ experiment had become uninteresting by the time we were able to perform it, the stringent requirements which it imposed on the design have been very important for the versatility and productivity of the spectrometer facility. In return, our debt to meson spectroscopy was amply repaid by E-268, a comprehensive comparison of K^* and ρ production, and a precise measurement of ρ-ω interference.

Early on in our design studies we began to collaborate closely with members of Drasko Jovanovic's Wire Spark Chamber Group, which also included Dave Rust and Charlie Ward. Their experience with online computer data acquisition systems, spark chamber construction and operation, and with the previous generation of spectrometers at the ZGS, provided much of the technical foundation upon which the EMS was built. As I mention a few of the names of those who played major roles in the early

days of the EMS, I realize that I am leaving a lot of people out.
In an attempt to rectify this, I have tried to list in Table I many of
those who contributed to the EMS at one time or another.

Table I: Major Contributors to EMS Construction

Physicists	HEP Technical	ARF Technical
I. Ambats	J. Dawson	R. Blaskie
M. Arenton	R. Diaz	R. Bouie
D. Ayres	R. Ely	C. Brzegowy
D. Cohen	J. Falout	J. Dvorak
R. Diebold	L. Filips	D. Gacek
A. Greene	C. Klindworth	R. Giugler
D. Jovanovic	R. Laird	E. Heyn
S. Kramer	R. Rivetna	R. Klem
A. Lesnik	L. Ruppert	C. Kreiger
E. May	E. Walschon	L. Marek
W. T. Meyer		D. R. Moffett
A. Pawlicki		A. Passi
T. Romanowski		W. Siljander
D. Rust		
J. Sauer		
E. Swallow		
C. Ward		
A. B. Wicklund		

Plus:
ZGS Accelerator and Radiation Safety Personnel
EPOG Technicians, Riggers, Surveyors
EMS Group Summer Students
ARF Instrument Shop and SHELF Group
HEF Magnet Measuring Group
ARF Superconducting Magnet Group
HEP Secretarial and Administrative Support
Equipment loaned from:
    U. of Chicago
    Elmhurst College
    Fermilab
    Indiana U.
    Lawrence Berkeley Lab.
    Ohio State U.

Figure 1 shows the assembled members of the two groups which
built the EMS, shortly after the start of operations in the summer
of 1971. The photograph also reveals (in the background) another
reason why the EMS has been so versatile - each group had its own
hydrogen target cart assembly mounted on a set of rails, so that the
targets and surrounding detectors could be easily interchanged. The
versatility of the Spectrometer as a facility in future years was in
large part due to the fact that it was originally designed to accom-
modate two rather different experiments: the Diebold Group's vector-meson production studies, and the Jovanovic Group's investigations of associated production.

124

Fig. 1. Builders of the original EMS. From left to right: (rear row) D. Jovanovic, E. Walschon, D. Ayres, W. T. Meyer, C. Ward, R. Rivetna, L. Ruppert, A. B. Wicklund, A. Greene, A. Lesnik; (front row) L. Filips, R. Diaz, R. Diebold, I. Ambats, S. Kramer, D. Rust. The spectrometer magnet can be seen in the center background, the Jovanovic-Group hydrogen target cart is in place in the beamline (to the left of the magnet), and the Diebold-Group target cart is pushed out of the beam-line (far left). The beam direction is from left to right.

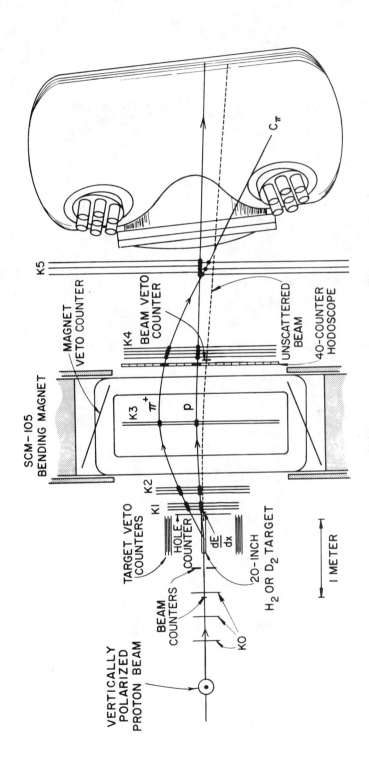

Fig. 2: Plan view of the EMS as it was used in E-339 for a study of $p_\uparrow p \to \Delta^{++} n$ with the polarized proton beam. K0, K1, K2, K3, K4, K5 are sets of magnetostrictive-readout spark chambers. The large downstream Cerenkov C_π distinguished between p and π^+.

126

Fig. 3. A photograph of the EMS as it was used for E-346, to study $\pi^- p \to K^0 \Lambda$. The Jovanovic-Group target cart, with the cylindrical spark chambers, is installed in the beamline, and the Diebold-Group target cart, with its array of target veto counters, is rolled out to the right of the beam. The beam enters from the right, and the large downstream Čerenkov is at the far left.

EVOLUTION OF THE EMS

In order to describe the evolution of the EMS, I will first need to review briefly some of the basic features of its construction[1]. Figure 2 shows a plan view of the EMS, essentially a large aperture dipole magnet (11 kG-m) surrounded by magnetostrictive-readout spark chambers. The basic elements of the trigger were the 40-counter hodoscope which counted particles emerging from the magnet, a beam veto counter, and veto counters surrounding the magnet aperture and hydrogen target. Experiments studying K^0 and Λ production used pulse height counters to define both ends of a decay space downstream of the target. The large downstream Cerenkov counter in Fig. 2 separated final-state pions from protons. It was borrowed from the University of Chicago for E-325, a study of K^+K^- production, to veto events with final-state pions. Figure 3 is a photograph of the EMS showing the cylindrical spark chambers which were used in E-331 and E-346 to detect recoil protons, Λ's, and Δ's. In the most recent configuration of the spectrometer, we have constructed a "Vertex Detector" from cylindrical proportional-wire chambers to detect recoil particles and to permit triggering on charged recoil multiplicity.

Over the years the original EMS detectors were gradually upgraded or replaced, and its versatility and reliability steadily improved. Although the K1 and K2 chamber sets have withstood the entire 8 years of operation, all other chamber sets were replaced at one time or another. Table II outlines a few of the major milestones in the EMS evolution. Table III summarizes a few numbers

Table II: Milestones in the Evolution of the EMS

1969 - 1971:	Design and Construction
1972:	Large Downstream Cerenkov Counter Installed
1973:	Cylindrical Spark Chambers Operational
1973:	First Polarized Beam Experiments
1975:	12 GeV/c Superconducting Beamline Commissioned
1978:	Vertex Detector and PDP-1134 Installed
1979:	Spin Rotating Solenoids Provide Longitudinal Beam Polarization

Table III: EMS Experiment Statistics

Years of Operation (May, 1971 - August, 1979):	8.3 years
Number of Submitted Proposals:	23 proposals
Number of Completed Experiments:	21 experiments
Number of Institutions on Proposals:	9 institutions
Number of Physicists on Proposals:	56 physicists
Total Number of Triggers on Tape:	4×10^8 triggers
Total Number of Shifts Charged:	1925 = 32 months
Published Journal Articles to Date:	29 publications

128

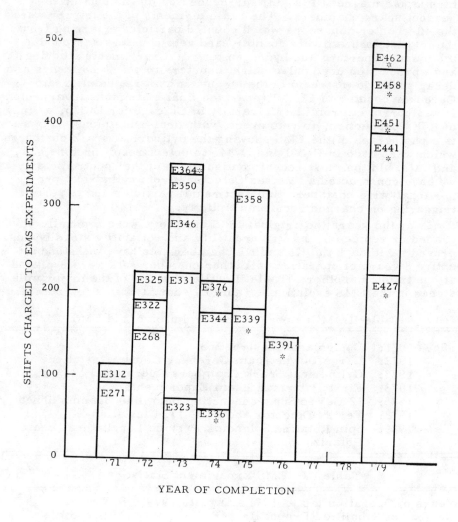

Fig. 4: Time distribution of EMS experiments. All shifts for each experiment are shown in the year when the experiment was completed. Asterisks indicate polarized beam experiments.

which are some measure of the spectrometer's achievements in the past eight years. Of course the number of publications from the EMS will increase considerably as we dig into the large volume of new polarized beam data recorded during the last eight months of operation in 1979. Figure 4 gives an idea of the time distribution of EMS experiments, which are listed in detail in Table IV. During the gap in operations in 1977 and 1978, the beamline to the EMS was rebuilt for the third time, to allow the installation of spin-rotating solenoids for longitudinal beam polarization, and the Vertex Detector was built and installed.

Table IV: List of EMS Experiments

E-268 Proposal to Study Meson Production with a Wire-Spark-Chamber Spectrometer (ANL), R. Diebold
Proposal - Jan. 1970, Completed - Jan. 1972, 155 shifts

E-271 A Proposal to Study the Reactions $\pi^- p \to K^0 \Lambda^0$ and $\pi^- p \to K^0 \Sigma^0$ in the Range 3.0 to 6.0 GeV/c (ANL), C. E. W. Ward
Proposal - Feb. 1970, Completed - Dec. 1971, 89 shifts

E-312 Proposal to Measure Elastic Scattering at Small Angles (ANL), D. R. Rust
Proposal - Sept. 1971, Completed - Oct. 1971, 23 shifts

E-322 Study of Λ^0 Production in 6 GeV/c Proton-Proton Interactions (U. of Chicago, ANL, Ohio State U.) T. M. Knasel
Proposal - Jan. 1972, Completed - Feb. 1972, 35 shifts

E-323 A Measurement of the Three Pion Mass Spectrum in the Reaction $\pi^- C^{12} \to \pi^+ \pi^- \pi^- C^{12*}$ (4.45 MeV) (U. of Illinois), U. E. Kruse
Proposal - Feb. 1972, Completed - Jan. 1973, 73 shifts

E-325 Proposal to Study $f^0 - A_2^0$ Interference in $\pi^- p \to K^+ K^- n$ (ANL), D. S. Ayres
Proposal - March 1972, Completed - April 1972, 31 shifts

E-331 Proposal to Study the Reactions $K^+ p \to K^0 \Delta^{++}$ and $K^- n \to \bar{K}^0 \Delta^-$ Using the Effective Mass Spectrometer (ANL, Calif. Inst. Tech.), B. Musgrave
Proposal - March 1972, Completed - Dec. 1973, 148 shifts

P-332 Study of Multiple Pion Productions, Associated with Nuclear Gamma Transitions, (USCD), O. Piccioni
Proposal - March 1972, (withdrawn)

Table IV (continued): List of EMS Experiments

E-336 Measurement of the Polarization of Λ's Produced by a Polarized Proton Beam (U. of Chicago, Ohio State U., ANL), R. Winston
Proposal - May 1972, Completed - May 1974, 59 shifts

E-339 Study of Resonance Production with a Polarized Proton Beam Using the Effective Mass Spectrometer (ANL), A. B. Wicklund
Proposal - May 1972, Completed - April 1975, 180 shifts

E-344 Neutral Meson Spectrum Near 1000 MeV (Iowa State U.), E. C. Peterson
Proposal - July 1972, Completed - April 1974, 116 shifts

E-346 A Proposal to Measure Λ^o Polarization with High Statistics in $\pi^- p \rightarrow K^o \Lambda^o$ at 5 GeV/c (ANL), C. E. W. Ward
Proposal - July 1972, Completed - Dec. 1973, 71 shifts

E-350 Further Study of Forward Elastic Scattering (ANL), R. Diebold
Proposal - Oct. 1972, Completed - Feb. 1973, 47 shifts

E-358 Study of the $K\bar{K}$ System Using the Effective Mass Spectrometer and a Deuterium Target (ANL), A. B. Wicklund
Proposal - April 1973, Completed - May 1975, 140 shifts

E-364 Proposal to Measure the pp Elastic Scattering Polarization Parameter Using the Effective Mass Spectrometer and the Polarized Beam (ANL, Indiana U., Ohio State U., U. of Chicago), D. R. Rust
Proposal - July 1973, Completed - Oct. 1973, 6 shifts

E-376 Measurement of the Polarization Parameter for pn Elastic Scattering from 2 to 6 GeV/c (ANL), R. Diebold
Proposal - May 1974, Completed - Oct. 1974, 35 shifts

E-391 Measurement of Polarization Effects Using the Effective Mass Spectrometer and Polarized Proton Beam at 9 and 12 GeV/c (ANL), S. L. Kramer
Proposal - March 1975, Completed - Dec. 1976, 143 shifts

Table IV (continued): List of EMS Experiments

P-394 Measurement of Spin-Rotation Parameters in Λ^o Production at 12 GeV/c Using the Polarized Proton Beam and the Effective Mass Spectrometer (Ohio State U., U. of Chicago, ANL), A. Lesnik
Proposal - June 1975, (withdrawn)

E-427 Proposal to Study Exclusive Λ-Production Reactions with the ZGS Polarized Proton Beam (ANL, Elmhurst College, Ohio State U.), D. Ayres
Proposal - Jan. 1977, Completed - July 1979, 205 shifts

E-441 Proposal to Study the Reaction $p_\downarrow p \rightarrow p\pi^+\pi^- p$ with the ZGS Polarized Proton Beam (ANL), A. B. Wicklund
Proposal - Oct. 1977, Completed - April 1979, 181 shifts

E-451 Proposal to Study Longitudinal Spin Correlations in the Reaction $pp \rightarrow p\pi^+ n$ at 6 and 12 GeV/c (ANL), A. B. Wicklund
Proposal - Sept. 1978, Completed - June 1979, 41 shifts

E-458 Proposal to Study pK^+K^-p and $\bar{p}pp\bar{p}$ Final States Using the Effective Mass Spectrometer (ANL, Elmhurst College), M. Arenton
Proposal - Feb. 1979, Completed - May 1979, 48 shifts

E-462 Proposal to Study $pp \rightarrow pn\pi^+$ Between 1.5 and 2 GeV/c (ANL, Elmhurst College, Rice U.) A. B. Wicklund
Proposal - July 1979, Completed - August 1979, 24 shifts

AN EMS PHYSICS SAMPLER

I could not possibly give a talk on the EMS without showing some data. Since time is short, I have simply chosen eight of my favorite figures to give you a sampling of the physics which has been produced so far. These are shown as Figs. 5 through 12. Because we've concentrated on recording as much data as possible before the ZGS shutdown next month, polarized beam results are somewhat under-represented in this collection. Within the next few years, however, these recent EMS experiments can be expected to add significantly to our knowledge: E-451 extended the study of Δ^{++} production at 6 and 12 GeV/c to include measurements with a longitudinally polarized beam, to help separate

natural and unnatural parity-exchange production; E-462 continued
these studies down into the threshold region with the help of the
Vertex Detector, and holds the additional interest of perhaps detecting dibaryon resonances; E-427 was a high statistics study of
$pN \rightarrow \Lambda K^+ N$, where amplitude analyses and baryon spectroscopy
will be constrained by the Λ spin information, and where the use
of proton and neutron targets isolates isospin dependence; E-441 is
expected to yield detailed information on baryon spectroscopy in
$pp \rightarrow p\pi^+\pi^-p$ at 6 and 12 GeV/c, with longitudinal beam polarization
providing discrimination between production-mechanism asymme-
tries and resonance-decay asymmetries; E-458 was a measurement
of $pp \rightarrow pK^+K^-p$, $pp\bar{p}p$, providing information on production mechanisms and baryon spectroscopy at high masses, and yielding a
unique comparison of ϕ production and ω production (from $pp \rightarrow$
$p\pi^+\pi^-\pi^0 p$ in E-441) in pp interactions. Thus, although the
Effective Mass Spectrometer is now being dismantled, you can expect EMS physics results to continue to emerge for several years
to come.

THE EMS IN PERSPECTIVE

Now that the ZGS program is drawing to a close I cannot help
but ask what lessons there are to be learned from the amazing
productivity of our simple spectrometer. Although the EMS had
its share of startup "debugging" and routine equipment failures,
the actual accelerator time lost to such exercises was quite low
due to the high level of technical support in the ZGS complex, and
to the conservative design of the spectrometer itself. The relatively high reliability of the detectors was in large measure due to
the use of well-proven technology - scintillation counters,
Cerenkov counters, and spark chambers were already "old hat"
when the EMS was being designed. A second essential element
was that, despite numerous improvements and replacements, the
basic EMS changed remarkably little over the years. Targets and
detectors upstream and downstream of the magnet-spark-chamber
system were changed, but the basic apparatus for measuring fast
forward charged particles was always the same.

Throughout the first few years of EMS operation I found myself
worrying that the big, multipurpose spectrometers at other accelerators would put us out of business as soon as they came into
operation. The LASS spectrometer at SLAC, the CERN Omega,
and the Brookhaven MPS were all more expensive and sophisticated
devices than the EMS, but despite my concern, the EMS never did
come into direct competition with them. Due to their size and
complexity, these facilities took much longer to build, and thus
began operation several years after the EMS. When they finally
were ready to do physics, they often chose to take advantage of the
higher energy beams available to them, and to study the more
complex reactions which were accessible with their large acceptance and sophisticated triggering and tracking instrumentation.
Their more ambitious goals seemed to further delay the output
of physics results. Life at the ZGS turned out to be much simpler:
reactions with only a few final-state particles were easy to detect

and record, and had high cross sections at ZGS energies. Triggers were easy because the reactions of interest were a relatively large fraction of the total cross section, and reconstruction of two - or three-track events was immensely simpler than high multiplicity final states. Because the scintillation counters, wire spark chambers, and Cerenkov counters of the EMS were not at the frontier of technology, they turned on quickly and performed reliably year after year. The device was so simple to operate, that university groups were able to perform experiments with only a minimum of instruction by Argonne physicists.

The choice of experiments also had a lot to do with our success: high event rates made it easy to perform high statistics, systematic studies of related reactions over the whole range of ZGS energies. We utilized beams of $\pi^{\pm}$, $K^{\pm}$, protons and antiprotons, on both hydrogen and deuterium targets to isolate isospin dependence. Such comprehensive studies of related reactions with a single apparatus allowed us to perform comparisons with much smaller systematic errors than had been achieved before. It turned out that there were things to learn about even the simplest reactions. An excellent example of this is the first EMS experiment to be completed (E-312), a measurement of elastic scattering of $\pi^{\pm}$, $K^{\pm}$, p and $\bar{p}$ from protons, which measured the crossover values and their energy dependence very precisely. High statistics studies of meson spectroscopy and production mechanisms dominated the program in the early years, and this evolved naturally into a study of inelastic reactions using the polarized proton beam when it became available. With the polarized beam we were once again able to learn new things about simple reactions, a good example being the EMS measurement of the polarization parameter in proton-neutron elastic scattering (E-376).

The timely availability of the polarized proton beam was an important factor in the EMS productivity. As Fig. 4 illustrates, the polarized beam experiments joined smoothly onto the conclusion of the unpolarized beam program - had the polarized beam come a few years later, there might have been no EMS to make use of it. As it turned out, 10 of the 21 completed experiments utilized polarized protons.

ACKNOWLEDGEMENTS

I cannot end this talk without acknowledging the enormous contribution made by the ZGS staff, experimental area technicians, and support groups in ARF, AD, HEF, and HEP Divisions. Table I represents only the tip of a very large iceberg, and gives little hint of the long hours of hard work which were devoted to keeping the EMS on the air, usually by people who had to take the physicists' word for it that what we were doing made sense. It was the dedication of the people who worked here that ultimately made our work a success. If I may borrow the words of Bruce Cork, we were most fortunate to be working "at the right machine, at the right place, at the right time."

REFERENCES

1. D. S. Ayres, in Proceedings of the International Conference on Instrumentation for High Energy Physics, Frascati, Italy, 1973, edited by S. Stipcich (Laboratori Nazionali del Comitato Nazionale per l'Energia Nucleare, Frascati, Italy, 1973), p. 665.
2. I. Ambats et al., Phys. Rev. D9, 1179 (1974).
3. A. B. Wicklund et al., Phys. Rev. D17, 1197 (1978).
4. C. E. W. Ward et al., Phys. Rev. Lett. 31, 1149 (1973).
5. J. J. Phelan et al., Phys. Lett. 61B, 483 (1976).
6. D. S. Ayres et al., Phys. Rev. Lett. 32, 1463 (1974).
7. A. J. Pawlicki et al., Phys. Rev. D15, 3196 (1977).
8. R. Diebold et al., Phys. Rev. Lett. 35, 632 (1975).
9. S. L. Kramer et al., Phys. Rev. D17, 1709 (1978).
10. R. Diebold et al., in Proceedings of Orbis Scientiae 1977 on Deeper Pathways in High-Energy Physics, Coral Gables, Florida, 1977, edited by A. Perlmutter and L. F. Scott (Plenum Press, New York, 1977), p. 109.

SUPPORT

This work was performed under the auspices of:
    The U. S. Atomic Energy Commission,
    The U. S. Energy Research and Development Administration,
    The U. S. Department of Energy.

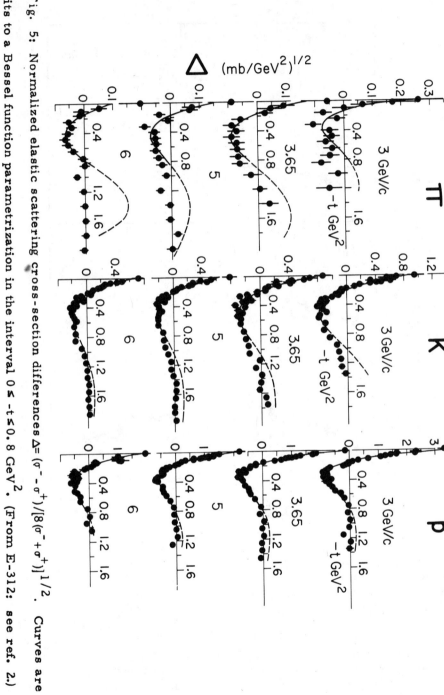

Fig. 5: Normalized elastic scattering cross-section differences $\Delta = (\sigma^- - \sigma^+)/[8(\sigma^- + \sigma^+)]^{1/2}$. Curves are fits to a Bessel function parametrization in the interval $0 \leq -t \leq 0.8$ GeV^2. (From E-312: see ref. 2.)

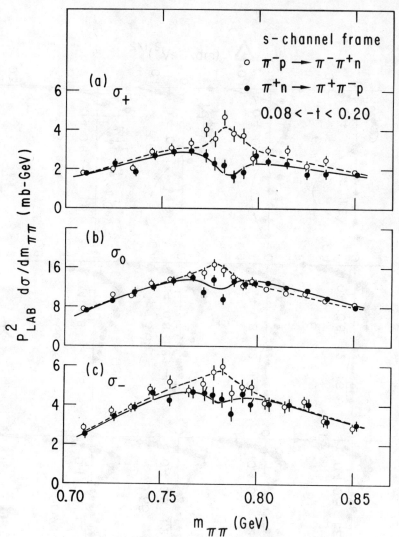

Fig. 6: Comparison of the dipion mass spectra in the ρ-ω inter-
ference region for the reactions $\pi^- p \to \pi^- \pi^+ n$ and $\pi^+ n \to \pi^+ \pi^- p$, where
the s-channel projections σ_0, σ_+, and σ_- are defined by
$\sigma_{ij} = P_{lab}^2 \rho_{ij} \frac{d\sigma}{dm}$. The data shown are for 4 GeV/c and
$0.08 \leq -t \leq 0.20$ GeV^2. This same experiment acquired precise data
on ρ^o production by pions and K^{*o}, $\overline{K}^{*o}$ production by kaons, and
permitted a systematic study of vector-meson production within the
framework of SU-(3) symmetric, strongly absorbed Regge-pole
models. New constraints to such amplitude analyses were provided
by the ρ-ω interference data and by ω-production data from the ZGS
Charged-Neutral Spectrometer. (From E-268: see ref. 3.)

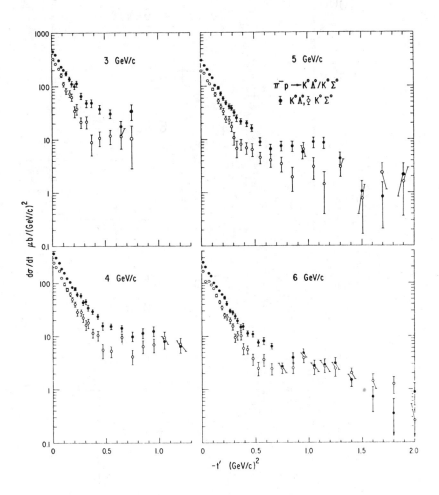

Fig. 7: Differential cross sections for $\pi^- p \to K^O \Lambda$ and $\pi^- p \to K^O \Sigma^O$ from 3 to 6 GeV/c. (From E-271: see ref. 4.)

138

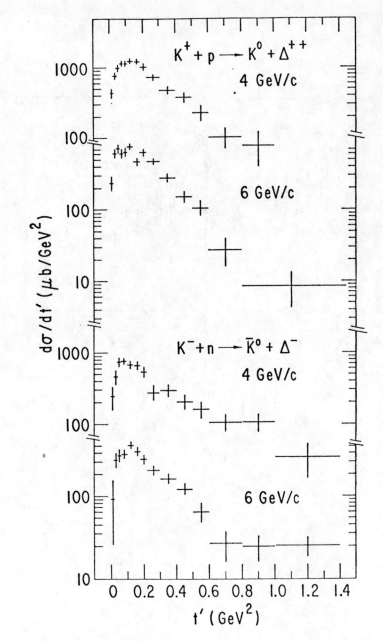

Fig. 8: Differential cross sections for line-reversal related reactions. The ratios of the $K^0\Delta^{++}$ to $\bar{K}^0\Delta^-$ cross sections at 4 and 6 GeV/c are 1.62±0.27 and 1.60±0.26 respectively. (From E-331: see ref. 5.)

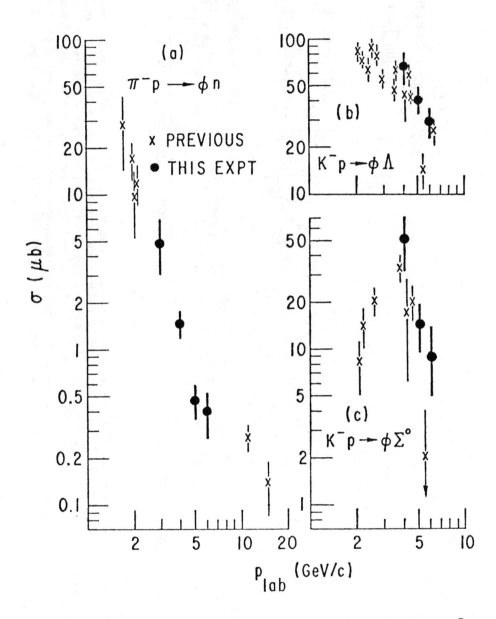

Fig. 9: Differential cross sections for φ-meson production in π⁻p and K⁻p interactions. In (a), cross sections have been integrated out to -t' = 0.9 GeV²; in (b) and (c), cross sections are integrated over the entire angular range. (From E-325: see ref. 6.)

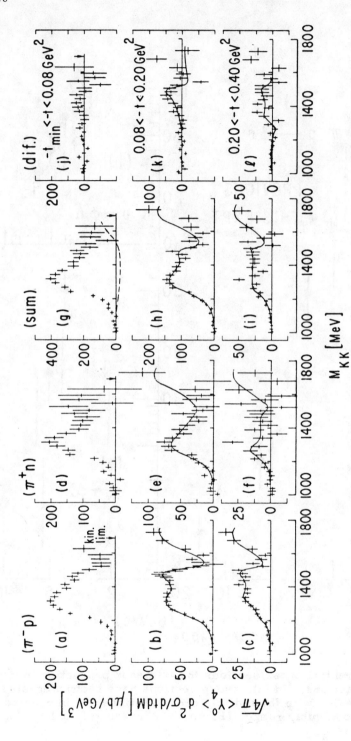

Fig. 10: The Y_4^0 moments of the K^+K^- decay angular distribution versus K^+K^- effective mass, for the reactions $\pi^- p \to K^- K^+ n$ and $\pi^+ n \to K^- K^+ p$ at 6 GeV/c, for the three t ranges given. Parts (g) through (l) show the sums and differences of the moments for the two reactions, and isolate the interference of isospin-0 and isospin-1 K^+K^- states. Production of S^*, ϕ, f, and A_2 mesons and their mutual interferences were clearly observed in this experiment, and f' production by pions was detected for the first time through the interference of the f and the f', as seen in the dramatic dropoff around 1500 MeV. (From E-358; see ref. 7.)

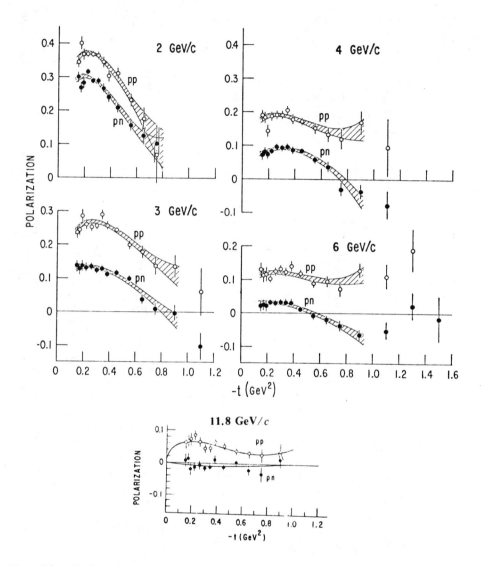

Fig. 11: Polarization asymmetries for pp and pn elastic scattering from 2 to 11.8 GeV/c. Fits to the data from 0.15 to 1.0 GeV^2 are shown as bands (±1 standard deviation). These data allowed for the first time the separation of isospin-0 and isospin-1 t-channel exchange contributions to the single-flip amplitude, and revealed a previously unsuspected rapid energy dependence in the isospin-0 amplitude. (From E-376 and E-391: see ref. 8, 9.)

142

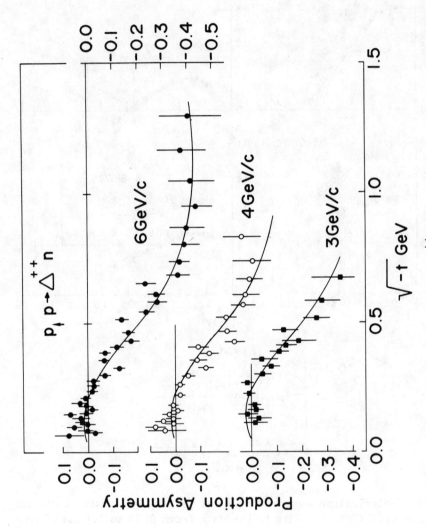

Fig. 12: Overall left-right asymmetries in $P_t p \rightarrow \Delta^{++} n$ at 3, 4, and 6 GeV/c. The curve which is shown with all three data sets is an eyeball interpolation of the 6 GeV/c data. (From E-339: see ref. 10.)

HISTORY OF THE POLARIZED BEAM[*]

Everette F. Parker
Argonne National Laboratory, Argonne, IL   60439

In 1973, the first high energy polarized proton beam was developed at the Argonne Zero Gradient Synchrotron (ZGS). It operated very successfully and productively until 1979 when the ZGS was shut down permanently. This report describes the development, characteristics, and operations of this facility.

In July 1973, the world's first high energy polarized proton beam became operational at the ZGS when a beam of $\sim 2 \times 10^7$ protons/pulse was provided to experiment E-324.[1] This marked the end of a development program that got its original start 15 years earlier. In the late 1950's, the first polarized ion sources were being put into operation and polarized targets were being developed. With sources and targets becoming available, the question of accelerating polarized protons in circular machines naturally arose at a number of labs. In 1959, Cohen and Burger[2] of the ZGS staff published the results of a study of the problem of acceleration in an injector linac having magnetic quadrupoles. They found that the small precessions of the spin as the protons pass through the quadrupole fields tend to average to zero net rotation because the betatron motion causes them to "see" as much field of one polarity as the opposite. In 1962, Cohen[3] published the results of an investigation of the problems of acceleration in the ZGS. Like Froissart and Stora[4] in their study of SATURNE, Cohen found that there is a series of depolarizing resonances which, if uncorrected, will depolarize the beam during the acceleration cycle. This depolarization is caused by the horizontal magnetic fields which provide the focussing forces for the beam. As with the linac, the spin rotation caused by these fields tends to average to zero. There are, however, very brief periods during the acceleration cycle when, for a few turns around the machine, these small rotations add coherently and the spin experiences a net rotation on average. These resonances occur when the energy of the particle is such that the following equality is correct.

$$G\gamma = [L \cdot K] \pm [m \cdot \nu_y] \pm [n \cdot \nu_x] \qquad (1)$$

where
G = 1.79 for protons
γ = the Lorentz factor
K = the sector number of the accelerator
ν_y, ν_x = the vertical and radial betatron tune numbers of the machine
L, m, n = integers

---

[*]Work supported by the U.S. Department of Energy.

Being inherent in the machine since they are caused by the fields which are required to keep the beam stably confined in the machine, these resonances are referred to as intrinsic resonances.

Horizontal fields can also be produced by errors in the magnetic alignment of the magnets. These field errors can produce what we refer to as imperfection depolarizing resonance, and they can occur when

$$G\gamma = \text{integer}$$

Cohen concluded that only the intrinsic resonances with m = 1, n = 0 would be particularly significant and proposed the use of pulsed quadrupoles to rapidly shift the tune of the machine through the resonance condition as a possible way to minimize their effect. This was, in fact, the technique we used.

These studies took place 20 years ago, four years before the ZGS became operational. With the press of completing the ZGS and starting up a conventional (i.e., unpolarized beam program), the subject of polarized protons was forgotten.

A second wave of interest developed in 1968 when Alan Krisch of the University of Michigan requested the Accelerator Division to consider the feasibility of accelerating polarized protons. The key result of this study was the conclusion that an acceptable polarized ion source (PIS) was commercially available but that its size and complexity would make switching back and forth between polarized and unpolarized beam operations very costly in terms of effort and lost beam time. Lost beam time was very significant at that time since the 12-ft bubble chamber program was soon to start, and this already represented a new mode of running; i.e., when the entire machine is dedicated to producing neutrinos for the neutrino program of the 12-ft. This second study of accelerating polarized protons at the ZGS, like the previous study, came to a quiet end primarily because the key outside user, A. Krisch, and the key Argonne National Laboratory (ANL) physicist, L. Ratner, went off to CERN to work on one of the first ISR experiments.

In 1970, again at the request of A. Krisch, the Accelerator Division considered the feasibility of developing a polarized proton beam facility at the ZGS. This study showed that a facility capable of intensities of the order of 10^9 protons/pulse and polarizations of at least 50% could be developed for a reasonable cost and with limited time loss associated with switching between polarized and unpolarized beam operations. No part of the approximately $500 K of equipment and materials and services funds required to design and construct this facility could be obtained as a special supplemental appropriation from the then Atomic Energy Commission. Bruce Cork, the Associate Laboratory Director for High Energy Physics at that time, felt the project should be carried out any way so he reprogrammed existing monies and told us to go ahead. We got approval in October 1971, and less than two years later the facility was operational.

A detailed technical description of the physics and hardware involved in producing the polarized beam at the ZGS has been given elsewhere[5] so only a very brief review will be given here. The key

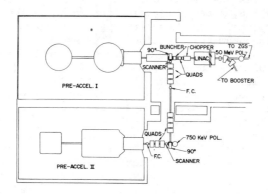

Fig. 1.   Layout of ZGS Preaccelerators.
Preaccelerator II houses the Polarized
Ion Source.

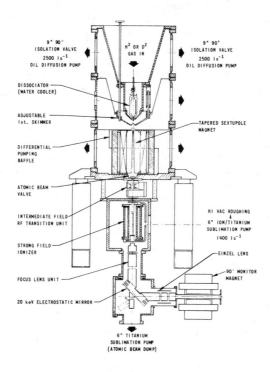

Fig. 2.   ANL Polarized Proton Ion
Source, Original Configuration.

equipment items that had to be added to the ZGS for the polarized beam facility were a PIS, a new preaccelerator to house the PIS, pulsed quadrupoles for jumping the depolarizing resonances, and polarimeters for measuring the beam polarization. A separate preaccelerator for the PIS was the solution to the interference problem between the polarized and unpolarized beam operations. Its location relative to the original preaccelerator is shown in Fig. 1. The enclosure was used as a storage room prior to its conversion. The floor was removed to give it the required vertical extent, and steel window screening was hung on the walls to make them electrically smooth. The high voltage power supply was constructed from spare parts from the original preaccelerator. The source terminal, 8 ft x 11 ft x 14.5 ft in size, was manufactured by a company that makes campers. The terminal was made using a rib and panel type construction with 6-in. radius corners. This was a bit of a gamble for a 750-kV terminal to be located in an enclosure that allowed only a 6-ft separation from the wall; however, cost and delivery considerations made the gamble necessary.

The ion source was purchased from the Auckland Nuclear Accessory Company (ANAC) of Auckland, New Zealand. Figure 2 is a diagram of the source as it looked originally. It is a ground state type source[6] with a strong field ionizer. Figure 3 shows the source in the ANAC shop in Auckland in its final stage of fabrication. The size and complexity of the ion source are awesome relative to the conventional duoplasmatron normally used on proton synchrotrons. In its present configuration, it weighs about 10,000 lbs and consumes about 35 kW of electrical power. It contains three rf systems; six magnets; six beam lens elements; and nine vacuum pumps including diffusion, ion, turbomolecular, sublimator, and mechanical; just to identify some of its components.

As delivered, the PIS produced a dc beam current of 6 μA with a proton polarization of about 75%. It had only one rf transition so the polarization direction could be reversed only by reversing the current in the ionizer solenoid magnet. Due to the finite life of high current contactors, the reversal could not be made very frequently; i.e., only a few times per hour. Also since the solenoid field effects the beam optical properties, spin reversal was also accompanied by a ZGS intensity charge. Originally, the PIS had several lens elements containing fine wire grids. These grids were supposed to last for thousands of hours; but in our operations, they lasted only a few. These grids, along with several other parts, gave the PIS an overall mean-time-to-failure of only a few days.

The two pulsed quadrupoles, outlined in Fig. 4, were machined from commercially available "C" cores. The power supply is a resurrected extraction bump magnet power supply. The quadrupoles have a 10 μsec rise time; a gradient of 50 G/in., which produces a tune shift, $\Delta\nu_y$ of ~ 0.01 at 12 GeV/c; and a decay time constant of 1.3 msec. The quadrupoles can be pulsed up to 12 times each machine cycle. The resonances are such that the pulses are of alternate polarity. The pulsing time, amplitude, and decay time of each pulse is independently controllable.

Originally we had two polarimeters, one at the end of the injector linac to monitor the beam polarization before it enters the ZGS

Fig. 3. ANL Polarized Ion Source in
the Final Stage of Construction.

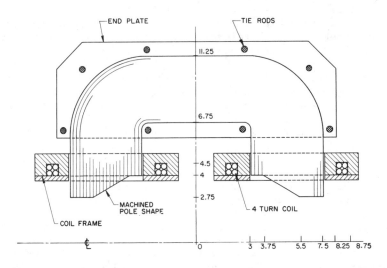

Fig. 4. Top Half of Pulsed Quadrupole Showing Computer
Designed Pole Profile and Copper Placement.

and one in EPB-I, beam 1 to measure the full energy polarization. The
50 MeV polarimeter, shown in Fig. 5, uses a fine carbon filament as a
target and looks at the left-right asymmetry in the elastically scattered protons with two triple coincidence scintillator telescopes.
The data rate from this device allows a 1% polarization measurement in
$\sim$ 10 min. The high energy polarimeter, Fig. 6, consists of two double
arm spectrometers for analyzing protons scattered from an LH_2 target.
Because of the small cross section and limited acceptance, this is a
very low data rate device. At 10^{10} protons/pulse on target, it takes
about 2 hours to make a 10% measurement at 6 GeV/c. In order to have
a faster indication of the beam polarization, a third polarimeter was
added in 1975. This device consists of two triple coincidence scintillation telescopes viewing recoils from a small CH_4 target mounted
in the extracted beam just down stream of the ZGS. Being uncon-
strained, the noise level in this polarimeter is high and an absolute
measurement is not possible. However, with an asymmetry about 3/4 of
the pure elastic value and a very high rate, relative polarization
measurements to the 1% level can be made in about 10 min. This device
has been a key item in allowing us to find and maintain the beam polarization in an efficient and rapid manner.

Along with these new equipment devices, there was also a significant machine physics, both theoretical and experimental, program required. Although the theoretical ground work had been laid years
earlier, detailed studies were required to make sure all of the possible effects were being considered and to more precisely identify which
resonances we had to worry about and what kind of tune shifts would be
required to avoid losing significant polarization. In the experimental area, there was the question of just how to set the machine tunes
and how to locate and correct for the depolarizing resonances. From
Eq. (1), the energy at which the resonance occurs is known to within
the uncertainty of one's knowledge of the tune value. Unfortunately,
the acceleration cycle timing is based only on a relative measurement
of the magnet field; therefore, there is no absolute relationship
between what is measured to be the magnet field and the particle energy. From experience, we felt we knew the relationship between the
measured magnetic field and the particle energy or γ to within about
±1%. This meant that the correct quadrupole pulsing time for the
first strong resonance was expected to lie somewhere within a 100 to
200 G region or 5 to 10 msec period in the acceleration cycle. This
does not now seem like a big window, but in July 1973 it seemed to be
infinite.

The exercise of locating the first resonance turned out to be
rather difficult not only because of the large window to be searched,
but also because of the very low initial beam intensity, the small
acceptance of the high energy polarimeter, and the fact that the
strength of the resonances (i.e., the amount of depolarization) was
twice as much as the theorists had predicted. The extracted beam
intensity was of the order of 2 x 10^7 protons/pulse during the first
few weeks the system was operational. This yielded a data rate in the
polarimeter of only a few events per pulse with which to find and map
the resonances. Also as originally constructed, the spin direction
could be reversed only once or twice per hour so the systematic error

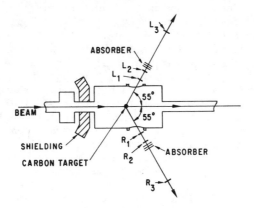

Fig. 5. 50 MeV Polarimeter Showing the
Two Symmetric Three-Counter Scintillator
Telescopes.

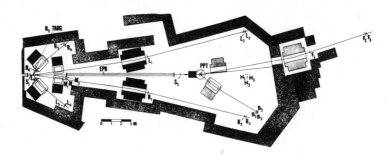

Fig. 6. ANL High Energy Polarimeter.

cancellation we have enjoyed with pulse-to-pulse reversal since 1974 was not available. Thus, poor statistical precision was compounded by systematic effects. Trying to find that first resonance was an almost hopeless task. Fortunately, after many hours of blindly searching, a data point was taken where the spin actually reverses. These reversal points are adjacent to the resonances and occur because the tune shift being too early actually enhances the strength of the resonance to the point a significant fraction of the spins actually reverse. (See Fig. 7.) This data point, taken where the spin is partially reversed, stood out even with the statistical noise of the data relative to the adjacent point where the spin is maximal in the proper direction. With this hint, we then concentrated on this region and discovered that we had located and jumped the first resonance. With the first resonance located, the others were relatively easy to find and jump since the γ vs. measured magnetic field curve was now normalized. Figure 7 shows a plot of the $G\gamma = 8 - \nu_y$ resonance. The spin reversal region at the leading edge is clearly demonstrated. With the mapping of the first resonance and finding it to be considerably stronger than predicted, the theorists were quick to find the necessary factor of two to make the theory and experiment agree very well. This, of course, is an often seen phenomena in physics.

With the beam thus polarized up to 6 GeV/c, the experimental program was turned on only ten days after the first polarized protons were injected into the ZGS. The momentum was restricted to 6 GeV/c and below for the first three years since low momentum running requires much less electrical power and is, therefore, cheaper than high momentum operations. This in turn allowed us to get much more running time per year from a limited budget. This, of course, was in no way a detriment to the physics program since any energy above 50 MeV was virgin territory for polarized beam experiments with a primary polarized beam. In February 1976, the momentum was raised to 11.75 GeV/c when the physics program requirements dictated this increase. The peak momentum was held to 11.75 GeV/c because a very strong resonance exists at 12.0 GeV/c, and the expected loss of polarization in crossing this resonance was deemed too much for such a small increase in momentum up to the nominal ZGS momentum of 12.3 GeV/c. Locating and jumping the resonances between 6 GeV/c and 11.75 GeV/c were straightforward, as expected. However, the resulting polarization was only about 45%, well below the expected 65 to 70%. After many days of investigating this unexpected loss of polarization, the causes were finally tracked down to two sources, one physics and one engineering. The physics source was the imperfection resonances discussed earlier; i.e., resonances caused by vertical bumps in the equilibrium orbits. Such resonances had been predicted[7] and indeed had been experimentally studied at the ZGS[8] but only by greatly enhancing their effects to make them detectable. Their apparent weakness meant that under normal circumstances, they would have negligible effect. However, after a very exhaustive and complicated study, it became clear that a number of the imperfection resonances which occur at the rate of about 2 per GeV/c were each costing a difficult to detect few percent loss in polarization. The net result of these many small losses was a very noticeable loss. These resonances were avoided by introducing

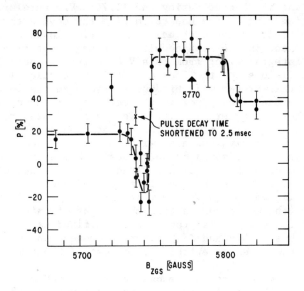

Fig. 7.   Beam Polarization Vs. Quadrupole Timing.

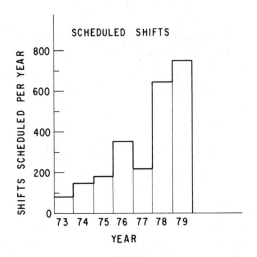

Fig. 8.   Shifts Scheduled Per Year.

vertical orbit bumps via the ZGS pulsed pole face winding system which removed the resonant condition. Eventually a total of 22 such corrective pulses was introduced during the 11.75-GeV/c acceleration cycle.

The engineering source had to do with the device that measures the magnetic field of the machine. This device looks at the output of an inductive loop in the machine with a voltage controlled oscillator (VCO). The integral count from this VCO is then proportional to the total magnet field since the loop output is proportional to dB/dt. Any noise in this system will produce spurious counts in this integrator which produces an error in the measured magnetic field. This means that there is a certain jitter in the field measurement and, therefore, in the firing time of the pulsed quadrupoles. This integration error obviously increases with energy, and the width of the correction for these resonances decreases with energy. This in turn means that the quadrupoles are not fired at the correct time on some machine cycles, and this shows up in a loss of average polarization. This problem is compounded if a "front porch" is used since the integrator "sees" the noise counts through out the "front porch" even though dB/dt is zero. This problem was minimized as much as possible by using the best VCO's and integrators; however, the problem still remains with us, at an acceptable level most of the time but not always.

In 1979, the very strong resonance at 12.1 GeV/c was jumped and the machine operated at 12.75 GeV/c for experiment E-452. In spite of our previous worries about this resonance, it was easily jumped and the 12.75 GeV/c turned out to be most successful.

In the fall of 1978, we produced and operated for two months the world's first and only high energy polarized deuteron beam. Conceptually, this is much easier than accelerating polarized protons since the G factor for the deuteron is so much smaller than that of the proton, -0.14 vs. 1.79. From Eq. (1), it can be seen that, with a tune value $\nu_y = 0.8$, only the L = 0 resonance exists within the energy range at the ZGS; i.e., at 10.4 GeV/c. All others are beyond the energy of the ZGS. In actuality, the production of this beam turned out to be less than straightforward for several reasons;[9] however, the run turned out to be quite successful. The press of the proton program in the face of the October 1979 shutdown precluded any further deuteron operations although there were requesters.

Up until the time the ZGS became a dedicated polarized proton facility, the polarized beam was used for physics about three months per year or about one third of the available time. These were usually one month runs separated by several months. Although this schedule was dictated by the overall physics requirements and funding limitations, it would have been necessary in any event because of the time required for maintenance and repair of the PIS plus the time needed to make the modifications and improvements required to keep up with the ever expanding requirements of the experimental program. Since becoming a dedicated facility in March 1978, the facility has run almost continually except for the month of July 1978. This has been possible primarily due to very significant improvements in the reliability of the PIS.

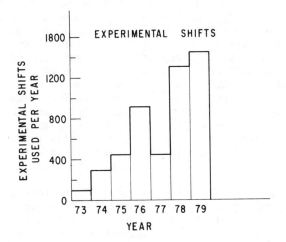

Fig. 9.   Experimental Shifts Used Per Year.

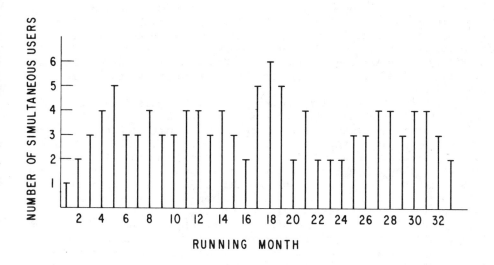

Fig. 10.   Simultaneous Users as a Function of Operating Month.

The best way of showing how the facility developed and how productive it was is to show certain descriptive parameters as a function of time. The first is shown in Fig. 8, which shows the number of 8 hour shifts scheduled for physics per year. Figure 9 shows the number of experimental shifts used per year; i.e., the number of shifts the beam was available times the number of experimenters using the beam. Figure 10 shows the number of simultaneous users we had during each month the system was in operation. The facility, as originally envisaged, would be able to support one, possibly two, experiments at a time. From this figure, it is clear that we rapidly outgrew this expected limit. For the first run, only beam 1 had a polarized beam user. By the second run, the Effective Mass Spectrometer had polarized beam users. Beam 22 then followed. The high momentum secondary beam in EPB-I, beam 5, was then modified for polarized beam use. Beam 2 was then modified for the parity experiment;[10] and then in 1978, beam 23 was added along with a new polarized target. In 1974, the 12-ft bubble chamber joined the users for a 150 K picture exposure.

Unlike unpolarized operations where the protons are mostly used to produce secondary particles, polarized beam experiments use the protons directly. Thus, energy dependence studies require frequent changes in the machine energy and agreement among the users as to the particular energy to be used. Generally, one could get agreement for two energies so frequent use was made of front porches. This worked very well after EPB-I was modified to include programmable magnets and power supplies so that multiple energies could be efficiently transmitted the length of EPB-I on the same machine cycle. In summary, we started off with one experimental beam site and the ability to transmit one energy per machine cycle. We had two N-type polarized targets, one with only an He^3 cryostat, and could produce only beams with the spin in the vertical plane. We will terminate this facility with the ability to transmit multiple energies each machine cycle to six experimental sites. We have three polarized targets, one of which is an R&A type target, and all three have He^3 cryostats; and longitudinally polarized beam at any energy is available to two of the sites, beams 21 and 22.

Over 40 different energies were used at the polarized beam facility during its six year life. These ranged from as low as 350 MeV to slightly over the design energy of the ZGS. A significant part of the time was spent at energies now available at TRIUMF, LAMPF, and SIN. Not only were numerous energies used, but also many energies were required during a single month run. In August 1977, we changed energies 11 times. That we were able to change energies so frequently and still have time for a viable physics program speaks well of our machine operators and experimental area people.

Figure 11 shows the intensity as a function of time. This shows the dramatic growth in the beam intensity that has taken place over the years, a growth that made the high utilization levels described above possible. The original system's goals called for 3×10^9 protons/pulse, and this to some of us seemed impossibly high. You can see this goal was surpassed by the 25th week of running. This almost three orders of magnitude increase in intensity since July 1973 was

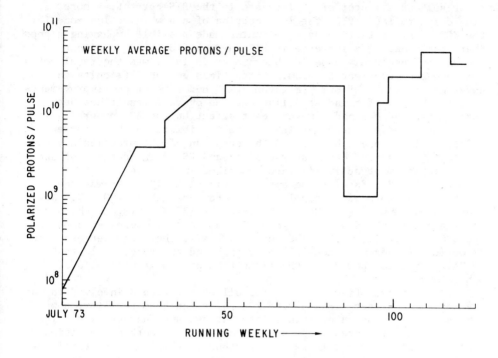

Fig. 11.   Intensity Profile of the ZGS Polarized Proton Beam.

the result of a factor of 15 increase in the PIS current, a more
stable and reliable PIS, the introduction of a new injection mode for
the ZGS,[11] and effective machine tuning made possible by having a good
operating crew. The increase in PIS current from 6 μA to almost
100 μA, an absolutely unbelievable number in 1973, was the result of a
continuous development program. Often times we had difficulty in re-
covering from some of our "improvements," and some of our improvements
were the result of blind stumbling but the overall result has been
outstanding and has had a significant effect on the ZGS program. It
has also had effect at other labs. The possibility of high current
polarized ion sources has raised the question of the practicality of
developing polarized beams at the AGS[12] and PS.[13] The details of the
source improvement activities are described in Ref. 14.

When the ZGS facility became operational in 1973, it was the
first machine to produce a polarized proton beam above 100 MeV. Now
beams in the hundreds of MeV range exist at TRIUMF, SIN, LAMPF, and
Indiana and in the few GeV range at Saclay; energy ranges encompassed by
the ZGS. Hopefully within the next few years, the range between 3 and
12 GeV/c will become accessible once more and the range of 12 to
25 GeV/c will be opened for the first time at Brookhaven National
Laboratory.

I am not qualified to speak for all of the people involved in the
development and operations of the ZGS polarized beam facility; however,
for my part, it was an exciting, interesting, and fun activity. The
pressure was often great, the hours long, and the working conditions
less than pleasant; but it was all somehow worthwhile. The thrill
never abated for seeing an asymmetry develop in the event rates of
two spectrometers consisting of 25-ton magnets operated at kilogauss
levels when making one or two gauss changes in a 10-lb magnet operating at 150 G. The beauty of the physics that allows one to make
microvolt changes in the energy levels of an atom travelling with a
$\beta = 10^{-4}$ and change the interaction probabilities at GeV energies by
60% never loses its shine.

This is supposed to be an historical review of this facility;
and as such, it should include the names and contributions of the
people involved. A complete list would be too long to manage here for
it would include just about everyone at the ZGS Complex plus some outsiders. However, the contribution of some must be recorded for they
made the key efforts, the key designs, or donated the most blood,
sweat, and tears. The entire Injector Group was involved in the design, construction, and operation of the preaccelerator and PIS; but
key among these were J. Madsen, E. Parker, N. Sesol, R. Stockley, and
R. Timm for design and construction and E. Parker, P. Schultz,
N. Sesol, and T. Singleterry for its operation and development.
C. Potts and R. Zolecki were the primary movers behind the pulsed
quadrupole system. L. Ratner provided the yeoman effort behind the
accelerator studies required to find and keep the polarization; and
in his imperfection resonance studies, A. Wright provided the key
weapons, pulsed pole face winding modifications. The machine theory
effort was provided by T. Khoe and the polarimetry by A. Krisch of
the University of Michigan; J. Roberts, then from Michigan but now at
Rice University; and J. O'Fallon, then of St. Louis University but

now of the Argonne Universities Association at ANL. Of course, none of this would have been possible without the guidance and support of our Associate Laboratory Director, B. Cork, and the ARF Division Director, R. Martin.

REFERENCES

1. E. F. Parker et al., Phys. Rev. Letters 31, 783 (1973).
2. D. Cohen and A. Burger, Rev. Sci. Instr. 30, 1134 (1959).
3. D. Cohen, Rev. Sci. Instr. 33, 161 (1962).
4. M. Froissart and R. Stora, Nucl. Instr. Methods 7, 297 (1960).
5. T. Khoe et al., Particle Accelerators 6, 213 (1975).
6. H. Glavish, Polarization Phenomena in Nuclear Reactions (University of Wisconsin Press, Madison, 1971), 267.
7. O. Barbalat et al., Proceedings of the Summer Studies on High Energy Physics with Polarized Beams, ANL Report ANL/HEP 75-02, XIV-1 (July 1974).
8. Y. Cho et al., High Energy Physics with Polarized Beams and Targets, AIP Conference Proceedings 35 (American Institute of Physics, NY, 1976), 396.
9. E. F. Parker et al., IEEE Trans. on Nucl. Sci. NS-26, 3200 (1979).
10. D. Nagle et al., High Energy Physics with Polarized Beams and Targets, AIP Conference Proceedings 35 (American Institute of Physics, NY, 1976), 231.
11. Y. Cho et al., IEEE Trans. on Nucl. Sci. NS-24, 1426 (1977).
12. D. G. Crabb et al., IEEE Trans. on Nucl. Sci. NS-26, 3203 (1979).
13. M. Bell et al., High Energy Physics with Polarized Beams and Targets, AIP Conference Proceedings 35 (American Institute of Physics, NY, 1976), 405.
14. E. F. Parker, High Energy Physics with Polarized Beams and Targets, AIP Conference Proceedings 35 (American Institute of Physics, NY, 1976), 382.

THE ZGS POLARIZED BEAM PROGRAM

J. B. Roberts, Jr.
Physics Department and T. W. Bonner Nuclear Laboratories
Rice University, Houston, Texas 77001

I will probably not be contradicted in saying that the polarized beam program has produced many of the highlights of the entire ZGS physics program. The scope of the discoveries has extended far beyond any expectations that existed at the time of the initial acceleration of the polarized beam in July of 1973. Some statistics about the polarized beam program are listed in Table I. (I apologize in advance to anyone whose data is not discussed. It is impossible to show all the high quality data which has been taken with the polarized beam in a talk of this length.)

TABLE I

POLARIZED BEAM PROGRAM

Beam Parameters: momentum: 1–12.8 GeV/c

polarization: protons 71%; deuterons 55%

intensity: up to 9×10^{10} circulating protons with 3×10^{10} extracted

beam lines: up to 6 fed simultaneously, 3 with polarized targets

Beam Lines	Primary Physics Studies
1	PPT V: Elastic scattering
2	Parity violation
5	Elastic and Inelastic scattering
21	Effective Mass Spectrometer (EMS)
22	PPT III: Elastic scattering, Total Cross Sections
23	PPT VI: Elastic Scattering, Total Cross Sections

Between July 1973 and September 1979, 43 polarized proton and four polarized deuteron experiments were completed.

Let us begin by remembering one original motivation for accelerating the polarized proton beam to high energies; viz., to study the dynamics of nucleon–nucleon scattering at small t. Measurements of a sufficient number of parameters for a complete amplitude analysis of elastic scattering and studies of the dynamics of Δ and N^* production were the desired goals which in fact have been partially achieved.

ISSN:0094-243X/80/600158-11$1.50 Copyright 1980 American Institute of Physics

The five complex amplitudes parameterizing p-p elastic scattering are listed in Table IIA (ϕ's), and linear combinations of these which correspond asymptotically to t-channel exchanges of definite naturality (N,U) are also listed. The subscripts give net helicity flip.[1] The spin parameters which are most sensitive to the various amplitudes are listed in Table IIB (n direction is normal to the scattering plane, ℓ is along the incident beam momentum, and $\vec{s} = \vec{n} \times \vec{\ell}$). Many of these measurements have been made at a variety of energies at the ZGS, the largest number by the Yokosawa Group. A sufficient amount of data has been collected for a reasonably accurate smplitude analysis at small t at 6 and 11.8 GeV/c. The result of such an analysis at 6 GeV/c at $-t = .3$ done by Berger and Sorensen[2] is shown in Fig. 1. The dotted lines give the errors on the amplitudes. A similar analysis will be done at 11.8 GeV/c when the Yokosawa Group has completed their data analysis. Among the interesting conclusions is evidence for an A_1-like exchange (U_0).

TABLE II

A. Amplitudes for Elastic Scattering (Helicity Basis)

$$\phi_1 = \langle ++|++\rangle \qquad N_0 = 1/2(\phi_1 + \phi_3)$$
$$\phi_2 = \langle --|++\rangle \qquad N_1 = \phi_5$$
$$\phi_3 = \langle +-|+-\rangle \qquad N_2 = 1/2(\phi_4 - \phi_2)$$
$$\phi_4 = \langle +-|-+\rangle \qquad U_0 = 1/2(\phi_1 - \phi_3)$$
$$\phi_5 = \langle ++|+-\rangle \qquad U_2 = 1/2(\phi_2 + \phi_4)$$

B. Spin parameters sensitive to various amplitudes:

$$N_0 \sim d\sigma/d\Omega, \rho = \frac{ReN_0}{ImN_0} \qquad ReU_0 \sim H_{\ell sn}$$
$$ReN_1 \sim A_n (P) \qquad ImU_0 \sim A_{\ell\ell} (C_{\ell\ell}) (\Delta\sigma_\ell)$$
$$ImN_1 \sim D_{ss} \qquad ReU_2 \sim H_{sns}$$
$$ReN_2 \sim H_{nss} \qquad ImU_2 \sim A_{ss} (C_{ss}) (\Delta\sigma_T)$$
$$ImN_2 \sim A_{nn} (C_{nn})$$

In addition to the p-p elastic measurements, the polarized beam has simplified the measurement of a certain number of n-p parameters by scattering from a liquid deuterium target. The measurements of the n-p polarization made by the Effective Mass Spectrometer (EMS) Group are shown in Fig. 2. The anomolously rapid energy dependence of the I=0 single flip amplitude was quite unexpected, and has led theorists[3] to postulate the existence of a new I=0 trajectory (ε) whose exchange contribution falls off very rapidly with energy.

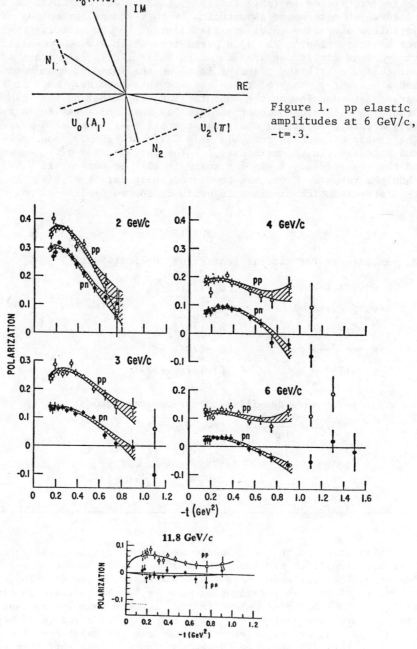

Figure 1. pp elastic
amplitudes at 6 GeV/c,
−t=.3.

Figure 2. pp and np elastic polarization. Curves
at 11.8 GeV/c are predictions based on the lower
energy data.

There are many nice examples of Δ and N* production processes studied by the EMS Group for which the polarized beam provides an excellent handle which illustrates the dynamics involved. However, explanation of any of these effects is too lengthy for this paper. Rather, I'll discuss two examples from the data of the Minnesota-Rice-Argonne collaboration in Beam 5. Polarization asymmetries in inclusive pion production at large x are shown in Fig. 3. The shaded area encompasses all the data for x>.7 for both 6 and 12 GeV/c. The data points are the $\pi^{\pm}p$ backward elastic polarization measurements at 6 GeV/c. Note that the inclusive π^+ and π^- asymmetries are both quite large and quite different from one another, but are similar in magnitude and crossing zeros to the backward elastic data. This supports the idea that the dynamics of high-x meson production can be described in terms of baryon exchange in a similar way to backward meson-nucleon elastic scattering. A second example is illustrated

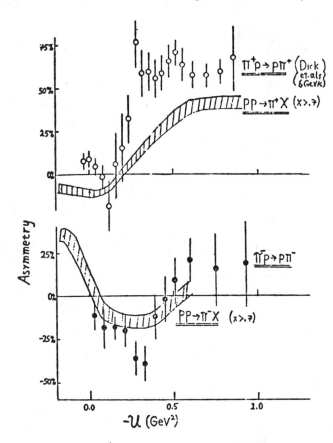

Figure 3. The $\pi^{\pm}$ asymmetry for large x at 11.8 GeV/c compared with the 6 GeV/c polarization data of Dick, et al for backward $\pi^{\pm}$ p elastic scattering.

162

by the data in Fig. 4. Figure 4a is a plot of the missing mass recoiling from the fast forward proton in p+p→p+X in the region where
$M_x \cong M_{\Delta+}$. Note the clear Δ^+ peak, with Δ/background $\cong 1/1$. Figure 4b
is a plot of the depolarization parameter D_{nn} of the fast forward
proton whose spin is analyzed by scattering from a second liquid
hydrogen target. Note that D_{nn} for elastic scattering and for missing
masses higher than the Δ^+ is about +1, but drops sharply to about
zero at M_Δ. Since Δ/background$\cong 1/1$, this gives D_{nn} for pure pp→pΔ^+
of about -1, which is the value expected for purely pion exchange.
This illustrates very strikingly that π-exchange is the dominant
dynamical mechanism for Δ production at small t.

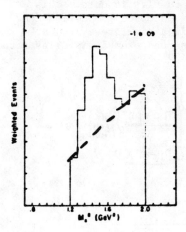

Fig. 4a. Missing mass spectrum
recoiling from the scattered
proton near the $\Delta(1236)$.

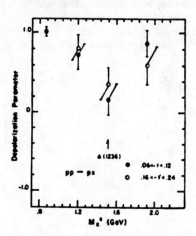

Fig. 4b. Depolarization in
p+p→p+X for small values of
M_X^2.

Many large and striking spin effects, some of which were quite
unexpected, have been found in small-t nucleon-nucleon scattering
using the ZGS polarized proton beam, as the above examples illustrate.
These discoveries have shown that the dynamics of small-t hadron
scattering in this energy range is much richer and much more interesting than had been previously thought.

A very nice experiment has been performed in Beam 2 in several
stages over the past five years by collaborators from Los Alamos,
Chicago, Illinois, Argonne, and Ohio State, searching for a parity
violation in p-nucleus scattering at 6 GeV/c. The beam polarization
was rotated into the longitudinal direction, and a search was made
for a helicity dependence of the p-nucleus total cross section. Although no effect was found, an upper limit was set on the helicity
dependent asymmetry $(\sigma_+ - \sigma_-)/(\sigma_+ + \sigma_-) < 7\times 10^{-7}$, which is certainly an
experimental tour de force. Several nice experiments studying spin
dependence in nuclear physics have been performed by the Minnesota-
UCLA-Argonne collaboration, which we unfortunately do not have space
to describe here.

Although we have described quite a rich physics program above, much of the emphasis of the polarized beam program for the past few years has been on other measurements, because surprising phenomena have been discovered which were completely unexpected in 1973. These phenomena fall into two categories: 1) Low energy data (1-3 GeV/c), especially pure-spin-state total cross sections. Very striking structure has been found in these spin-dependent cross section measurements, which is absent in the spin-averaged p-p and n-p total cross sections. This structure has been interpreted by several authors[4] as evidence for nucleon-nucleon resonances, which previously were thought not to exist; 2) Elastic scattering at high-transverse momentum. Strong spin dependence has been found in high-t pp and np elastic scattering which may be caused by the spin dependence of the scattering of the constituents of the nucleon.

There are two possible pure-spin-state total cross section differences: 1)$\Delta\sigma_\ell \equiv \sigma_{++} - \sigma_{+-} = 4\pi/k \ \text{Im}(\phi_1 - \phi_3)$, for spins in helicity states: and 2) $\Delta\sigma_T \equiv \sigma\uparrow\downarrow - \sigma\uparrow\uparrow = 4\pi/k \ \text{Im}\phi_2$, for spins transverse to the beam momentum, where the helicity amplitudes are evaluated at t=0. Data have been recorded for both p-p and p-n (via p-d) scattering, but the p-n data is very preliminary and will not be discussed here. The $\Delta\sigma_\ell$ measurements of the Yokosawa Group are shown in Fig. 5. The striking structure between 1 and 2 GeV/c has been interpreted by several authors[3] as being due to the formation of s-channel nucleon-nucleon resonances, called dibaryons.

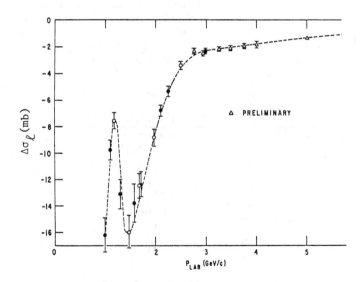

Fig. 5. $\Delta\sigma_\ell$ for p-p scattering.

The $\Delta\sigma_T$ measurements, taken in several stages, largely by our Rice Group and collaborators, are shown in Fig. 6. The more recent measurements (1978 and 1979) show more structure at the lowest momenta than the earlier data, more similar to the $\Delta\sigma_\ell$ measurements.

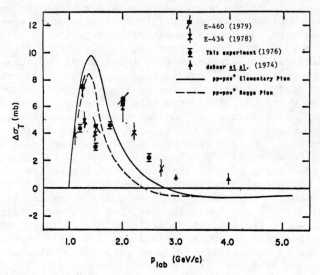

Fig. 6. $\Delta\sigma_T$ for p-p scattering.

Although these results are very preliminary, and a detailed discussion of systematic errors is inappropriate here, I believe that these recent measurements are more nearly correct. Phase shift and amplitude analyses done using these and other data, such as the $C_{\ell\ell}$ measurements of the Yokosawa Group shown in Fig. 7, have indicated the possible formation of three dibaryon resonances corresponding roughly to incident beam momenta of 1.2, 1.5, and 2.0 GeV/c. The

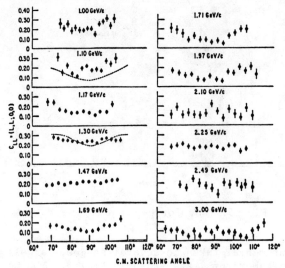

Fig. 7. $C_{\ell\ell}$ (or $A_{\ell\ell}$) for p-p elastic scattering.

question of the existence of such particles can only be settled by making measurements of many spin parameters for both elastic and inelastic channels, perhaps at LAMPF or Saturne II, but the first strong evidence for their existence was found using the ZGS polarized proton beam. The existence of dibaryons would have a fundamental impact on the theory of matter; viz., how quarks bind to one another to form hadrons.

The second and perhaps most surprising category of results found using the polarized beam involves scattering at large transverse momentum, where very large spin effects have been found up to the highest $p_\perp$ kinematically available at the ZGS. Figure 8 is a plot of the overall production left-right asymmetry for $pp \to \Delta^{++}n$ measured by the EMS Group. Note that this asymmetry A_n is independent of energy

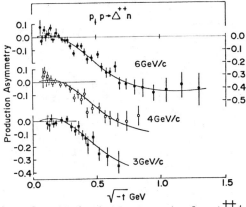

Fig. 8. Analyzing power, A, for Δ^{++} production.

between 3 and 12 GeV/c and attains a value of about -.5 at a $p_\perp$ of 1-2 GeV/c, similar except for sign to that in np charge exchange. Since $A_n=0$ for pion exchange, another amplitude of the correct phase interfering must exist to produce these large effects. Barry Wicklund[5] has shown that the unpolarized density matrix elements are related very well by SU_3 to those in vector meson production $K^+n \to K^{*0}p$. It has thus been suggested[6] that there may be a strongly spin dependent quark scattering amplitude which interferes with the pion exchange to produce the large values of A_n. If this is so, it is remarkable that direct scattering of constituents plays a major role at these relatively modest values of $p_\perp \sim 1-2$ GeV/c.

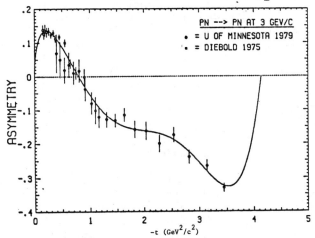

Fig. 9a. Analyzing power, A, for n-p elastic scattering.

The large angle n-p polarization data of the Minnesota-Rice-Argonne collaboration are shown in Fig. 9 (the smaller angle data from the EMS are in good agreement with these measurements at 3 GeV/c but there is some disagreement at 6 GeV/c, although the errors are large). Note that at large t the polarization is large (20-40%) and, if anything, is increasing with energy.

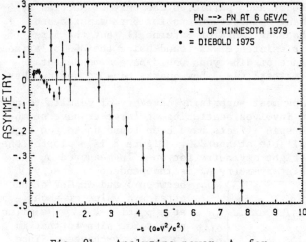

Fig. 9b. Analyzing power, A, for n-p elastic scattering.

Some $A_{nn}(C_{nn})$ measurements taken by the Michigan-Argonne collaboration using the polarized deuteron beam as a source of polarized neutrons are plotted in Fig.10. A_{nn} is sizeable ($\sim 20\%$) and quite different from that in p-p scattering even at this modest $p_\perp \sim 1 GeV/c$. Unfortunately, studies of spin effects in large-t n-p scattering were not very extensive due to the termination of the ZGS program, although the limited number of results obtained seemed quite interesting.

Certainly one of the most striking spin effects which has ever been observed was that of the Michigan-Argonne collaboration headed by Alan Krisch, which studied p-p elastic scattering at large angles. Their results at 12 GeV/c are plotted in Fig. 11 as the ratio of spin parallel to spin anti-parallel elastic cross sections. At $p_\perp \sim 2$ GeV/c the ratio increases dramatically with increasing $p_\perp$. This same effect is seen in Fig. 12 where A_{nn} at 90° C.M. is plotted vs. momentum. At an incident beam momentum of about 9 GeV/c, where $p_\perp \cong 2$ GeV/c for 90° elastic scattering, A_{nn} begins to increase, reaching a value of about .6 at 12 GeV/c. Thus A_{nn}, or equivalently the ratio of spin parallel to spin anti-parallel differential cross sections, begins to increase dramatically at a $p_\perp$ of 2 GeV/c, independent of incident energy.

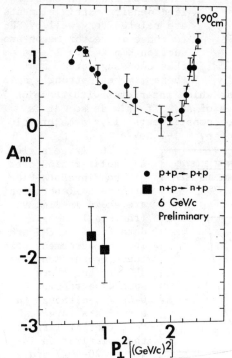

Fig. 10. Spin-spin correlation parameter, A_{nn}.

In Fig. 13, the pure-spin-state differential cross sections at 12 GeV/c from the ZGS and the spin-averaged cross section from the ISR at s=2800 (GeV)^2 are plotted vs. the

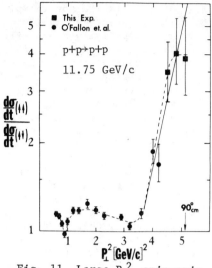

Fig. 11. Large $P_\perp^2$ spin-spin effects.

the <u>scaled</u> variable $\rho_\perp^2$. Note that use of this scaled variable removes the shrinkage in the diffraction peak. Moreover, the slope of the ZGS data in the large-t region for spins parallel is the same as that of the ISR data, while for spins antiparallel the cross section continues to fall off much faster as at lower t. This has led to speculations, such as that hard scattering occurs only with spins parallel, and that the threshold for this behavior is at $p_\perp \cong 2$ GeV/c.[7] There have also been attempts to calculate the large A_{nn} values in terms of quark scattering models such as the Constituent Interchange Model.[8] Whether or not any of these conjectures are correct, it certainly seems very compelling to investigate these large $p_\perp$ spin effects more thoroughly and at higher energies where higher $p_\perp$ values can be attained.

The very interesting and exciting results discovered during the ZGS polarized beam program have played a major role in the revitalization of strong interaction dynamics as a field which is both rich and exciting and at the same time amenable to a theoretical understanding. Although the lower energy work in the "dibaryon resonance" region can be continued at such facilities as LAMPF and Saturne II, the shutdown of the ZGS precludes any further study of high energy spin effects, especially the high $p_\perp$ scattering.

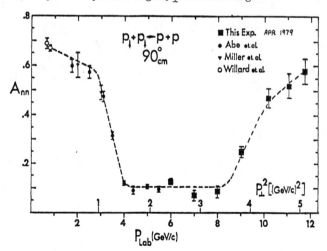

Figure 12. A_{nn} in p-p elastic scattering at 90° C.M. vs. incident momentum.

168

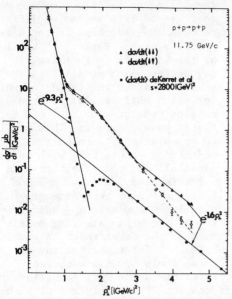

p+p→p+p

11.75 GeV/c

▲ dσ/dt(↓↓)
○ dσ/dt(↑↑)

• ⟨dσ/dt⟩ deKerret et al
s = 2800(GeV)²

$e^{-9.3p_\perp^2}$

$e^{-1.6p_\perp^2}$

$\frac{d\sigma}{dt} \left[\frac{\mu b}{(GeV/c)^2}\right]$

$p_\perp^2 [(GeV/c)^2]$

Fig. 13. Pure spin cross sections
in proton proton elastic
scattering.

Let me make a special plea
to the high energy physics community, especially since Bill
Wallenmeyer is in the audience,
for the continuation of this
work at higher energies via the
implementation of polarized
beams at Brookhaven and Fermilab. Finally, let me reiterate: The most exciting results
from the ZGS polarized beam program have been unexpected surprises. Although there are
compelling reasons for continuing the polarization studies
to higher energies based on
present data, there also may
well be surprises which turn
out to be very exciting.

REFERENCES

[1] F. Halzen and G. H. Thomas, Phys. Rev. D10, 344, 1974.
[2] A. C. Irving, et al., High Energy Physics with Polarized Beams
and Targets-1978, AIP Conf.Proc. No. 51, p. 445.
[3] E. L. Berger, A. C. Irving, and C. Sorensen, Phys.Rev. D17, 2971,
1978.
[4] H. Spinka, High Energy Physics with Polarized Beams and Targets-
1978, AIP Conf.Proc. No. 51, p. 382, and references therein.
N. Hoshizaki, Ibid., p. 399, and references therein.
[5] A. B. Wicklund, ANL/HEP 75-02, Section III; R. E. Diebold, High
Energy Physics with Polarized Beams and Targets-1976, AIP Conf.
Proc. No. 35, p. 92.
[6] F. E. Low, Higher Energy Polarized Proton Beams-1977, AIP Conf.
Proc. No. 42, p. 35.
[7] Alan Krisch and Stan Brodsky, for example, have expressed these
views in many places.
[8] D. Sivers, High Energy Physics with Polarized Beams and Polarized
Targets-1978, and references therein.

PANEL ON THE TECHNICAL HISTORY OF THE ZGS

E. G. Pewitt, Moderator (Argonne National Laboratory)

F. F. Bacci (ANL)	R. L. Kustom (ANL)
R. Bouie (ANL)	R. J. Prien (ANL)
J. A. Bywater (ANL)	C. W. Potts (ANL)

INTRODUCTION

E. Gale Pewitt

The performance of high energy physics requires the participation of a great number of people: designers, builders and operators. There are beam lines to design and cables to pull, meters
to watch, and detectors to design, build and operate. Of course,
many disciplines are required to carry out this effort. At the ZGS,
we are pleased that all this activity has been carried on with a
great deal of teamwork, and that it has worked well. We've had
good, even excellent productivity, although in the lean times, this
was best exemplified by the grafitti out by the 12' Bubble Chamber
Camera building. This read, "We've been doing more for less for so
long, that we can now do anything with nothing!"
     Our panel members are representative of the many talents needed
to carry out high energy physics. We've asked them to address the
question, "What makes high energy physics work at Argonne?" Several
of the members will mention some of the significant events, projects
and happenings that occurred during their stay here. All of them
are long-term employees, so in the process, we'll get a technical
history of the effort at Argonne.
     If there is time, we'll ask them to give their favorite User's
Story. This is when a particular user has exhibited exceptional
judgement, resourcefulness, and above all, tactfulness. We had a
"dry run" of this session. Our Chairman, Alan Krisch, was with us
and it seemed he was getting mentioned so many times, he felt there
probably wouldn't be enough time for all this!

THE ZGS AS SEEN FROM THE MIDNIGHT SHIFT

F. F. Bacci*
Argonne National Laboratory, Argonne, IL 60439

For those of you who do not know who I am, my name is Frank Bacci. I am one of the Zero Gradient Synchrotron (ZGS) Assistant Chief of Operations (ACO). As an ACO it is my responsibility to operate the ZGS. This means keeping the machine operating and beam extracted for the users. This also means shift work. I prefer working the midnight shift. It gives an ACO more time to tune for intensity or extraction efficiency. Midnights also mean you are the only one tuning. During the day shift, many well-meaning people come in the control room and start adjusting their respective systems. Well, the control room is rather large. Many days I have been on one side of the room wondering why what I am doing is not producing the usual response, only to find someone adjusting timing or voltage on the other side of the room. Machine intensity is closely coupled to the telephone. If intensity is increased without readjusting timing, extraction efficiency goes down and the telephone rings. Midnights also mean more of a chance to speak to the many different university groups who have used the ZGS. Now, let me go back to earlier days, my own and the ZGS.

After two years at DeVry Technical Institute I went to work for Webor Inc. repairing and building test equipment, then spent two years in the Army at White Sands Missile Range working on small short range missiles. After Webor and the Army, I joined the old Particle Accelerator Division (PAD) in December 1959. My first impression of Argonne National Laboratory (ANL), fresh out of the Army, was that somehow I had reenlisted. Have you ever crested the hill on Bluff Road and thought how much the east area looks like an army post? At that time, all of PAD was housed in building 818. My first job, as a linac technician, was to help in linac design and construction. We tested 200 MHz cavities, built waveguides and constructed a 300-kV power supply for source development. In those days when we had an arc-down we would take out line fuses all over the building. We tested the old preaccelerator column voltage gradients in a salt water tank and, in the meantime, anxiously watched the buildings for the ZGS go up.

After what seemed like forever, we moved in. The linac components started to arrive. All eleven copper-clad cavity sections forming the linear accelerator were received. They were then trimmed to their resonant frequency of 200 MHz with copper bars. After the cavity was assembled, water headers for cooling were installed. Cavity temperature must be maintained to close tolerances in order to maintain resonant frequency. Vacuum manifolds were installed for the drift tube stem boxes. Then the drift tubes with their quadrupole magnets were installed. At this point cavity tune measurements

*Work supported by the U. S. Department of Energy.

started. This involved pulling a metal bead down the bore of the drift tubes and adjusting each gap to frequency.

During the design phase, trials were run in order to decide which type of ion source to use: a rf source, where a burst of rf energy is used to form the plasma, or the VanArden duoplasmatron. It was decided to use the High Voltage Engineering VanArden source for the ZGS. At this point, design and construction of the necessary source electronics was started. A mock-up of the source enclosure was made, components were positioned and wiring harnesses were assembled.

The 800-kV Cockcroft-Walton preaccelerator power supply, built by Haefely in Basel, Switzerland, started to arrive at ANL in April 1961. My first major job in the Linac Group was to work on the power supply. This meant I had to go over all the Haefely prints, which were in German, and work out an interconnecting scheme, plan the reassembly and conduct high voltage and regulation tests. Finally, the supply and column held 750 kV with 0.1% regulation. The source and its electronics were installed and in December 1961, 750-keV protons were extracted.

Continental Electronics of Dallas, Texas, had been chosen as the supplier for the 200-MHz, 5-MW linac cavity rf system. Shipment from Texas for the transmitter started in October 1961 and reassembly at ANL was done by Continental. By March 1962, the transmitter had been installed and testing was underway. Radio-frequency power to the linac cavity was first applied in August of 1962. Cavity matching and tuning started. Fifty-MeV protons were first detected in October of 1962.

We then started emittance studies for ZGS injection. In August of 1963, protons were coasted around the ZGS ring. Then at 0335 hours on September 18, 1963, protons were accelerated to 12.7 BeV for the first time. Machine operation was intermittent at first while we worked out the bugs. Protons were first extracted from the ZGS in April 1964.

My time from this point until October 1968 was spent in linac operation and maintenance. In October 1968, I joined the Operations Group as an ACO for the ZGS. As of February 1969, the machine had targeted extraction to the Meson Area and to both external proton lines. To the ACO this meant dealing with as many as ten simultaneous users. Beam was extracted to three bubble chambers, the 30-inch and 40-inch chambers in the Meson Area and the 12-foot chamber in External Proton Beam II (EPB-II). In 1970, the octant 2 main coil failed. Almost a year later, the octant 3 main coil failed. By March 1972, we were accelerating 3.0×10^{12} protons per pulse.

Then from May to September 1972, the new titanium vacuum chambers were installed. By March 1973, our machine operating efficiency had climbed to 97%. In July 1973, the ZGS became the first and only accelerator in the world able to provide high energy polarized protons for our users. Negative ion and low B injection came along and we hit our peak beam of 7.55×10^{12} unpolarized protons per

pulse. Then came simultaneous injection sharing, H⁻ beam to the
Rapid Cycling Synchrotron and polarized beam to the ZGS.

High energy physics at the ZGS works because we have ACO's who
try to improve, or at least match, the preceding shift in beam
intensity, extraction, and operating efficiency; also, because we
have users who, for the most part, understand some of our problems.
One of the many things I have been called in my life is "a biological
feedback loop." That is, as intensity or efficiency decreases, the
loop closes and the appropriate action is taken to bring one or both
back up. There have been times when only two machine pulses have
passed with little or no protons and that old telephone coupling
takes over.

Ten years ago we were scratching for protons and our users
wanted more. In the later years of the ZGS our users have, at times,
called to ask that we reduce beam intensity because their target was
"too hot." I think my most memorable event as an ACO was watching
two users at 3 a.m. arguing about what seemed individual protons.

A DESIGN ENGINEER'S VIEW OF THE ZGS*

Panel on the Technical History of the ZGS

J. Bywater

Argonne National Laboratory, Argonne, IL 60439

I have been requested to review my long association with the Zero Gradient Synchrotron (ZGS) and some of the major projects that have been completed during this time. My present position is Associate Group Leader for the Engineering Group of the Accelerator Research Facilities Division here at Argonne. I was first assigned to the ZGS from Central Shops during the summer of 1957 as a draftsman and have been associated with the accelerator ever since, officially transferring to the Division in March 1973.

During the past 22 years I have worked on a variety of assignments for design and engineering of equipment necessary to make high energy physics work at the ZGS. Working closely with the physicists it is the draftsman and engineer's responsibility to put on paper what the physicist has in his head so that the machine-shop can produce a workable part.

To me the most interesting and challenging part of an engineering assignment at any accelerator is the many disciplines one must be knowledgeable in, such as vacuum, hydraulics, magnetics, electrical, and cryogenics. In addition to these the engineer obtains a large amount of experience in working and communicating with the outside vendors on contracts with both large and small dollar amounts. Needless to say, working with the low bidder has not always been totally successful; however, the majority of times the vendors have performed satisfactorily. Fifty percent of the equipment required by the physicist cannot be purchased off the shelf. New and innovative designs and fabrication methods are often required, in many cases for a one-of-a-kind piece of equipment which prohibits extensive research and development prior to fabrication. Many times it has been necessary to stick your neck out and go ahead with a job relying on calculations and gut feelings that it's going to work.

In addition to the challenging work at the accelerator, the ability and willingness of all the personnel to rise to the occasion in times of crisis has been outstanding. The classic example of this occurred on January 9, 1971, when the Ring Magnet Coil in Octant 3 failed. Since the only spare coil had been used to replace the coil that failed in Octant 2 on April 21, 1970, the ZGS was out of business totally unless the Octant 3 coil could be repaired.[1]

Assessment of the damage began immediately and the decision to repair the coil in place was soon made. Teams of physicists, engineers and technicians were assigned to handle various phases of the repair, such as octant disassembly, tooling and repair method development, actual replacement of damaged sections of turns, insulation repair, testing of repaired sections, both mechanical and electrical and final reimpregnation of the repaired section.

---

*Work supported by the U.S. Department of Energy

ISSN:0094-243X/80/600173-05$1.50 Copyright 1980 American Institute of Physics

Several unique problems had to be overcome in order to insure a completely successful repair. One problem was the brazing in of a one meter long conductor section, 2.5 cm x 3.3 cm in cross section, to replace the melted conductor caused by the failure. The second joint had to be made with the ends of the conductor rigidly fixed at a distance of approximately 2.2 m on each side of the joint. The actual silver brazing caused the area of the joint to expand 3 mm which if not allowed for would have exceeded the compressive strength of the copper conductor. In addition the brazing had to be done away from the unaffected turns of the coil to avoid damaging the good insulation behind the joint area.

A fixture was developed which contained the free ends of the conductor to be joined parallel to each other and the coil assembly but at a distance of 7.5 cm out from the main body of the coil assembly.

Fig. 1. ZGS ring magnet coil repair fixture.

The free ends to be joined were spaced apart to allow for expansion when brazed. The additional conductor length required to position the joint away from the assembly was obtained by strip heaters placed between the fixture and conductor fixed point. Water-cooled chill blocks were used between the fixture and strip heaters to protect the heaters from the brazing heat. These heaters were also used to hold the completed joint out from the assembly (3 cm) to allow for cleanup and testing of the joint. The fixture was employed to feed the joint and coil back towards the assembly as it cooled after brazing.

After joint cleaning and testing the joined conductors were allowed to cool fully and thereby pull into its place in the assembly with the aid of about 100-kg tension built into the joined conductor.

Many more problems were encountered and solved during the coil repair but time does not permit me to go into detail in this session. However, with the total efforts of everyone in the complex the coil repair and reassembly of the magnet iron was completed in little more than three (3) months with pulsing and beam accelerating resuming on April 27, 1971.

Several other projects are worthy of note, such as the magnet iron for the 12-foot Hydrogen Bubble Chamber (HBC),[2] the superconducting beam line referred to as 21S installed in Building 366[3] and the two Polarized Proton Targets V and VI. The 12-ft H_2BC magnet iron formed the poles and return frame for the magnetic circuit generated by the superconducting coil.

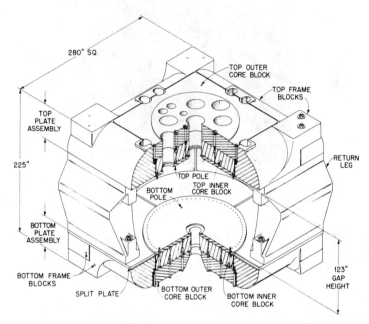

Fig. 2. 12-ft H_2BC magnet iron assembly.

A number of geometries and fabricating techniques were evaluated during the design stage. The use of rolled plate, forgings and castings were compared. Steel castings were chosen for the final design. The design flexibility offered by cast construction techniques permitted simultaneous achievement of maximum material utilization efficiency, maximum experimental access and maximum structural rigidity. An equivalent "all forged and machined" structure would have been 300 tons heavier than the "cast" design chosen and considerably more expensive. The final "cast" design weighed a total of 1600 tons.

The superconducting beamline transports 12-GeV/c polarized protons a distance of 60 meters to the Effective Mass Spectrometer (EMS).

176

Fig. 3.   Installation of three superconducting magnet cryostats
as viewed downstream with the EMS visible just beyond the last
cryostat.

It consists of ten superconducting dipoles and two superconducting
quadrupoles and several other conventional magnets.   The superconducting magnets are contained in four cryostats; three 3-dipole
cryostats, each 3.5 meters long; and one quadrupole-dipole-quadru-
pole (QDQ) cryostat 2.7 meters long.
     The Polarized Proton Targets consist of three subsystems: the
insulating shells, the common stable isotope helium^4 (He^4) cryostat,
and the rare stable isotope helium^3 (He^3) cryostat.

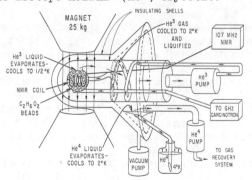

Fig. 4.   Schematic diagram of a typical polarized proton target.

The insulating shells act as a thermal insulator to isolate the two cryostats from their thermal surroundings. The He^4 cryostat fits around the outside of the He^3 cryostat and intercepts the heat load from the outside and cools the warm side of the He^3 refrigerator. In the He^3 cryostat, gas is compressed, cooled, expanded, and liquefied in thermal contact with the He^4 cryostat. The liquid He^3 is evaporated at the target by means of a closed loop vacuum pumping system in order to achieve the 0.4 to 0.5 K temperature required.

During my many years at the ZGS I have seen many people come and go from all job classifications. They have all had their own merits and contributed to the ZGS program. I am sure we all have gained valuable experience in how a team effort works. Now it is time to employ these lessons in new areas, and I know we all can be successful.

REFERENCES

1. J. Bywater, et al., "Repair of ZGS Ring Magnet Coil Damaged Conductor Sections," Proceedings of 4th International Conference on Magnet Technology, Brookhaven, p. 381, 1972.
2. J. Bywater, et al., "Design and Fabrication of Steel Core for the 12-foot Hydrogen Bubble Chamber Superconducting Magnet," Proceedings of 3rd International Conference on Magnet Technology, Hamburg, p. 655, 1970.
3. J. Bywater, et al., "Design and Operation of a Superconducting High Energy Beam Line at Argonne National Laboratory Zero Gradient Synchrotron," IEEE Transactions on Magnetics, Vol. MAG-13 No. 1, p. 294, January 1977.

EXPERIMENTAL AREA OPERATIONS

R. Bouie*

Argonne National Laboratory, Argonne, IL 60439

My name is Rudolph Bouie, and I have been employed by Argonne
National Laboratory (ANL) for 15 years. From the first day of my
employment, I started cleaning up management's problems. You see,
I started as a janitor. Today I am an Executive Assistant to the
Accelerator Research Facilities Division Director. However, during
the span of time between day one and now, I have furthered my formal
education and for the past seven years, I have been the Experimental
Area Manager of the ZGS. In that time, I have observed that Argonne
employees are basically of two types, regardless of job titles and
descriptions. The first type is called "staff" and consists of
scientists, engineers, administrators, and supervisors. The second
type is called "hourly" and consists of technicians, craftsmen, and
secretaries. After observing and being a part of each, I have
drawn some conclusions relating to personalities and attitudes of
staff and hourly employees.

The Experimental Area Manager has to place heavy emphasis on
interpersonal relations and getting work accomplished by motivating
rather than commanding. The manager is responsible not only for
seeing that the work gets done but also for establishing the best
environment for getting it done and making the most creative use
of the abilities of those who do it. A manager's success depends
in some measure on his ability to supervise without actually appearing to do so.

Staff persons basically all have bachelor, masters, and/or
doctorate degrees and receive their greatest satisfaction and
strongest motivation from achievement, responsibility, growth,
advancement, work itself, and earned recognition. Staff persons
are usually inner-directed, self-sufficient, and less subject to
influence by the environment. They enjoy work, strive for quality,
tend to overachieve, and benefit professionally from experience.
They realize great satisfaction from accomplishment and have
positive feelings toward work in general.

The hourly person, usually not a college graduate, tends to see
himself in a supportive role doing the unpleasant and uninteresting
tasks. Often hourly personnel realize little satisfaction from
accomplishment and exhibit cynicism regarding the virtues of work in
general. They may succeed on the job through sheer talent, but
seldom profit professionally from their accomplishments. The importance he gives to the work assigned supports the hourly person's
contention that he "gets stuck with the dirty work." The above
observations for hourly personnel are related to the experimental
area and not necessarily to the ZGS Complex as a whole or Argonne in
general.

The question to be answered now is, if that's the Experimental
Area Manager's opinion of his hourly personnel, how in the world

*Work supported by the U.S. Department of Energy

did the place ever operate? The factors I consider very important
to maintain a high level of operating efficiency are (1) goal
congruences and (2) the actual treatment of employees.

In the past seven years, it has become common practice to install
large experiments on very limited budgets with exceedingly short
installation time frames. The working conditions most of the time
are less than desirable, and the hours are long and seemingly endless. If the work didn't get you, the mosquitoes did. Who could
forget that green can of insect spray that fattened up the little
bugs and made their bite create a bump as big as a dime? And the
roofs leaked so bad that you would have to go outside during a
rain storm to get dry.

Of all the people responsible for a successful operation, the
technicians are the backbone of the experimental area. Technicians
in the experimental area are some of the most unique people I have
ever encountered. They have constantly produced excellent results
under the most severe circumstances. With the advent of the technicians union, I personally believe it made my job easier. I didn't
have to worry about salary adjustments, promotions, rates of pay,
or methods of pay. Instead, I could concentrate on establishing
short range goals and job assignments which would allow every
technician the opportunity to succeed. Since the achievement of a
goal ultimately lies more with the people at the bottom than at the
top, it was important that the goal itself and the individual
contribution to it be clearly understood. Our primary goal was to
keep the experimental area operations running smooth, safe, and
efficient.

The attitude of the people in the experimental area has proven
to be an eye opener to outsiders. We didn't care if the accelerator
generated quality beam or peanut butter, and we couldn't care less
if the experimenters' equipment were operational; we only cared
about the beam transport system. It's been my experience that when
each group worried about the conditions of the equipment that's not
in the realm of its responsibility, it has the same effect as a
dog chasing a car; once the dog catches the car, all he can do is
bark at it and become a nuisance.

The motivation of the technicians never seemed to be a real
problem for me. I regularly scheduled personnel meetings to hear
problems, discuss solutions, and request suggestions to accomplish
predetermined goals. People were treated as individuals and not as
a class. This method was laughingly called "Rudy's rule." This
rule allowed for special consideration to be given for sickness,
vacation, overtime, picnics, Christmas, and G jobs (activity not
necessarily covered by the ANL policy manual). We had a lunch
room complete with stove, refrigerator, vending machine, and a
television. Who could forget those fresh Polish sausage luncheons;
once you got past the smell, you could eat the food. The television
added to our efficiency; we have been known to fix major problems
during a commercial break or halftime of the Bears game. However,
the most important element to the motivation of technicians is
money. There are technicians that will move the ZGS to the East

Area if there is overtime involved. They are all basically young, aggressive, adaptable to change, and money oriented. They don't believe that money is the most important thing; but when we don't pay, they don't stay.

It is very important for the experimental area personnel to work sympathetically with the scientific people. Our working relationship with them has been fantastic over the years. This is primarily because we understand their situation and we realize that we are a service group and not their peers. We understand that experimenters generate many changes during the course of a success- ful run and if everything they did was right the first time or as expected, they wouldn't do it. However, superior service, honesty in dealing with them, and being seen and not heard have proven to be the necessary ingredients to enhance a solid working relationship.

Since all good things must come to an end, the ZGS is shutting down September 30, 1979. I am going to take this opportunity to thank everyone associated with the experimental area for a job well done. I guess I have mixed emotions about the closing of the experimental area. You see, it's like seeing your nagging mother- in-law drive over a cliff in your new Cadillac. You hate to lose the car, but the future looks promising.

AN ENGINEER'S VIEW OF THE ZGS

Robert L. Kustom
Argonne National Laboratory, Argonne, Illinois 60439

I came to Argonne in 1958 from an industrial research laboratory that specialized in developing protective equipment for
lightning and transients on power transmission lines. The group I
joined in the Physics Division was destined to be the charter
group of the High Energy Physics Division. In the first few years
in the High Energy Physics Division, I concentrated first on
developing high field magnets (20 T) for a spin precession experiment at the Cosmotron and, later on, magnetic spark chamber spectrometers for an experiment on the PS at CERN in Geneva,
Switzerland.

I was a member of the spark chamber neutrino experiment
(E1) on the ZGS. My activities included developing the large
spark chambers, the control systems, the film-scanning tables,
and the digital track measuring equipment.

In the midsixties, I took an academic leave from Argonne
to pursue a Ph.D. in the Electrical Engineering Department at
the University of Wisconsin, Madison. After my return from the
University, I joined the Particle Accelerator Division, working
on electrostatic particle separators, rf traveling wave and superconducting separators, and particle track devices. I also participated in some of the early research on the resistive wall instability and the use of the vertical damper system.

I became Group Leader of the ZGS Operations (ZSO) Group
in late 1971; Associate Division Director of the Accelerator
Research Facilities Division in October 1973; Deputy Division
Director in March 1978; and Director in August 1979.

The single event that I will describe as part of the technical history of the ZGS is the installation of the titanium vacuum
chamber. When I became ZSO group leader in late 1971, the installation of the vacuum chamber was the major forthcoming event
in the group's activities.

Prior to joining the ZSO group, my activities were part
of tightly-knit research groups in which all members were closely
coupled to a monolithic goal. I would describe inheriting the
ZSO Group as a true educational experience into the mechanism
that makes high energy accelerators perform. Figure 1 shows the
ZSO Group organization in late 1971. The spectrum of talent
varies from engineers and physicists who spent years in formal
university education; and technicians and other specialists that
acquired their expertise through fewer years of formal education,
plus many years of experience. As can be seen from the organizational chart, there are control room personnel to adjust machine
parameters for a high level of performance, power house personnel
to keep the 15,000-hp, 100-MVA motor-generator system under control, a complete maintenance hierarchy to keep everything working
from high-pulsed power apparatus to delicate low-level digital and
analog control systems, proton line tuners to adjust the extracted

ISSN:0094-243X/80/600181-09$1.50 Copyright 1980 American Institute of Physics

182

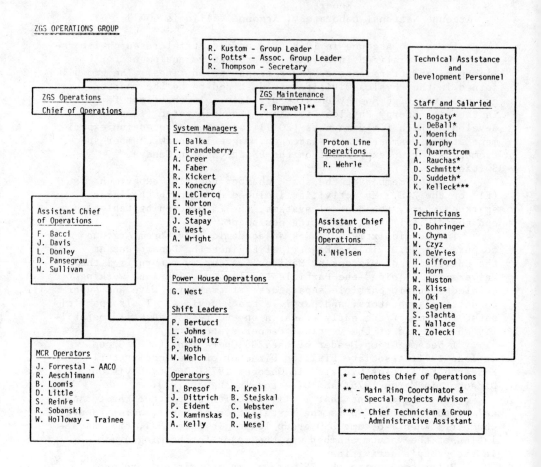

ZGS OPERATIONS GROUP

R. Kustom - Group Leader
C. Potts* - Assoc. Group Leader
R. Thompson - Secretary

Technical Assistance and Development Personnel

ZGS Operations
Chief of Operations

ZGS Maintenance
F. Brumwell**

Staff and Salaried

J. Bogaty*
L. DeBall*
J. Moenich
J. Murphy
T. Quarnstrom
A. Rauchas*
D. Schmitt*
D. Suddeth*
K. Kelleck***

System Managers

L. Balka
F. Brandeberry
A. Creer
M. Faber
R. Kickert
R. Konecny
W. LeClercq
E. Norton
D. Reigle
J. Stapay
G. West
A. Wright

Proton Line Operations

R. Wehrle

Technicians

D. Bohringer
W. Chyna
W. Czyz
K. DeVries
H. Gifford
W. Horn
W. Huston
R. Kliss
N. Oki
R. Seglem
S. Slachta
E. Wallace
R. Zolecki

Assistant Chief of Operations

F. Bacci
J. Davis
L. Donley
D. Pansegrau
W. Sullivan

Assistant Chief Proton Line Operations

R. Nielsen

Power House Operations
G. West

Shift Leaders

P. Bertucci
L. Johns
E. Kulovitz
P. Roth
W. Welch

MCR Operators

J. Forrestal - AACO
R. Aeschlimann
B. Loomis
D. Little
S. Reinke
R. Sobanski
W. Holloway - Trainee

Operators

I. Bresof   R. Krell
J. Dittrich   B. Stejskal
P. Eident   C. Webster
S. Kaminskas   D. Weis
A. Kelly   R. Wesel

* - Denotes Chief of Operations
** - Main Ring Coordinator & Special Projects Advisor
*** - Chief Technician & Group Administrative Assistant

Fig. 1.

beam and provide an interface to the downstream users, and support engineers and physicists to provide technical assistance to the operations and maintenance staff and execute hardware development generated from within or outside the group.

Within their own area of interests, individuals or groups of individuals had developed a high level of skill and understanding necessary to maintain and improve specialized apparatus. Further, the ZSO Group as a whole had to interact with other specialized groups within the Accelerator Division, an organizational chart of which is shown in the Figure 2.

The vacuum chamber installation was more than just a mechanical replacement. The chamber was fitted with 82 pole-face windings radially distributed above and below the vacuum space. **The** ampere-turn requirements for these windings were generated by physicists in the Accelerator Physics Group who rarely, if ever, approached the accelerator control room. Their concepts were developed into hardware designs by experimental physicists, accelerator engineers from ZSO, and other engineers from within the Accelerator Division or High Energy Physics Division. Finally, the hardware was built by technicians who had little or no knowledge of vector potentials, resonances, or betatron motion. Their interest centered on transistors, integrated-circuits, trimpots, volts and amperes per second. The total integration of all of these disciplines and the utilization of individual skills were needed to successfully accomplish the titanium vacuum project.

The conceptual design work was started in March 1964, and the construction project was approved in July 1966. The first vacuum chamber was delivered in November 1968. The ZGS was shut down on May 1, 1972, to begin the vacuum chamber installation. The installation activity extended from May 1, 1972, until September 18, 1972. The startup and retuning exercise lasted from September 18, 1972, until October 2, 1972. The high energy physics program was restarted five months after the installation started.

It is impossible in this short paper to document all of the individual contributions to this effort, nor is it possible to document the significant impact the installation had on simultaneous resonant extraction, $\nu_x = 1$ extraction, polarized proton operation, or ordinary operation. I have attached, as appendices to the paper, Frank Brumwell's installation schedule; some memos relating to schedules and personnel activities; and a summary of turn-on events maintained during late September. The information in the Appendices will provide an overview of the program, but not the dedication and hard work of the individuals involved.

There is really no way to describe the moments of frustration, confusion, understanding, or triumph experienced. Many hours were spent by physicists like Crosbie, Ratner, or Cho interpreting data provided by engineers Potts, Rauchas, and Bogaty with diagnostics equipment they designed and made work, while technicians and system managers Brandeberry, Wright, and Konecny kept

March 10, 1972

ACCELERATOR DIVISION

DIVISION MANAGEMENT

R. Martin - Director
G.W.Bryan
E. Crosbie
T. Khoe
M. Knott
R. Kustom
L. Lewis
A. Lindner
D. Manson
E. Parker
L. Ratner
J. Simpson

A. Garneski*
C. Herrs*
B. Jackan*
R. Thompson*
K. Volk*

ACCELERATOR PHYSICS

E. Crosbie - Leader
G. Concaildi
R. George
R. Lari
S. Marcowitz

*Clerical

ZGS OPERATIONS & MAINTENANCE

R. Kustom - Leader
C. Potts-Assoc.Leader
L. Balka
J. Bogaty
F. Brunwell
J. Creer
J. Davis
L. DeBall
L. Donley
H. Easton
K. Kelleck

A. Aeschliman
P. Bertucci
F. Brandeberry
I. Bresof
D. Bohringer
W. Chyna
W. Czyz
K. DeVries
J. Dittrich
P. Eident
M. Faber
J. Forrestal
H. Gifford
W. Holloway
W. Horn
L. Huston
L. Johns
S. Kaminskas
A. Kelly
R. Kliss
R. Konecny
R. Krell

R. Kickert
W. Leclercq
J. Moenich
R. Nielsen
D. Pansegrau
T. Quarnstrom
A. Rauchas
D. Schmitt
D. Suddeth
W. Sullivan
R. Wehrle
G. West

E. Kulovitz
D. Little
B. Loomis
J. Murphy
E. Norton
N. Oki
D. Reigle
S. Reinke
P. Roth
R. Seglem
S. Slachta
R. Sobanski
J. Stapay
B. Stejskal
E. Wallace
C. Webster
D. Weis
W. Welch
R. Wesel
A. Wright
R. Zolecki

COMPUTER DEV.

M. Knott - Leader
R. Timm
A. Valente
A. Brescia
R. Combs
G. Gunderson
K. Kellogg
R. Kmiec
H. Rabe
F. Toussaint
D. Voss

INSTRUMENT SERVICE

T. Carothers
G. Fouts
L. Davies
D. Emery
D. Moore
T. Saddler
L. Shirkey

BOOSTER DEV.

J. Simpson - Leader
M. Lieberg
K. Menefee
R. Sanders
J. Volk
D. Connor
V. Patrizi
J. Seeman

MAGNET MEAS.

E. Berrill
T. Hardek
J. Lewellen
S. Phillips
D. Piatak

INJECTOR DEV.

E. Parker - Leader
J. Abraham
J. Fasolo
A. Gorka
G. Marmer
J. Madsen
N. Sesol
V. Stipp
R. Stockley
A. Vanderflught
Q. Allen
R. Barner
R. Broholm
A. DeWitt
M. Erickson
K. Meacham
W. Richards
C. Saunders

Fig. 2.

the equipment operating that they had built. These hours can only be appreciated by them and all the other participants who cannot be named because so many contributed in just the same way.

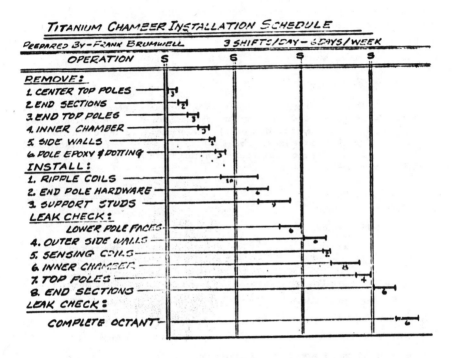

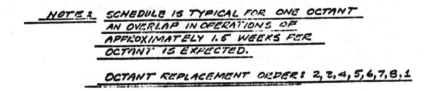

Fig. 3.

ARGONNE NATIONAL LABORATORY

August 4, 1972

To: R. L. Martin   AD

From: C. W. PottsCwp AD

The following items are essential for the startup of the ZGS. They should be completed and tested by the noted dates. The completion of these items will give us the capability of (a) conventional targeted extraction to EPB-I, EPB-II, and the meson areas (b) fast $\nu_x = 1$ resonant extraction to EPB-I or EPB-II (c) slow resonant extraction (with rf on) at $\nu_x = 1$ to EPB-I or EPB-II.

Item	Completion Date
Octant 7 Ti Chamber	August 7
Octant 8 Ti Chamber	August 20
Octant 1 Ti Chamber	September 4
Main Motor Repair	August 25
Main Motor Installation and Testing	September 11
S-4 BM-100 Installation	August 25
L-3 Bumper Magnet Installation	August 10
EPB-I Resonant Magnet Installation	August 17
Resonant Magnet Power Supply Installation	September 4
Pole Face Winding (Eddy Current Correction)	September 18
Ti Chamber Ground Detection System	August 30
Internal Meson Target Installation	September 1
All Straight Section Boxes Reassembled	September 13
Vertical Damper Modification	August 28
Closed Circuit TV System Reassembly	September 11
Main Ring Beam Diagnostic Checkout	September 18
L-2 and L-4 Diffusion Pump Installation	August 9

9700 South Cass Avenue, Argonne, Illinois 60439 · Telephone 312-739-7711 · TWX 910-258-3285 · WUX LB, Argonne, Illinois

Fig. 4 (a)

R. L. Martin
August 4, 1972
Page 2

       The following items are not essential for startup of the ZGS. They are items which we had hoped to complete at this shutdown, but due to manpower limitations these items have been delayed.

Item	Completion Date
Pole Face Winding-Tune Correction (Simultaneous resonant extraction)	September 15
Active Ripple System (To provide structure-free spill)	November 1
Digital Phase Control of RMPS	October 15
Main Programmer Modification	September 1
Additional EPB Diagnostic Construction	November 15
New Radial Damper	November 8
BM 201-1 Pulsed Supply Installation	October 15
RMPS Passive Filter Addition	August 25
Extraction Monitoring Inside ZGS (Test)	September 15
Booster Line Energy Spectrometer	August 28

CWP:rt

cc: R. L. Kustom
       F. R. Brumwell
       Division/CF

Fig. 4 (b)

A.   Titanium Chamber Installation Team - R. Kustom, F. Brumwell,
                                           C. Potts

Chamber Installation Shifts Supervisors

P. Bertucci   F. Bacci
L. Johns   D. Pansegrau
W. Welch

Chamber Installation Team

I. Bresof   R. Aeschlimann   W. Czyz
P. Eident   J. Murphy   B. Loomis
A. Kelly   R. Sobanski   S. Slachta
R. Krell   S. Reinke   3 HEF Techs
D. Weis   W. Huston
R. Wesel   R. Kmiec

Electrical Installation and Testing Team

E. Kulovitz   W. Sullivan
K. Kelleck   B. Stejskal

Pole Face Winding Amplifiers and Testing

D. Bohringer   A. Wright

Vacuum System Modifications

C. Webster   J. Dittrich
E. Wallace

B.   MCR Operation of Linac and H⁻Source for Booster

J. Davis
L. Donley
J. Forrestal

C.   General ZSO Activities

Support of System managers, general maintenance, development projects,
and general chamber installation.

F. Brandeberry   W. Chyna   K. DeVries
S. Kaminskas   R. Kliss   N. Oki
P. Roth   R. Seglem   R. Zolecki

Fig. 5.

September 18, 1972

DAILY SUMMARY OF ZGS TURN-ON AND RETUNING

First beam injected into ZGS at 0945. Approximately 28 - 30 mA of current observed at 701 toroid.

50 MeV was adjusted using L 1 - L 2 segmented Faraday cups. Beam spot looked good at both locations.

Energy ramp of linac and coarse tune measurement was made with L-3 screen. Some instability in beam location was noted. Most of the instability was eliminated by raising the rest field from 188 to 248 gauss. A few of the pole face winding amplifiers are unstable with an apparent 5 kc oscillation.

Preliminary coasting tune is near 0.86. Three turns around machine have been observed.

October 2, 1972 12:08

DAILY SUMMARY OF ZGS TURN-ON

Beam on and targeting on M4 target for Longo, E 317

Accelerated beam is approximately 1.9×10^{12} ppp average and peaks over 2.2×10^{12} ppp. Voltage on column still approximately 60 kv low to avoid sparking. Expect when problem fixed ZGS will accelerate and beams of same intensity as before shutdown.

Radial blow up problem still prevalent. Does not seriously effect targeting now, but must be solved for future conditions.

Simultaneous extractions on P2 and M4 was successful over the weekend. Single efficiency of near 40% was achieved in EPB2. Resonant extraction was also tried with some success. More effort required before this is operational.

Fig. 6.

A TECHNICIAN'S VIEW OF HIGH ENERGY PHYSICS

R. J. Prien
Argonne National Laboratory, Argonne IL   60439

For the past 14-1/2 years I have been a technician at Argonne
National Laboratory. I've worked in the ZGS Experimental Area for
11 years, coming from the now-defunct Reactor Engineering Division.
I started in the Experimental Area as an operator at the 40-in.
heavy liquid bubble chamber and then worked on high voltage separators
and vacuum systems. More recently, my work has been with Ion Beam
Fusion and superconducting magnets.

For the past 6 years I've had the honor of being elected President of the Argonne Technicians Union, representing the 380 technicians at Argonne, in some 22 different divisions.

With this experience behind me, I was asked to talk today about
the technician in the ZGS, especially the Experimental Area technician.
You can't study a group of people without having some idea of where
they came from. The technicians at the ZGS seem to have come from
all over the country. Argonne and the ZGS did quite a recruiting job
back in the sixties - going to various technical schools and col-
leges. Many others came out of the service, and the ZGS picked up
people from other divisions at Argonne. At the peak, the ZGS employed 260 technicians.

At that time, Argonne had quite a bit to offer. It was the
early sixties, and "Atoms for Peace" was the national cry. The United States was in a massive technical push and Argonne, one of the
world's leading atomic energy research centers, had a fantasy about
it that could make dreams come true for up-and-coming technical people. The lure here was tremendous, a far cry from the ever-growing
"anti-nuclear anything" movement of today.

People came here hunting interesting jobs, stature, and maybe
a chance to contribute to the country's technical advance. They also
came for more basic reasons - good pay compatible with the surrounding area; a good benefit package; a suburban job in a nice area to
live; and, for some the most important, a deferment from the military
draft.

It takes less people to run a large facility like the ZGS than
it takes to build it, so our numbers have diminished. People left
because they found the work too hard, too complicated, too hot
(thermal and nuclear), too cold, too wet, or the hours too long or
staggered. Others left for better jobs, of course. But what about
those who stayed? Were there reasons besides those monetary ones
that kept people here after the initial fantasy wore off?

In the Experimental Area, any results from their efforts were
hard for the technicians to see. Unless you were working at one of
the facilities which recorded its data on film (such as the bubble
chamber) results were just not visible. All we would see was a cart

full of computer read-out, with a mumble-jumble of numbers on it
which meant nothing to the layman.

There are probably very few technicians in the ZGS who could
explain any of the experiments and what they were trying to do,
unless it was basic - like hunting a neutrino. Their job was to get
the beam to the experimenter and maintain his support equipment.
When the experimenter was happy, so were the technicians. The
technicians maintained a variety of sophisticated equipment, many
one-of-a-kind things. They became good trouble shooters, electri-
cians, pipefitters and plumbers - and did it well. Their basic re-
ward was being proud of the job they were doing.

The ZGS management realized early in its history that it would
need this team of technicians with their expertise if the ZGS were to
keep running and produce results. There developed an understanding
of how to treat technicians which doesn't exist in all other divisions of the Laboratory. Too many other divisions put the value of
a man into how many years of schooling he has had. Fortunately for
the ZGS that attitude didn't exist here, and the physicists and technicians developed a mutual respect for one another. This allowed the
technician to devote his energy to the job for which he was hired and
did best. This kept the technician who liked this type of work at
peace with his job, and in no great hurry to leave. It also developed the camaraderie and team effort mentioned so often about this
facility and is one of the underlying reasons for the great success
of this facility.

It wouldn't be fair for me not to mention the Technicians Union
and why it was formed. As far as I know, we are the only organized
group in the world composed wholly of technicians. Let me just say
that in a research laboratory this large, with 22 different divisions,
or dynasties as I call them, you are bound to have administrative
problems. The technicians felt that they would be much more secure
against fluctuating budgets and management changes if they were to
band together and bargain as a group, rather than begging as individuals. Hopefully, the ZGS technicians have, or will realize the
importance of this as the ZGS closes down.

As we look to the future the technicians are hopeful of a new
project at the ZGS site which will be large enough to make use of the
expertise which they have developed with accelerators. The Department of Energy certainly has its hands full with the new anti-nuclear
movement, and a major public relations job seems to be in order.

For the time being, however, we must tear apart the ZGS and
destroy the work many of us have done. This may come hard for some,
since the ZGS has become a dear old friend to many of us.

ZERO GRADIENT SYNCHROTRON (ZGS) OPERATIONS - THE VINTAGE YEARS

C. W. Potts*
Argonne National Laboratory, Argonne, IL 60439

INTRODUCTION

My Argonne career started in June of 1959 in the Remote Control Division where I worked on control system electronics for the electrical master-slave force reflecting manipulators. I transferred to the Targetry Group of the Particle Accelerator Division (PAD) in early 1967. The Operations Group acquired my services in 1969.

While in this division, I have worked on special target systems, the External Proton Beam II magnet power supply systems, injection and circulating beam diagnostics, automatic tune measuring equipment, resonance extraction, the fast pulse quads for the polarized beam, and failures too numerous to mention.

I have been Operations Group Leader since the fall of 1973. My first official act as Group Leader was to reduce Frank Brumwell's working day to eighteen hours.

The resonance extraction spill control system has also occupied a good deal of my time. Figure 1 shows the smooth spill that we send out and the chopped up mess that the users often received. There must be a chopper somewhere in the beam line! These signal discrepancies have been noted with some emphasis by Jim Cronin, Tom Romanowski, Dave Ayres and others. We have **even** been accused of using electronic filters!

I would like to point out that Bob Wehrle, whom most of you know, and I have been coworkers in Remote Control, which closed down in 1969; the PAD Targetry Group, which disappeared in 1970; and in the Accelerator Research Facilities Division (ARF). We seem to be a jinx. Heaven help the project which gets us both.

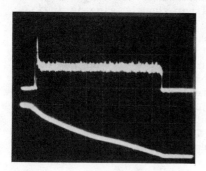 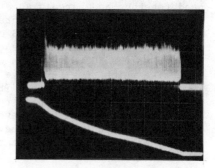

Operators View of Proton Beam Spill   Users View of Proton Beam Spill

FIG. 1.

*Work supported by the U. S. Department of Energy.

REVIEW OF OPERATIONS

Next, a few graphs that only Program Committee members normally get to see should brighten your day. Figure 2 is the ZGS operating efficiency since 1966. The ZGS has truly been a very reliable machine as the continued efficiency of 90% is shown here. Note also that it hasn't slackened the last couple of years. Figure 3 is the average proton intensity since 1966. You will note the rapid rise in beam intensity in 1966 and 1967, then the rather sluggish operation until 1975. Actually the titanium chambers and their attached pole face windings offered us many opportunities not enjoyed before their installation. Construction of numerous diagnostic devices, digital control of the pole face system and rf systems, introduction of an octupole mode on the pole face windings and ramped energy injection combined to give the great improvement noted the last few years. A reliable H⁻ source also came along to help us at this time. The advice and help of Yanglai Cho has been very significant in these improvements. One should bear in mind also that we were continually interrupting the operation for changes to polarized beam.

Figure 4 shows the fantastic improvement in the polarized beam these last five years. Most of this is due to ion source improvements; i.e., the current delivered to the ZGS has gone from 1 µA to 50 µA, but the factors improving conventional proton intensity,

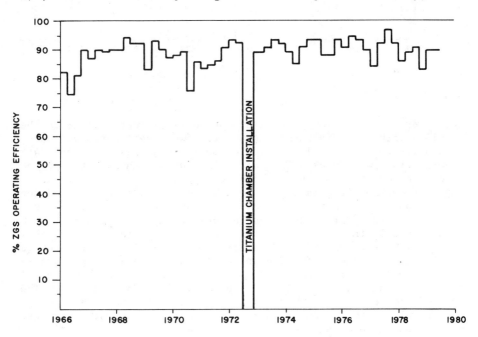

FIG. 2. ZGS Operating Efficiency.

194

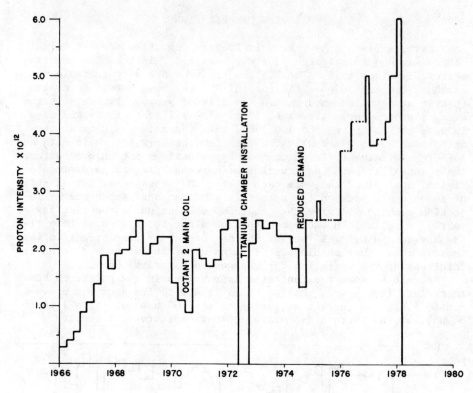

FIG. 3.   ZGS Proton Intensity.

which were just mentioned, have also been significant. Through the
tireless work of Larry Ratner and others, we have located and successfully crossed some 11 intrinsic depolarizing resources and 24
imperfection resonances. The fast pulse quad system and the pulse
pole face octant bump systems both deserve special plaudits. We
must not overlook the fact that in this period we also produced the
world's only high energy polarized deuteron beam. Another change of
pattern occurred during the lean budget years. The accelerator ran
with virtually the same 12.33 GeV/c operating program most of the
time in 1966 through 1972. Polarized proton physics was interesting
at numerous low energies. The Operations Group, using the diagnostics developed in 1972 and 1973, learned to rapidly tune up the
machine at any energy. Once we changed 11 times in one month. The
work of the beam line tuning group can't be neglected. They not
only developed their own techniques for rapid tuning during energy
changes but also developed a myriad of new beam line diagnostics for
the low signal levels present during polarized operation.

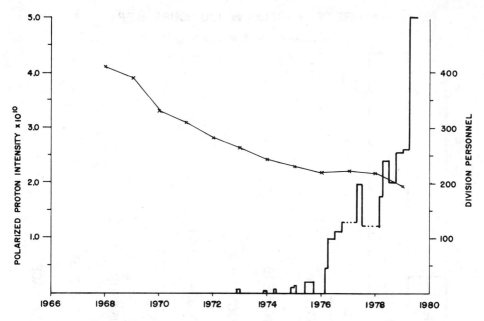

FIG. 4.   ZGS Polarized Proton Intensity and Divisional Personnel.

Another interesting curve is also shown on Fig. 4--number of personnel involved.   It is interesting to speculate what about fifty people could produce here since polarized proton intensity seems inversely proportional to the number of personnel.   Perhaps this is some comment on the quality and interest of those who stayed on.

MOTIVATION

The last paragraph is sort of a natural lead in to my next topic--motivation.   Specifically, what has made our reliability so good and what has made the last couple years of operation so good despite a discouraging future.   We basically use the carrot more than the stick.   Figure 5 shows how we compare the reliability of systems, thus, in effect, urging the system managers to compete.   Twenty-four hour graphs of beam intensity allow the machine operators to compare their performance to other operators and, of course, allow management to make comparisons of operator performance.   We have, on occasion, resorted to awarding bottles of liquor for establishing new intensity records.   This technique proved very effective in maintaining good performance during the last couple neutrino runs. It was obviously the prestige and not the value of the gift that perked things up.   By the way, the money for the liquor came from our own pockets, not from the HEP budget.

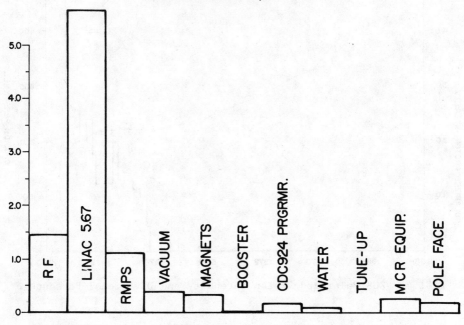

FIG. 5. ZGS System Reliability.

The most important factor, however, is the example set by staff personnel. If they perform with enthusiasm and interest, it seems to affect the whole operation. Despite the many "acid stomachs" that I have endured from dealing with physicists, I must confess that I think their drive infects the engineers who, in turn, infect many of the hourly personnel. Technicians often spend their own time when a problem develops. Many put up with midnight calls with no extra compensation except "job satisfaction." The people here in the technician bargaining unit do not, as a group, have the anti-management bias so often seen in factories.

The frequent parties thrown by the experimenters also help to show the operating personnel that their work is appreciated despite the crabbing about intensity stability, spill structure, etc., during the run. Another factor of significance during these last two years has been the growth of the peripheral programs such as the Intense Pulsed Neutron Source and Heavy Ion Fusion. They offer interesting job opportunities for many after the ZGS closedown.

NOSTALGIA

These new programs help to make the shutdown as painless as possible but, in a greater sense, it can't be painless for many of us; that big pile of iron has become a part of our lives. I personally will miss the Tuesday operations meetings which often took three hours to set up a schedule that would be changed in three minutes by a breakdown or by a user's complaint. Other great memories include being part of the committee that tried to convince Alan Krisch that he should have 75.6% of the beam, not 75.7%. That took about three hours also! Sharing Larry Ratner's satisfaction when we successfully jumped the imperfection resonances was a great moment. The tireless work of so many people has made our success; but I think the most important job of all was done by the five Assistant Chief of Operations, the machine operators. Their dedication, skill and resourcefulness have been the key to all those nice rising slopes on our intensity graphs. All the members of the ZGS Operations Group can remember the ZGS days with pride.

HISTORY OF THE SUPERCONDUCTING MAGNET BUBBLE CHAMBERS

M. Derrick, L. G. Hyman, and E. G. Pewitt
Argonne National Laboratory

"It has been said that although God cannot alter the past, historians can. It is perhaps because they can be useful to him in this respect that he tolerates their existence."

Erehwon
Samuel Butler

EARLY HISTORY

In the late 1950's, a group of physicists at Carnegie Tech[1] built a number of small bubble chambers for use with the 450-MeV synchro-cyclotron facility that was located at Saxonburg, PA. Since bubble chambers were the dominant technique in studies of resonances and since those studies required the use of higher energy beams, it was natural for this group to consider continuing the work at the new 12-GeV machine proposed for construction at Argonne. To this end, Tom Fields took leave from the University and came to Argonne in the fall of 1960 to work with Bob Thompson, who had joined the Laboratory from Indiana University. Thompson was charged with the responsibility of planning a bubble chamber construction program for the ZGS.

At that time the Berkeley 72" chamber was being completed and so Thompson was thinking of something larger, perhaps as big as a 5 m bubble chamber. The major technical problem foreseen was the difficulty of obtaining a suitable glass window. The physics rational for such a large hydrogen chamber was not completely clear, but Fields and Thompson carried out optimization studies to explore the costs associated with different choices of chamber size and magnetic field.[2] However, Thompson was aware of the possibilities of using a large bubble chamber to do neutrino physics as the following quotation[3] from a round table held at the 1960 Instrumentation Conference for High Energy Physics attests: ".... finally let me mention the possibility of using a large freon chamber for the neutrino experiments. The cost of a big magnet, the optical system, and the equipment to analyze the photographs is a substantial part of the total cost. The cost of the freon chamber would be only a small fraction. One has to remember that in a few years, neutrino physics may be one of the very hottest things, and a really large freon chamber detector for neutrinos would be a nice thing to have.... ." Shortly after this meeting, Thompson left Argonne for the University of Chicago. However, a cryogenic test building (829) had been provided, and the nucleus of a staff of

ISSN:0094-243X/80/600198-29$1.50 Copyright 1980 American Institute of Physics

technicians and engineers had been recruited.

In early 1960, the Bell Laboratory group published a paper[4] describing the discovery of continuing superconductivity of Nb Sn in a high magnetic field. As a result, the Argonne-CIT collaboration started the design of a small bubble chamber. The aim of the project was to build a device large enough to work usefully and competitively in the physics of elementary particles but small enough to have a reasonable chance of success. At best this promised to be a somewhat experimental device since no one could be sure that superconducting magnets would be practical, so the Laboratory, in collaboration with a number of users, proposed to copy the Brookhaven 80-inch chamber as the quickest way of providing a large bubble chamber facility. This proposal was not supported by the AEC. A year or two after this, there was also an informal proprosal from L. Alvarez to build two 5 m hydrogen chambers, one for SLAC and one for the ZGS. This idea was not greeted with enthusiasm by the Midwest university community.

SUPERCONDUCTING MAGNET DEVELOPMENT

Following the publication from Bell Labs, Westinghouse issued a press release describing the operation of a superconducting magnet made of niobium/zirconium wire. The ANL-CIT group, which at that time consisted of Tom Fields, John Fetkovich, Ken Martin, and Gale Pewitt, solicited proposals to build a high field magnet to fit around a 10-inch bubble chamber. Responses were received from Atomics International, General Electric, Magnion, Westinghouse, and Avco Everett. As a result of visits to all the vendors, and discussions with the people involved, Avco Everett was awarded a contract.[5] John Stekly from Avco made a great impression on the physicists in the group and seemed to understand many of the technical problems associated with such a venture.

Shortly after the contract was placed with Avco, Charles Laverick joined the group, and he and Tom Fields started a series of small experiments on superconductivity in order to build up some in-house expertise. They began making tests on short samples of wire and operated a number of small magnets that were wound in the lab. At that time, a liquid helium test facility was not the trivial thing it later became, and there was some period of learning how to design the cryogenic systems and transfer liquid helium without boiling it all away.

The conductor used in the Avco magnet was 10-mil Nb/Zr wire with a very thin layer of copper on the outside, and one of the problems encountered was in making satisfactory electrical joints. The joint resistance depended on the orientation with respect to the magnetic field. As a result of a development effort at Argonne, a double parallel joint was developed, using both a crimped copper tube and a Nb/Zr block. This technique gave a resistance of $< 1\,\mu\Omega$ at 30 kG when aligned parallel to the field.[6] We had a technician, called Joe Bozman, who became very good at making these joints, and we sent him to Avco to make the joints for the magnet.

The magnet, shown in Fig. 1, was made of a stack of pancakes;

200

Fig. 1 AVCO magnet coil package.

Fig. 2 Argonne magnet being withdrawn from the test dewar.

the most impressive thing was that it used a total of 40 miles of 10-mil wire, giving an inductance of 7,000 H. Each pancake was layer-wound with external shunts placed every second layer. Shorted copper rings were used between layers to control the time constant of current decay during a quench. Much of the energy was dissipated in the external shunts. [7]

After one or two of the eight pancakes were wound, the job slowed down at Avco, and no progress was being made. We later learned that Stekly had developed a copper-stabilized conductor, and the company had no enthusiasm for proceeding with our contract. There was a showdown between Charlie Laverick and Arthur Kantrowitz, the head of the Avco Laboratory; Stekly at the time was in Florida on vacation. To accelerate work on our magnet, Charlie threatened to send a truck to pick up the remaining pieces and said we would finish it ourselves. This bit of gamesmanship forced Kantrowitz to have Stekly come back from Florida and work full time on the project. The magnet was completed and tested at Avco. It carried 11A giving a central field of 32.8 kG. We used most of the helium on the eastern seaboard in the test.

By this time it was clear to everyone that developments of the art had made this prototype obsolete. The magnet was quite unstable, and so not suitable for use in a complicated experiment where reliability of every component was required.

As a result of the work on the Avco magnet and other ideas, it became clear in which direction to go in order to build a more reliable magnet. The quality control in the early superconducting material was not good, so the technique of using a single strand of conductor as the only path of the current meant that one small bad section could limit the current-carrying capacity of a whole pancake. Another problem came from the use of epoxy resin to hold the wires. The poor thermal conductivity of this material made it difficult to provide adequate cooling in the middle of the coil.

The solution of these difficulties was to make a composite conductor consisting of many parallel superconducting strands in intimate contact with a normal material. Charlie Laverick developed a conductor that was made by cabling several strands of superconductor and copper and impregnating the resulting cable with indium. George Lobell designed a machine to make the cables in the Argonne Central Shops. This simple technique allowed a wide variation in the fraction of copper and superconductor, and also meant that each superconducting strand followed a helical path; the importance of which we did not immediately realize. Good cooling was provided by using stainless steel mesh between layers of the magnet.

On the first test, the new magnet [8] gave a central field of 44 kG and behaved in a very stable manner although the strong inductive coupling between the different sections which were on separate power supplies challenged our understanding of Lenz' law. A liner of 7-inch bore was installed in the magnet for the first test in a helium dewar and when this inner section was energized, a field of 67 kG was obtained. Fig. 2 shows this magnet being withdrawn from the test dewar. It is now on display at the Smithsonian Institution in Washington.

It was then just a matter of installing the magnet in its cryostat and testing the whole system - not such a simple matter as the cryostat had a removable end plate with large indium seals and it took some time to make these leak-tight.

SMALL BUBBLE CHAMBER CONSTRUCTION

In parallel with the work on the magnet, the bubble chamber was being designed and built. At the time, we had anticipated a long program of physics, and so the chamber was designed to operate with hydrogen or deuterium as well as helium. This meant that the chamber, as shown in Fig. 3, ended up being a rather complicated device.

In order to conserve magnetic field volume, the flanges and windows of the chamber were placed beyond the magnet, resulting in a rather deep chamber. The chamber vessel itself was a thin-wall stainless steel cylinder machined eccentric to have a thicker wall section at the top than the bottom. Cooling channels were machined in this top section and capped with a copper plate. There was considerable difficulty in fabricating this piece as the cylinder always distorted when the copper insert was brazed onto the wall. We finally had a second chamber body made in which the copper was electroformed onto the steel.

The illumination of a long cylinder for dark field photography presented a special problem. It is desirable to have the light from the source focus at the camera plane so that the scattered intensity to each camera from bubbles anywhere in the chamber is roughly constant. Fig. 4 shows the illumination system that was designed by John Fetkovich. There were two cylindrical reflectors (one of which was the wall of the chamber itself), that produced an intermediate vertical ring source, which was then imaged by the plastic lens window of the chamber at the camera plane.

The expansions were made by using the condensing lens as a moving wall of the chamber. The Plexiglas lens was sealed to the chamber body by a stainless steel omega bellows. Because of the large expansion coefficient of plastic, a special plastic-metal seal had to be developed that was later patented.[9] A complete expansion-recompression cycle took 15 msec. The cryogenic systems were complicated by the fact that both liquid nitrogen and liquid hydrogen heat intercepts were used on all support members to reduce the heat input to helium temperature. Two separate helium systems were used. Since the chamber operated at about $3.2^\circ K$, the chamber cooling was done using helium under reduced pressure. The magnet was cooled by helium slightly above atmospheric pressure.

In August 1963, a proposal was submitted to the ZGS Program Committee. The title page of what became ZGS Experiment 12 is shown as Fig. 5.[10] It is notable that Tom Fields was the only physicist on the proposal with a tenure job. In the proposal, we emphasized the importance of good momentum resolution. For low momentum particles, the relevant parameter is $B^2 \ell$, where ℓ is some typical track length, and in this parameter, the chamber was superior to any other chamber then operating with the exception of the Berkeley 72" bubble chamber.

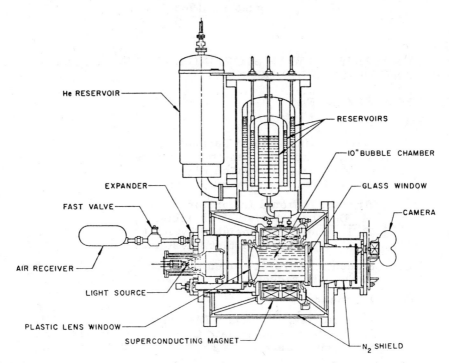

He RESERVOIR

RESERVOIRS

10" BUBBLE CHAMBER

GLASS WINDOW

EXPANDER

FAST VALVE

CAMERA

AIR RECEIVER

LIGHT SOURCE

PLASTIC LENS WINDOW

SUPERCONDUCTING MAGNET

N$_2$ SHIELD

Fig. 3 Superconducting magnet bubble chamber assembly.

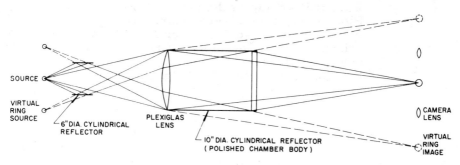

SOURCE

VIRTUAL
RING
SOURCE

6"DIA. CYLINDRICAL
REFLECTOR

PLEXIGLAS
LENS

10" DIA. CYLINDRICAL REFLECTOR
(POLISHED CHAMBER BODY)

CAMERA
LENS

VIRTUAL
RING
IMAGE

Fig. 4 Ray diagram of illumination optics.

ARGONNE NATIONAL LABORATORY
Argonne, Illinois
August 30, 1963

1. EXPERIMENTS WITH STOPPING KAONS IN HELIUM AND
   ANTIPROTONS IN DEUTERIUM

2. EXPERIMENTERS

   A. Ph. D's:

   Carnegie Institute of Technology
   J. G. Fetkovich, Assistant Professor
   J. McKenzie, Research Physicist
   E. G. Pewitt, Assistant Professor

   Argonne National Laboratory
   M. Derrick, Assistant Physicist
   T. H. Fields, Associate Physicist
   L. G. Hyman, Assistant Physicist

   B. Students: Several students at Carnegie Tech will help
      in the data reduction.

   C. Engineers:

   C. Laverick, Associate Electrical Engineer, ANL
   K. B. Martin, Associate Cryogenic Engineer, ANL
   S. Stasak, Electronics Engineer, CIT

   D. Technicians:

   J. Santori, Technical Assistant, ANL
   C. Barnes, Senior Technician, ANL
   J. Sheppard, Senior Technician, ANL
   F. Piotrowski, Senior Technician, ANL
   P. Arvidson, Research Technician, ANL
   J. Griffith, Senior Technician, CIT
   J. Sadecky, Laboratory Technician, CIT
   C. Petrigni, Laboratory Technician, CIT

Fig. 5 Proposal for E12.

A separated, low momentum kaon beam was needed in order to provide a source of stopping K⁻. The beam that was built is shown in Fig. 6. It used conventional two-stage optics, each stage having a 10-ft long electrostatic separator. Positive particles were taken off at 28° from an internal target, whereas negative particles could be used at smaller production angles depending on the target position inside the ZGS octant.

At ZGS turnon, the meson shield wall had a simple unseparated beam at 30° that was used for studies of hypernuclei and K^o decay by the Northwestern-Argonne and Illinois groups, respectively. The shielding round the three beams coming from the internal target was fairly complex and so there was some reluctance on the part of the management of the Particle Accelerator Division to change the arrangement in the meson cap as it was called, but finally in late 1965, the change was made and the separated beam installed. The beam was tuned and behaved as expected although the intensity was less than that predicted.

There was some difficulty in the beginning of the experiment in centering the beam in the chamber, and the last quadrupole had to be moved by about 1/4-inch to steer the beam horizontally. It later turned out that bending magnet B2 in the beam was misplaced by one inch from the nominal position. Fig. 7 shows the helium chamber installed at the end of the 28° beam in the meson building before the camera plate was put on.

This beam was also used later to provide particles for the Michigan-Argonne heavy liquid chamber and the Northwestern helium chamber.

Just before our run took place, there was a proposal from the Northwestern group to use their 20" x 10" x 10" helium chamber instead of the supermagnet chamber to carry out the physics program of K⁻ in helium, and, indeed, throughout our exposure, the Northwestern chamber was on standby at liquid nitrogen temperature waiting for us to break down! Fortunately, the problems anticipated by our competitors did not materialize and in the period March 5-29, 1966, 476,800 photographs were taken. The run summary is given in Table I. It was compiled by Gary Keyes, a student who did a Ph.D. thesis studying the properties of hypernuclei produced in K⁻ He interactions.

A competition developed between the physicists on the different shifts to see who could take the most pictures. This placed a premium on devising an optimal strategy of transferring helium from the 1000-liter storage dewars into the chamber reservoir. Several physicists were also known to have advanced the cameras manually when the automatic film advance failed.

We took some pictures of stopping K^+ as we wished to use the $K^+ \rightarrow \pi^+ \pi^+ \pi^-$ decay as a calibrator of the magnetic field and also a few pictures of stopping antiprotons to see if there was any physics in p̄He reactions. The latter had not been approved, and the management of the Particle Accelerator Division had a strong reaction to this minor infringement of the rules.

Table I Summary of 10-Inch Helium SBC Run
March 5 - 29, 1966

ANL-SBC-75
G. Keyes
March 13, 1967

Stopping K⁻ Film:

Good film:	61 rolls (No. 44-69, No. 91-125)	195, 200 pictures
Fair film:	25 rolls (No. 19-43)	80, 000

In-flight K⁻ Film:

280 MeV/c:	10 rolls (No. 81-90)	32, 000
370 MeV/c:	11 rolls (No. 70-80)	35, 200
400 MeV/c:	12 rolls (No. 126-137)	38, 400

Stopping K⁺ Film:	4 rolls (No. 146-149)	12, 800
Antiproton Film:	2 rolls (No. 138-139)	6, 400

π⁺ Film

~ 500 MeV/c:	6 rolls (No. 140-145)	19, 200

π⁻ Film
(tuning beam

& chamber):	18 rolls (No. 1-18)	57, 600

Total number of
K⁻ pictures: 380, 800

Total number of
pictures: 476, 800

(See Table 1 for detailed film summary.)

Averages: Rolls 45-149   VPT-1 231 mm. Hg.
(See Table 2 for detailed summary.)   VPT-2 245 mm. Hg.
Overpressure: 305 mm. Hg.
ZGS Intensity: 4. 3 x 10" protons/pulse

Liquid Consumption:

Helium:	420 liters/day	
Magnet:	270	"
Chamber:	150	"
Hydrogen:	270	"

Magnet Current Regulation:

Maximum fluctuation:	0. 45 amps = 0. 09%
Standard deviation:	0. 16 amps = 0. 03%

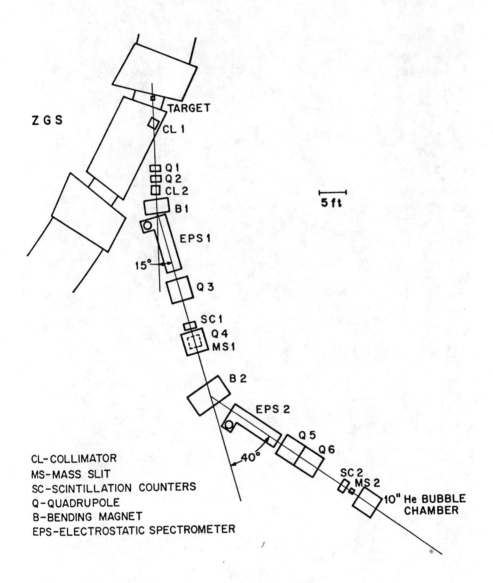

ZGS

TARGET
CL 1

Q 1
Q 2
CL 2
B 1

EPS 1

15°

Q 3

SC 1
Q 4
MS 1

B 2

EPS 2

40°

Q 5
Q 6

SC 2
MS 2
10" He BUBBLE
CHAMBER

5 ft

CL-COLLIMATOR
MS-MASS SLIT
SC-SCINTILLATION COUNTERS
Q-QUADRUPOLE
B-BENDING MAGNET
EPS-ELECTROSTATIC SPECTROMETER

Fig. 6 28° low momentum separated beam.

Fig. 7 Bubble chamber setup in low momentum beam.

HIGH ENERGY PHYSICS RESULTS FROM
THE SMALL CHAMBER

We immediately set to work to do the physics analysis. In addition to the K^- pictures, we had also taken some film of low energy pions for a Kansas group, and the IIT group approached us about doing some independent analysis using the K^- film, so that with the Argonne and Carnegie efforts there were four groups working on the analysis.

The chronology of the program and, in particular, the physics publications, is shown in Table II. A total of six Letter publications and 10 articles resulted from the work, and the total program from first discussions to final publication extended over about 15 years. It is not appropriate to discuss the physics in detail, but it was notable that many experiments were done that were not anticipated in the proposal.

An early result of some importance was a measurement of the lifetime of the hypernucleus $_\Lambda H^3$ where we showed that the lifetime was in accord with theory and in disagreement with an earlier measurement. We also did a number of other studies of the light hypernuclei $_\Lambda H^3$, $_\Lambda H^4$, and $_\Lambda He^4$ and many of the results were presented at an International Conference on Hypernuclear Physics held at Argonne in 1969 and organized by A. Bodmer and L. Hyman.

The chamber also proved to be a rather accurate device. We made the best measurement of the range-energy relation in helium, and this precise calibration, combined with the high magnetic field, allowed us to make good measurements of the Σ^+ and Λ masses and give an improved upper limit on the ν_μ mass. In addition, we made systematic studies of many reactions resulting from K^- interactions in helium - something that had not been done previously.

EARLY HISTORY OF THE TWELVE-FOOT
BUBBLE CHAMBER PROJECT

In late 1963, a group from Madison led by Jack Fry and Ugo Camerini proposed to build a large heavy liquid bubble chamber to be used to study neutrino physics at the ZGS. Although somewhat smaller, the parameters were very similar to the Gargamelle chamber that was later so successfully operated at CERN. Wilson Powell worked with the group and built a 2-ft model chamber to test various parameters of the final project. After the construction of the 12-foot hydrogen chamber was approved for Argonne and as a result of the progress of the Gargamelle project, the Wisconsin group proposed in 1966 to locate their heavy liquid chamber at the AGS, but this was not accepted by the Brookhaven management. The magnet for the chamber was built and later used as the basis of the Multiparticle Spectrometer at Brookhaven. The chamber project was then terminated. The decision not to continue with the Wisconsin chamber was influenced by the realization that a hydrogen bubble chamber could be operated with neon and so do much of the physics accessible to a heavy liquid chamber.

In January 1964, Roger Hildebrand asked Gale Pewitt if he would go to Brookhaven to attend a High Energy Discussion Group

210

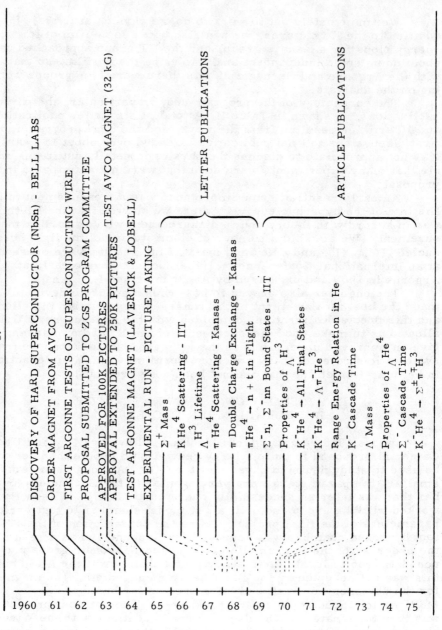

Table II Chronology of E12 K⁻ in Helium

(HEDG) meeting at which large bubble chamber proposals would be discussed. Two proposals were presented at that meeting. Bob Palmer described a proposed sixteen-foot hydrogen bubble chamber and a number of considerations were presented, which Palmer felt illustrated that a chamber was feasible on this large a scale. The other proposal was by Mel Schwartz and Jack Steinberger of Columbia for a scaled-up version of the 30" hydrogen bubble chamber built at Nevis. The Columbia proposal appeared to be very preliminary and it was not clear, for example, if any magnetic field calculations had been made. The proposal was perhaps intended to ensure the Columbia group would be a lead user of a large chamber if it were built.

Later that year an AUI Committee was appointed under the chairmanship of Dave Frisch of MIT to resolve the conflict. We met with this committee in New York to discuss the 12-ft chamber project and combined this business with an interesting visit to the World's Fair. The outcome of the Frisch committee was to scale down the BNL proposal to 14 ft and to insist that it be a facility available to all users on the basis of competitive proposals.

Following the HEDG meeting and after discussions between the Argonne high energy physicists, Gale Pewitt was asked by Roger Hildebrand to develop a proposal for a large chamber to be used at the ZGS. A small proposal team was assembled, and we hired Peter VanderArend as a cryogenic consultant. We were assigned a designer and started putting together some preliminary ideas.

The major question was the optics of the system. Both the BNL 14-ft chamber proposal and our plans called for the use of small windows, and these depended on being able to use Scotchlite as a retrodirector for the light. The Wisconsin group had earlier chosen this solution for their proposed heavy liquid chamber. The technique had been used successfully in small heavy liquid chambers, but it was not known if it would work in liquid hydrogen. Two different types of Scotchlite were obtained (the blue open-bead type and the green covered-bead variety) and tested at Carnegie-Tech in a six-inch hydrogen chamber. Some Cobalt-60 Compton electrons were photographed in liquid hydrogen and pictures of these tracks were shown in our proposal as a proof that the system was practical. The proposal was completed in June 1964 and submitted to the AEC. The title page is shown in Fig. 8.

The Laboratory also had an FY1966 proposal jointly prepared by the MURA and Argonne accelerator groups to build a 200-MeV linac as a new injector for the ZGS. In July 1964, the Users' Advisory Committee met at Argonne to discuss the relative priority of the bubble chamber and the linac. The committee included, among other people, Bill Walker and Ned Goldwasser. They failed to agree, and the Argonne management gave first priority to the bubble chamber and the new external proton area (Building 375). It was interesting to note that the Brookhaven users recommended that their linac system be put first in priority and the bubble chamber second.

There were some vigorous discussions about where the project would be located within the ZGS Complex. The proposal work had all been done by physicists and engineers within the High Energy Physics Division under the general direction of Tom Fields, the Division Director, but most of the engineering talent was in the Particle

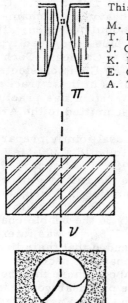

PROPOSAL
FOR THE CONSTRUCTION
OF A 12 FOOT HYDROGEN
BUBBLE CHAMBER

p

π

ν

June 1964

This proposal was prepared by:

M. Derrick, Argonne National Laboratory
T. H. Fields, ANL and Northwestern University
J. G. Fetkovich, Carnegie Institute of Technology
K. B. Martin, Argonne National Laboratory
E. G. Pewitt, ANL and Carnegie Inst. of Technology
A. Tamosaitis, Argonne National Laboratory

Fig. 8

Argonne National Laboratory

Accelerator Division. Bob Sachs decided that the HEP Division would build the chamber with engineering support from PAD, and Lee Teng assigned some very good people to the project.

When the President's budget for FY1966 was announced on January 20, 1965, the 12-foot bubble chamber was included as a line item. In addition to the bubble chamber, the extraction system and experimental area which became EPB-II was also funded for a total of about $17M. For Brookhaven, a 200 MeV linac and the AGS improvement program was supported at about $50M, but not the 14-ft bubble chamber.

We don't know all the political factors which resulted in the 12-foot chamber being included in the President s budget, but it is significant that MURA was phasing out at this time, and there was strong Midwest political pressure for facilities that would be competitive with the East and West coasts.

Before we knew about the construction approval, two of us went to California to discuss the proposal with Luis Alvarez. He listened to us for about thirty minutes and then told us that in projects like this, you've got to work on the politics as well. He thought we had already put too much work into the technical side, and that we should concentrate on getting political support for the chamber. He also said we ought to see Bill Brobeck and get him to do a cost estimate of the chamber. Alvarez felt that the AEC would request an independent cost estimate, and that we should beat them to it. We took this advice, and before we left the building, we called Bill Brobeck, made an appointment, and left a copy of the proposal with him. Some time later, Brobeck visited Argonne. He put the proposal down and said, "Our estimate is that, based on the proposal and our discussions, it will take ten million dollars to build, but it will cost you ten thousand dollars to get this in writing!"

We wrote a contract and Brobeck assigned Carl Skabalsky to the job of confirming our cost estimate. Skabalsky wrote a number of letters and received quotes that added up to less than our estimate, so we supplied him with more information which made his estimate come much closer to ours.

TECHNICAL DECISIONS

In the proposal we had suggested the use of a large diaphragm to expand the chamber. When we got down to detailed design, the difficulty, if not impracticality of this, became clear. We decided to go to a larger version of the omega bellows that we had used for the 10-inch chamber, but this time located at the bottom of the chamber with photography done from the top. There was some controversy over whether having the cameras looking directly at the expansion system was advisable. Since no one had experience in the thermal design of large chambers, we also had real concern about photographing through many feet of liquid hydrogen.

To minimize these problems, we decided that the cooling of the chamber would be done behind a shroud, which would be located in the chamber and spaced off from the wall by a few inches. This shroud prevented the cold liquid from the heat exchanger from falling into the photographed volume.

Cooling loops were installed to intercept all the conduction heat leaks to the chamber. The heat added, due to irreversible expansion, was to be removed at the top. The design of this copper-finned heat exchanger underwent many changes before it was finally built and installed.

The cameras were to be located at some radius from the center of the chamber, whether at the full diameter or at some intermediate diameter, was left undecided. A point that we quickly realized was that flat windows would not work. Because of the dispersion of the liquid hydrogen if the light rays entered the glass at other than normal incidence, the image would be smeared. This smearing would amount to as much as 5 mm for a 45° angle of incidence. This problem could be overcome either by going to a very small bandwidth on the light source or by using a hemispherical fisheye window. We chose the latter solution.

It was decided to view the chamber volume with four cameras arranged on a square. The exact location of the camera windows received considerable attention. There was a strong desire to locate the cameras near the axis of the chamber, thereby greatly enhancing the ease of scanning the film, in that the stereo views would look very similar. On the other hand, there were arguments for placing these cameras at the edge of the chamber, so as to maximize the stereo angle, improve the measurement precision, and maximize the volume of liquid covered. Ray Ammar wrote a program to generate events as seen through cameras placed at different locations. The resulting film was scanned by a number of physicists. As a result of this exercise, we located the cameras at a compromise location about half way out from the center.

We decided to build a model of the chamber in order to test the optical turbulence problems and to measure the specific dynamic heat load. The latter was not well known as the heat load of the existing chambers was dominated by boiling round the window seals and at the expander. The model was two feet in diameter but had many features in common with the full-size device.

Work proceeded in building the model, and Lyle Genens developed a procedure for forming the large bellows. The procedure utilized some of the features employed by the Badger Fire Extinguisher Company of Lowell, Massachusetts, who made bellows for expansion joints. We modeled the bellows formation on a two-foot diameter scale, and formed a number of bellows that were used for other applications, including the two-foot chamber.

Jack Frolich was the engineer in charge of putting the two-foot chamber together. It was assembled first in Building 362, and then was moved to Building 829 and operated there. Stan Stoy and Mike Morgan also worked on the job. The results of these tests convinced us that the dynamic heat load on the 12-ft chamber would be manageable and that thermal turbulence would not dominate the precision of reconstruction.

The expansion system that we built for the two-foot chamber was a state-of-the-art system. We contracted with an English engineer by the name of Brian Wolfington. He lived in Billerea, Massachusetts, and had his shop in a house built in the 1700's.

Tony Tamosaitis was the Chief Engineer of the bubble chamber, and it was through him that experts in the relevant technology areas were located. The turbulence problem was examined, and we talked with a professor from the Computing Department of Purdue University to ask him if he would study the problem. He said it would take enormous computing time to calculate the shift of a spot due to heat flux through 12-feet of hydrogen. He also pointed out to us that we didn't really want consulting; we wanted laborers, and that was not suitable to a university professor. Tony also contacted Dave Fultz from the Geophysics Department of the University of Chicago. We visited him and described the problem. We told him that we were building a model chamber, and he asked the dimensions. We told him a two-foot model, and he said, "Do you mean to tell me that you want to build a twelve-foot chamber and you will model it on a two-foot scale?" "I'm used to scaling by factors of 10^5 or 10^6; sounds like you people are going to make an actual measurement!" We were much encouraged by his comments.

One of the pleasures of the whole job was developing new systems. We chose the expander to be a resonant hydraulic system. The calculations were checked by Larry Turner and Richard Hoglund, a professor of hydraulics at Purdue University. The final design for the expansion system was made by Jim Simpson with help from Lyle Genens and Norm Majeski. Simpson's approach was that everything had a simple solution; that sort of confidence was certainly highly beneficial.

We had John Purcell visit for a couple of weeks of consulting to look at the transient problems in the superconducting magnet for the 10-inch chamber. He attacked these problems quite well, so we hired him for the 12-ft project.

He was assigned to work in Charles Laverick's laboratory and was also charged with the design of the 12-ft superconducting magnet. John Purcell, who had worked on cryogenic magnets, quickly appreciated the concept of cryogenically stable superconductors that had resulted from the work on the 10-inch chamber. In parallel, a conventional magnet coil was designed by Ray Krizek, an engineer in the Particle Accelerator Division. The iron yoke of the magnet was designed by Jim Bywater with help from Jack Nolan and Pete Marston of Magnetic Engineering Associates.

In June of 1965, we had to decide whether to go with a conventional magnet or with superconducting coils. The magnet iron was designed to be compatible with either. This was a very exciting question. The cost estimates of the two alternates were almost identical. The saving was in the operating cost of the superconducting magnet. We received a lot of advice, and probably the pivotal aspect of the whole matter was that the superconducting magnet was such a challenging frontier device it attracted high quality people. As a result, it was easy to see that we could do a better job on the superconducting magnet than we could on the conventional magnet. In the event the decision turned out to be correct and the superconducting magnet provided the practicality of the new technology on the largest scale.[11] Fig. 9 shows the magnet coils under construction.

Fig. 9　Assembly of Superconducting Coils

TABLE III 12 FT HBC CONSTRUCTION SCHEDULE

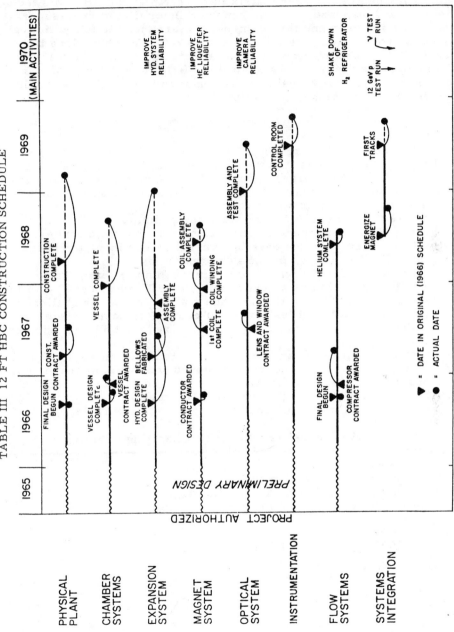

The conductor used in the magnet consisted of six strands of Nb Ti extruded in a copper matrix into a strip 2" x 0.1". This technique is now standard, but many more thinner filaments of superconductor are used and the filaments are arranged to follow helical paths to avoid eddy current in the substate that can have long time constants. These problems did not affect the magnet of the 12-ft chamber since the iron yoke shaped the field to be solenoidal. In the 10-inch chamber by cabling the conductor, we also avoided this problem of persistent eddy currents.

CHAMBER CONSTRUCTION

The construction schedule is shown in Table III as it was originally projected, and as it actually turned out.[12] It is notable that one of the longest slippages occurred in the conventional construction of the physical plant, although there was also a similar delay in the fabrication of the chamber vessel.

Lyle Genens and Bob Kleb were the principal mechanical engineers in the design and construction of the bubble chamber. Fig. 10 is a picture of the chamber vessel before insertion in the magnet. The 19,000-lb. piston is shown in Fig. 11.

The cryogenic design of the chamber and the refrigeration systems was provided by a consultant, Peter VanderArend, with the assistance of Stan Stoy of the Argonne staff. The major challenge was to intercept all heat leaks and, of course, to have a chamber with a design of the internals which minimized the boiling during expansion.

The system, as it was finally built, is shown in Figs. 12 and 13. The reconstruction accuracy achieved was 300μ in space, corresponding to $\Delta p/p \sim 1\%$ for all momenta up to 12 GeV/c. A photograph of a typical $\nu p \rightarrow \mu^- p\pi^+$ neutrino interaction is shown in Fig. 14. Some typical operating conditions are listed in Table IV.

Table IV Typical 12-Foot HBC Operating Conditions

$T_H = 25.7^{\circ}K$
$VP = 3.72$ atm. abs. (= 40 psig)
Static Pressure = 4.8 atm. abs. (= 55 psig)
Expanded Pressure = 1.2 atm. abs. (= 33 psig)
$\Delta V/V = 0.67\%$
Piston Stroke = 19 mm
$\oint PdV = 3000$ J/pulse ≈ 0.1 J/liter per pulse

Pressure Wave Form

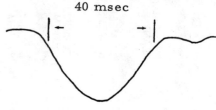

40 msec

Flash Delay After Beam = 4.6 msec.

Fig. 11 Expansion piston.

Fig. 10 Chamber vessel ready for installation.

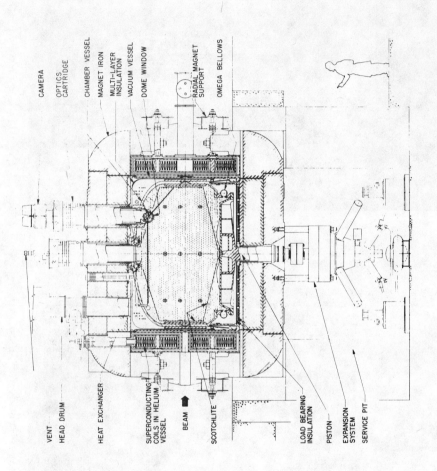

CAMERA

OPTICS
CARTRIDGE

CHAMBER VESSEL

MAGNET IRON

MULTI-LAYER
INSULATION

VACUUM VESSEL

DOME WINDOW

RADIAL MAGNET
SUPPORT

OMEGA BELLOWS

VENT

HEAD DRUM

HEAT EXCHANGER

SUPERCONDUCTING
COILS IN HELIUM
VESSEL

BEAM

SCOTCHLITE

LOAD BEARING
INSULATION

PISTON

EXPANSION
SYSTEM

SERVICE PIT

Fig. 12 Cross section of chamber and magnet.

Fig. 13 View of assembled 12-foot bubble chamber system.

222

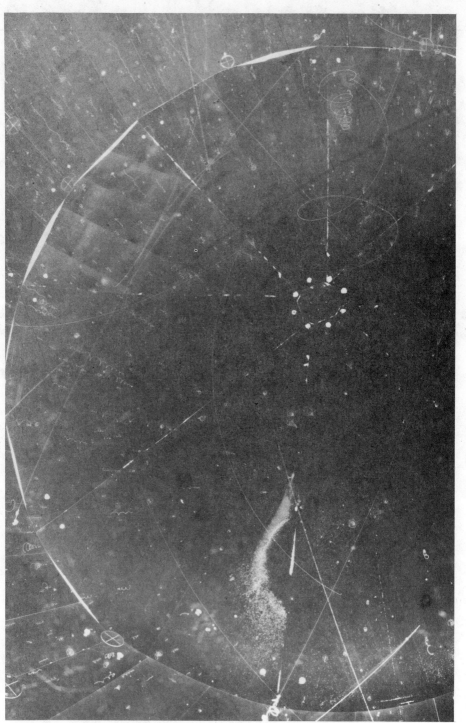

Fig. 14 $\nu p \rightarrow \mu^- p \pi^+$ event.

CHAMBER OPERATION

Bob Picha was in charge of hiring and training an operating crew. Tony Tamosaitis later took over this responsibility with Klaus Jaeger as the Deputy Group Leader. The superconducting magnet was installed in the fall of 1968, prior to the installation of the bubble chamber. During tests of the magnet, it experienced two quenches. A plumbing modification was made, such that cold gas was returned at four points around the top of the circumference of the magnet cryostat rather than the one location that was in the original design. We had no further problem with the magnet after this change.

The chamber was first operated with liquid nitrogen in July, 1969. We were aided in this run by Paul Williams of the Rutherford Laboratory. The first hydrogen run was in October of 1969, when tracks were photographed. The major problems pointed up in this run was the lack of reliability in the expansion and the optics systems. The hydrogen compressor also lacked reliability and the helium refrigerator needed some improvement. These systems were worked on and another run was held in April, 1970, which showed some improvements were needed in the neutrino shield. In the summer of 1970, these improvements were made, and in the fall, the first physics run was made with the chamber.

Some milestones in the first ten years after approval in 1964 are given in Table V. The experiments carried out in the chamber[13] are listed in Table VI.

TABLE V. MILESTONES OF 12-FT BUBBLE CHAMBER PROJECT.

HISTORY OF CHAMBER	
JUNE 1964	PROPOSAL SUBMITTED.
JANUARY 20, 1965	PRESIDENT'S BUDGET FY66.
MAY 1965	DESIGN STARTED.
JUNE 1965	SUPERMAGNET SELECTED.
JUNE 1966	CONSTRUCTION STARTED.
DECEMBER 1968	SUPERCONDUCTING MAGNET TESTED.
AUGUST 1969	FIRST CHAMBER COOLDOWN WITH LIQUID NITROGEN.
OCTOBER 1969	FIRST CHAMBER COOLDOWN WITH LIQUID HYDROGEN.
SEPTEMBER 1970	FIRST EXPERIMENTAL PICTURES TAKEN.
NOVEMBER 1970	FIRST NEUTRINO EVENT IN LIQUID HYDROGEN.
OCTOBER 1971	FIRST CHAMBER COOLDOWN WITH LIQUID DEUTERIUM.
OCTOBER 1971	FIRST NEUTRINO EVENT IN LIQUID DEUTERIUM.
MARCH 1972	1,000,000 PHYSICS PICTURES TAKEN.
MARCH 1972	TRACK SENSITIVE EXPANSIONS IN THE DOUBLE AND TRIPLE PULSE MODE AT INTERVALS AS CLOSE AS 150 MS.
NOVEMBER 1973	FIRST TRACK SENSITIVE TARGET (TST) RUN WITH HYDROGEN. TRACK SENSITIVE INSIDE AND OUTSIDE TST AT 27.5°K.
DECEMBER 1973	2,000,000 PHYSICS PICTURES TAKEN.
DECEMBER 1973	EXPANSION SYSTEM REACHED 3,000,000 PULSES WITH THE SAME TOROIDAL BELLOWS.

Table VI 12-Foot Bubble Chamber Experiments

Experiment Number	Experiment	No. of Pictures Taken in 1,000's	Last Pictures Taken	Collaborating Institutions
E267	p+p, 12 GeV/c	115	Oct. '71	ANL, IIT, Concordia Teachers College
E234	ν+d	1,154	Dec. '74	ANL, Purdue University, CMU
E234	ν+p	381	Sept. '71	ANL
E349 E308	p+p; 7.1, 8, 0, 9.0, 11.0 GeV/c	357	Feb. '73	Tohoku University
E244	K_L^0, 500 MeV/c	350	Sept. '73	ANL, CMU
E367	$\bar{p}$+p, 6 GeV/c	157	June '74	Case-Western Reserve, CMU
E303	$\bar{p}$+p, 800 MeV/c	291	June '75	Wisconsin University, ANL
E367	$\bar{p}$+p, 12 GeV/c	210	Feb. '76	Case-Western Reserve, CMU
E289/292	K^-+p, 6.5 GeV/c	1,115	Mar. '76	ANL, Michigan State University, Kansas University, Tufts University, Brussels
E413	$\bar{p}$+d, 800 MeV/c	242	April '76	ANL, Wisconsin University, Irvine, Michigan University, University of Tennessee
E371	p+p; 11 GeV/c TST 38% mole neon-hydrogen	25	Dec. '75	ANL
E370	π^-p, 8 GeV/c TST 38% mole neon-H_2	8	Dec. '75	Notre Dame University
E368	$\bar{p}$, stopping TST 38% mole neon-H_2	18	Dec. '75	Tohoku University
E383	$\bar{p}$, 4.1 GeV/c TST 38% mole neon-H_2	11	Dec. '75	Rutherford Laboratory, Melbourne University, CMU, ANL
E409	$\bar{p}$, 600 MeV/c 38% mole neon-H_2	39	Jan. '76	Tohoku University
E368	$\bar{p}$, 800 MeV/c 35-37% mole neon-H_2 TST	309	May '77	Tohoku University
E429	$\bar{p}$, 4.1 GeV/c TST 35% mole neon-H_2	227	July '77	Notre Dame University, ANL, Melbourne University, University of Tennessee
E412	ν d without plates	353	Oct. '76	ANL, Purdue University, CMU
E412	ν d with plates	1,246	Feb. '78	ANL, Purdue University, CMU

During the nine years it was in operation, a total of 6,948,000 physics pictures were taken for 17 different experiments, and the overall picture-taking efficiency, including all downtimes for the chamber, beam, and ZGS was 29%. The yearly summary is given in Table VII. As can be seen, the chamber was operated with hydrogen, deuterium, a neon-hydrogen mixture, with a track-sensitive target and with a system of tantalum-plates for γ-conversion.

Table VII Yearly Summary of 12-Foot Chamber Operation

Year	No. of Hadron Pictures (K)	No. of Neutrino Pictures (K)	Total K
1970	24	65	89
1971	156	675	831
1972	185	0	185
1973	868	419	1,287
1974	344	376	720
1975	465	0	465
1976	1,217	353	1,570
1977	555	843	1,398
1978	0	403	403
Total			6,948,000

$$\text{Overall Watt Factor} = \frac{\text{No. of Pictures Taken}}{\text{Maximum Possible}}$$

$$= 29\%$$

Including accelerator, beam, and chamber inefficiencies.

PHYSICS HIGHLIGHTS

The chamber was planned with both neutrino physics and strong interaction studies in mind. The neutrino experiment was the first such experiment to be done using a liquid hydrogen or deuterium target, and the data obtained on the sample charged-current final states as $\mu^- p$, $\mu^- p\pi^+$, $\mu^- n\pi^+$, $\mu^- p\pi^0$ are still the best available. A notable early picture showed an example of associated production by the neutral current $\nu n \rightarrow \nu \Lambda K^0$. A successful experiment to study the interactions and decays of low energy K^0 mesons was also carried out.

In the strong interactions, the first experiment studying 12-GeV pp interactions started a minor industry of π^0 spectroscopy, using the relative good γ conversion probability of the ~ 2 m γ-ray flight path available.

This work was followed up by similar studies using both high energy protons and low energy antiprotons in a track-sensitive target. Paul Kenney and his collaborators from Notre Dame were particularly active in the early planning for the use of this technique.[14] The TST program was done in a collaborative effort between Argonne, CERN, and the Rutherford Laboratory under the general direction of Klaus Jaeger.

REFERENCES

1.   M. Derrick, J. Fetkovich, T. H. Fields, R. L. McIlwain, and G. Yodh.

2.   R. Thompson and T. Fields, Bull. Am. Phys. Soc. $\underline{6}$, 24 (1961).

3.   Proceedings of an International Conference on Instrumentation for High Energy Physics, p. 156, Interscience, New York (1960).

4.   J. Kunzler, E. Buehler, F. Hsu, and J. Wernick, Phys. Rev. Lett. $\underline{6}$, 89 (1961).

5.   The proposal from Avco Everett is dated April 1962 and had an estimated cost of $143,771.

6.   E. J. Lucas, Z. J. J. Stekly, C. Laverick, and E. G. Pewitt, International Advances in Cryogenic Engineering, p. 113 (1964).

7.   Z. J. J. Stekly, Advances in Cryogenic Engineering $\underline{8}$, 585 (1962).

8.   C. Laverick and G. Lobell, Rev. Sci. Instr. $\underline{36}$, 825 (1965).

9.   K. B. Martin, T. H. Fields, E. G. Pewitt, and J. G. Fetkovich, Advances in Cryogenic Engineering $\underline{9}$, 146 (1963), U. S. Patent No. 3238574.

10. The proposed antiproton exposure was dropped since similar runs were made at Brookhaven and CERN in the period between submitting our proposal and the experimental run.

11. J. Purcell, "The Superconducting Magnet System for the 12-Foot Bubble Chamber" - Design Report, ANL/HEP 6813; Colloques Internationaux Du Centre National De La Recherche Scientifique (Grenoble 1966), No. 166, p. 261, Editions Du Centre National De La Recherche Scientifique.

12. The people involved in the chamber design wrote a total of 167 internal Argonne technical memos called BBC Notes that described various phases of the project. Some general references in the open literature that describe the 12-ft chamber are: M. Derrick, "Bubble Chambers 1964-1966", p. 431, Proceedings of the 1966 Instrumentation Conference on High Energy Physics.
     T. Fields, "Status of the Huge Liquid Hydrogen Bubble Chamber", Proceedings of the 1970 Dubna Conference on Instrumentation for High Energy Physics.
     There are also a number of papers in the Proceedings of the International Conference on Bubble Chamber Technology, Ed. M. Derrick, Argonne 1970.

13. K. Jaeger, "Parameters and Performance Characteristics of the 12-Foot Bubble Chamber", ANL/HEP 7210.

14. K. Jaeger, "Performance of the Argonne 12-Foot Bubble Chamber", and H. Leutz, "Track Sensitive Target Construction and Operating Conditions", ANL/HEP 7414.

NON-HEP PROGRAMS AT THE ZGS - PAST, PRESENT AND FUTURE

R. L. Martin
Argonne National Laboratory, Argonne, IL   60439

Several years ago, when high energy physics budgets were first beginning to come under fire, John Teem asked us to identify spin-offs of the high energy physics program. John was Director of the Division of Physical Research of the AEC (later ERDA, now DOE) and was looking for justification of the program that the lay person would understand. At the time superconductivity, with which high field magnets had been developed to the practical level almost exclusively in the high energy physics field, appeared to have the most promise for practical applications. But we couldn't say which of the many potential applications was really going to lead to a commercial use. So it was a bit difficult to justify the high energy physics program apart from what we all believed to be the huge intrinsic long-range value of fundamental knowledge in basic science.

Today the situation is quite different and if John Teem were here, I think he might be pleased. There was not a conscious effort at the ZGS to head in the direction of non-HEP programs. Those of us involved with the accelerator were motivated to make the ZGS an outstanding tool for high energy physics. In carrying out the work we recognized important applications of the technology and of some techniques. So everything I will cover grew in one way or another from the high energy physics effort and it will not be possible to separate the origins from the high energy physics programs. The applications are spin-offs and I recount here how some of the programs originated.

The list of non-HEP programs at the ZGS, shown in Table I, is quite large, and perhaps it isn't even complete. I have omitted several applications of HEP technology that I think are quite important and which were considered, but which, for one reason or another, were not pursued at the ZGS. These include intense pulsed muon beams from our Booster-II synchrotron; K meson factories with intense beams from somewhat higher energy synchrotrons; medical therapy with beams of neutrons, protons, pions, and heavy ions; electron storage rings for synchrotron radiation; and weather modifications by initiating lightning strokes with proton beams in the atmosphere.

ARTIFICIAL KIDNEY

The artificial kidney was the brainchild of Finley Markley. He was head of our plastics group and had become quite expert in bonding techniques of all kinds for the HEP program. He applied this expertise to bonding plastics such that there could be a large number of parallel paths for blood flow to exchange the wastes in the blood with saline solution. The lowered resistance to flow made it possible for the heart to pump the blood through, eliminating the expensive and complex apparatus of the normal artificial kidney

ISSN;0094-243X/80/600227-15$1.50

machine. After carrying on the development at quite a low level for about two years, Finley left Argonne to try to bring the development to the stage that it would be accepted and could be put on the market. I believe his units are now being manufactured. An IR-100 Award was given for this development in 1970. This is an award given by the Industrial Research Magazine which annually presents awards to the "100 most significant new technical products of the year."

TABLE I. NON-HEP PROGRAMS AT THE ZGS

Artificial Kidney
Thermonuclear Fusion (Magnetic)
Ocean Thermal Energy Conversion
Solar Collectors
Superconductivity
H^- Charge Exchange Injection
Intense Pulsed Neutron Sources
Proton Radiography
Heavy Ion Fusion (Inertial)

THERMONUCLEAR FUSION (MAGNETIC)

It has always been clear that the technology involved in the high energy accelerator field is closely related to that required for controlled thermonuclear fusion. In particular, this applies to high power coils, magnets, power supplies, superconductivity, and ion sources. So it is not unusual that the Accelerator Division would get involved with Argonne's CTR program.

APEX is a small tokamak, built here in one year by Jim Norem and his associates. The use of some ZGS spare parts, notably the 7-MW power supply, which was formerly used with the magnet test set of the ZGS, made the construction relatively inexpensive. The machine is now operating and is designed to do two rather specific experiments better than can be done on existing machines. So it should make a contribution to the magnetic fusion program.

Walter Praeg's expertise in high-power power supplies is widely recognized around the country and he contributes to many programs, including protection circuitry for powerful neutral beam injectors at the large TFTR tokamak at Princeton.

Ray Fuja and Bob Kustom have demonstrated rapid switching of energy in and out of a superconducting magnet with very little loss at a remarkable rate of 11 tesla per second. A special superconducting cable was developed for the purpose and also a special cryostat. The unique cable balances the need for electrical contact between the superconducting alloy and surrounding copper in order to shunt current into the copper, and the need to limit that same electrical contact to prevent eddy current losses. The cryostat is plastic, since eddy currents would produce excessive losses in normal stainless steel cryostats. It is composed of two tanks of fiberglass-reinforced polyester with 100 layers of super-insulating mylar between the tanks. The cable and cryostat were developed by Bert Wang and

Suk Kim of the superconducting magnet group. The magnet, which stored
1.5 megajoules of energy, won an IR-100 Award in 1978. Larger versions
of this magnet, with stored energies of 100 megajoules or more, could
be very important for ohmic heating coils in magnetic fusion reactors.

HOPE is an acronym for homopolar generator experiment. It is a
clever idea originated by Bob Kustom and Bob Wehrle. The experiment
consists of several counter-rotating cylinders making up a homopolar
generator. A small working model has been operated. It is an energy
storage device very much like a capacitor, but with much higher density, and much cheaper storage.

OCEAN THERMAL ENERGY CONVERSION

Lloyd Lewis, who developed the ZGS computer control system, was
instrumental in planning the use of ZGS facilities for the Ocean
Thermal Energy Conversion program. It is the large plumbing apparatus in the Meson Area. The idea of OTEC is to remove the energy
stored in the relatively small temperature difference between water
at the surface of the ocean and that at a deeper level. Ammonia is
used as the working fluid with a phase transition between gas and
liquid in this temperature range at the right working pressure. We
had available the necessary space, power, cooling, and pool of operating technicians to man the sporadic testing of commercial heat ex-
changers using ammonia. As a result, our involvement has cut one
year out of the national OTEC schedule and saved the program a lot
of money in this initial phase. After the Argonne program is completed in about one and a half years, the OTEC program will need
much larger facilities. The present plan is to move into the ocean
off Hawaii. I presume some of the people here would like to take
part in the continued development of the OTEC program there.

SOLAR COLLECTORS

Most people at this symposium are familiar with the clever idea
of the solar collectors originated by Roland Winston of the University
of Chicago. Roland developed these collectors to optimally detect
the weak Cerenkov light in a large chamber for his high energy physics
experiments on the ZGS. They are nonfocussing collectors with a wide
acceptance angle. Consequently, as solar collectors they do not have
to track the sun and can give a multiplication of the energy per unit
area of a factor of 3-10 over that of flat plate collectors. The ZGS
talent in plastics and materials was useful in development of the
collectors. John Martin of the ZGS, in collaboration with Roland
Winston, initially set up a program for this purpose at Argonne.
Argonne's solar program has not gotten very large because DOE has
made Sandia Laboratories in the Southwest the lead laboratory for
solar energy concentration systems and has established a Solar Energy
Research Institute (SERI) in Golden, Colorado. Nevertheless, two
commercial firms have been licensed to produce the Winston collectors
and I believe they will have a big impact on heating and cooling of
buildings in the future. Work in this area at Argonne has led to two

IR-100 awards, one in 1976 for the dielectric concentrator used with
photovoltaic cells and the other in 1977 for the nonimaging designs
used with tubular evacuated receivers.

SUPERCONDUCTIVITY

Superconducting coils for high field magnets were developed
specifically with high energy physics uses in mind and the extension to non-HEP programs occurred much later, at least at Argonne.
When I first arrived at Argonne in 1962, Charlie Laverick was working on superconductivity in the laboratory. He, and independently
John Stekley at AVCO, originated the idea of stabilized superconducting cable. This was superconducting wire surrounded by enough
copper to shunt current around a section which had gone normal long
enough for it to cool back down to the superconducting state without
the entire magnet going normal. The concept was a breakthrough in
the practical applications of superconductors and made the construction of large magnets with a great deal of stored energy practical.
An initial superconducting magnet for a 10" heavy liquid bubble chamber was built, followed by the magnet for the 12' hydrogen bubble
chamber at the ZGS (which won the first of the ZGS Complex's five
IR-100 Awards in 1969), and later the magnet for the Fermilab 15'
hydrogen bubble chamber. Gale Pewitt was in charge of the construction of the 12' HBC and John Purcell was deeply involved in both of
these large superconducting bubble chamber magnets.

Other applications of superconductivity to the accelerator field
were developed more slowly. It was clear that one could make higher
magnetic fields with superconducting coils than was easily possible
with iron magnets. There was some enthusiasm to build high field
quadrupoles and dipoles. But it was not clear that they would be useful for our beam lines because higher magnetic fields were not required. If such magnets were not particularly advantageous and not
used consistently their value and reliability could not be established.
Consequently, we and other laboratories made some false starts.

Pulsed superconducting coils have an inherent energy loss related to field penetration into the superconductor. To some extent this
problem is relaxed when the superconducting filaments are made very
small. However, since any such fundamental loss occurs at 4°K it is
not likely that superconducting synchrotrons with repetition rates
of one pulse per 2 to 4 seconds could be competitive with conventional
synchrotrons. However, superconducting storage rings with dc fields
or slow cycling synchrotrons with a long full field duration looked
very attractive. It was with these ideas in mind that the Superconducting Stretcher Ring for the ZGS was proposed in 1971.

The Superconducting Storage Ring (SSR) was to be a dc ring in
the ZGS tunnel to store 12-GeV protons. It would not only demonstrate the operation of a superconducting ring for the first time,
but would also significantly add to the capability of the ZGS for
high energy physics. It would, at the same power cost, increase the
average ZGS intensity by a factor of two (because one could operate
the ZGS at twice the repetition rate without the flattop). At the
same time it would provide beam nearly 100% of the time for high

energy physics (compared to a normal of 25% duty cycle). Alternatively, one could operate the ZGS at any slower rate desired and save money on the power bill, still retaining the 100% duty cycle for physics beams.

To demonstrate that we had the technology to construct the SSR, John Purcell and his group designed and constructed 10 superconducting dipoles and 5 quadrupoles (128 dipoles and 64 quadrupoles were required for the complete ring). They all worked as advertised. When the proposal was turned down these magnets were installed in a polarized proton beam line to the Effective Mass Spectrometer. They have operated reliably for many years.

The SSR was, in my opinion, an excellent and well-thought-out proposal to the AEC. Since the magnetic field required was only 30 kG, it would accomplish very significant operational and physics goals without challenging the state of the art in high field superconducting magnets at the same time. Such a demonstration is lacking even today, many years later. At the time, however, the future of the ZGS was in doubt (or at least a couple of years later when decisions on this proposal were being made). There was pressure from the President's Office of Managment and Budget to turn the ZGS off, and the AEC did not believe it could justify a new improvement program under those conditions. For these reasons we had to turn to non-HEP applications with our superconductivity program.

It was rather clear that there were applications requiring very high field magnets in both the thermonuclear program (CTR) and in magnetohydrodynamics (MHD). We were not supported in a bid to develop high field magnets in the CTR program, and as a result we lost John Purcell to General Atomics; Bert Wang took over the superconducting group and has carried the work on quite successfully. A large saddle coil dipole was built, the largest of its kind in the world. It was shipped in 1976 to the Soviet Union as part of a joint US-USSR program in MHD. The USSR presently has the only operating MHD facility, a 25-MW unit in Moscow. The Argonne magnet has operated successfully in about 12 tests in a 3-MW bypass loop of the facility over the past three years. The superconducting group is presently building a very much larger saddle coil dipole for an MHD facility being constructed at the University of Tennessee Space Institute. (This is one of the two MHD facilities under construction in the U.S.; the other is in Montana.) The magnet will have an 80-cm diameter circular bore at the entrance, expanding to more than a meter at the exit after a length of 5 m. It will have a field of 60 kG resulting in 80 MJ of stored energy. That is larger than the stored energy in the ZGS. It is being built in the HEP high bay area and the total cost will be about $8 million.

H⁻ CHARGE EXCHANGE INJECTION

The development of charge exchange injection into a practical and operational technique has had a very major impact in the high energy physics field as well as in several other fields. The stimulation for Argonne's program might be traced back nearly to the beginning of the ZGS. At the time we were involved in a collaboration with the former MURA group in Madison, Wisconsin, to examine several alternatives for upgrading the ZGS. The study considered as a new injector an FFAG machine, a rapid-cycling booster, and a 200-MeV linac. One did not then know how to achieve high intensity in a small rapid-cycling booster, so the most straightforward choice was a 200-MeV linac. A detailed design was carried out and submitted to the AEC as a major accelerator improvement project for the ZGS. In 1964, we were requesting two other major improvement projects, a 12-foot hydrogen bubble chamber and a second external proton beam experimental area. Placing these three projects in order of priority was a responsibility of Bob Sachs, shortly after he took over as Associate Laboratory Director for High Energy Physics. He opted for the 12-foot HBC and EPB-II, and we eventually were authorized to build both of these facilities. This decision, and one a few years later by Bruce Cork to invest all of our accelerator improvement funds in one year to an extension of the first proton area, set the tone of the ZGS operation; it was to be facility- and user-oriented. Brookhaven, on the other hand, chose a major upgrade in the capability of their alternating gradient synchrotron (AGS).

I agreed with the Argonne decision to press for a user-oriented operation. However, we were concerned about how we might keep the ZGS competitive in the future.

On my first trip to the Soviet Union in 1968, I met G. Dimov in Novosibirsk. He had a 15-mA H$^-$ source operating and was doing charge exchange injection at 1.5 MeV into a small tabletop synchrotron. His was the most intense H$^-$ ion source in the world. It would solve the injection problem to reach high intensity in a small rapid-cycling synchrotron if charge exchange injection could be made an operationally practical and reliable technique at 50 MeV. Argonne's development program was initiated immediately upon my return.

Dimov's group of about 30 people have led the world in the development of H$^-$ ion sources. Their sources, which have exceeded 100 mA at high repetition rate, have been copied at many places in this country.

At Argonne, we first did H$^-$ charge exchange injection into the ZGS in 1969. It was quite successful. However, since we only had 1/4 mA of H$^-$ at 50 MeV at the time, the intensity in the ZGS was not particularly high. The main effort was to be on a rapid-cycling booster injector, which, because of its higher injection energy, could raise the ZGS intensity above 10^{13} protons/pulse. A proposal was submitted to DOE (then AEC) in 1969 to build a booster injector for the ZGS at a cost of $3.5M. To get a head start on developing the charge exchange injection technique we obtained the Cornell 2-GeV electron synchrotron when its operation was discontinued in the fall of 1969. This became Booster I and it did, indeed, demonstrate that charge exchange injection could be done in a operationally reliable way with 50-MeV injection. Proton radiography and prototype pulsed neutron work were also carried out on Booster I. Jim Simpson had the major responsibility for Booster I and later for the design, construction, and initial operation of Booster II.

Eventually, Booster II, with a design not very different than that of the original proposal, was built. It was done piecemeal,

however, at a bare miminum cost with the magnet authorized one year
and the rf system the next. It was not completed in time to be used
as an injector into the ZGS. For the critical last five-month run
on the neutrino experiment, which began in October 1977, we instead
used H⁻ charge exchange injection into the ZGS at 50 MeV. It was a
remarkable success. Record ZGS intensities of 7 x 10^{12} protons/
pulse were achieved and the pulse-to-pulse stability of the beam was
significantly enhanced. Fermilab has since adopted charge exchange
injection into their booster as their normal mode of operation and
Brookhaven has embarked on such a program to improve AGS beam quality
for injection into ISABELLE. Charge exchange injection has, therefore, had a significant impact on the high energy field as well as
spawning the many spin-offs which I will discuss. I don't think
we've heard the last of it yet.

INTENSE PULSED NEUTRON SOURCE

When we first started thinking about the booster it was clear
that the circulating beam current, optimistically estimated to be
8 A at 500 MeV for Booster II, was quite large. It occurred to us
that such high currents might be quite useful for other applications
as well as injectors. One of the ideas was as a source for subcriti-
cal reactors - to exceed criticality for a short period of time, and,
therefore, produce an intense burst of neutrons. As soon as we suggested it, however, we were made aware that many other people had
previously considered this possibility, including people from Los
Alamos. But, if we were allowed to build the rapid-cycling synchrotron we would be the first to have such a proton source. At about
this time, Jack Carpenter suggested to us that more neutrons could
be obtained per unit of heat deposited in a target from spallation
on heavy metals than from fission. This was very important because
research reactors, the most powerful of which is represented by the
ILL reactor at Grenoble, France, have pretty much reached an upper
limit on available flux from reactors, whereas spallation sources
might go beyond that limit even for thermal neutrons. For epithermal
neutrons of energies up to 1 electron volt, spallation sources exceed
reactors by several orders of magnitude. It was at this time that the
proposal to produce intense pulsed neutron sources became quite serious and prototype work was begun under Jack Carpenter's direction
using Booster I. We initially called it the ZING project, representing the ZGS Intense Neutron Generator, in recognition of the Canadian
proposal to build ING ten years earlier. The ING project was much
more ambitious, with a linac of 300 mA average proton beam at 1 GeV.
It was the first proposal for the so-called electric breeder linac.
Eventually, Argonne's later developments were renamed IPNS for
Intense Pulsed Neutron Source.

Let me here inject my own enthusiasm for this kind of application of the efforts of myself and others like me in the accelerator
field. I firmly believe that one can trace, ultimately, the progress
of science to the development of instruments. Whenever a new instrument makes possible measurements an order of magnitude better (or faster or more precise) than had been possible previously or measurements

that had not been possible at all in the past, then whole new areas
of science are opened up. Couple this with another opinion of mine,
that the ultimate limitation of all technology comes down to the
properties of materials, and one can believe that any new instrument for materials studies or solid state science will have a major
impact on technology in the future. Intense pulsed neutron sources,
as well as accelerator-based synchrotron radiation, qualify as such
new instruments.

We proposed that if one designed a rapid-cycling synchrotron
using H^- charge exchange injection specifically for intense pulsed
neutron sources that one could exceed the capabilities of Booster II
by an order of magnitude. This proposal, with the neutron target
required, is called IPNS II. The program has not yet been authorized by DOE, although a facility quite similar, called the SNS, is
under construction in England at the Rutherford Laboratories. Also,
a storage ring is under construction at Los Alamos. It will utilize
H^- charge exchange injection to produce high current proton beams for
the same purpose. Germany and Japan have expressed a serious interest in such neutron sources so the development is almost certain to
be quite widespread.

PROTON RADIOGRAPHY

A neuropathologist from the University of Chicago, Dr. V. W.
Steward, visited me one day in 1973 to inquire about the availability
of 200-MeV protons at Argonne for proton radiography. Dr. Steward
had been doing proton radiography with Andy Koehler of Harvard using
the Harvard 160-MeV cyclotron. They had published several articles
on their results. Dr. Steward wanted to extend the range with higher
energy protons and so was approaching us to allow him to carry on the
program at Argonne. He pointed out the great sensitivity of proton
range measurements to detect the total mass of material penetrated.
In general, cancerous tissue has a different density than normal tissue and so could be detected by such range measurements.

We, of course, would be happy to provide them with whatever help
we could, just as we would any user. Any significant commitment of
experimental facilities or effort would have to be approved by the
ZGS Program Committee. But of what practical value would proton
radiography be if it had to be carried out at a high energy physics
facility?

It was while Dr. Steward was still in the office that the concept
of the Proton Diagnostic Accelerator (PDA) occurred to me. Since such
weak proton beams were required ($\sim 10^8$ per pulse) a 200-MeV synchrotron could be designed which would be simple, inexpensive, and reliable
compared to physics machines we were familiar with. The PDA could have
a small aperture magnet, use single turn injection from a low voltage
H^- source, and have slow acceleration. Stripping the H^- ions would
be used for ejection rather than injection and would not only provide
inexpensive slow extraction, but preserve the extremely high circulating beam quality as well. This would make possible scanning the extracted beam in one second over the area desired much like a TV scan.
Therefore, it seemed that the PDA might be suitable for use in a

hospital or clinic setting and make possible the realization of the very high potential diagnostic value of proton radiography.

Such ideas never quite emerge from a vacuum. Similar thoughts about small inexpensive synchrotrons had occurred to me in 1956. After working on the Cornell 1-GeV electron synchrotron, I thought about scaling them down to desk top size of 100 MeV energy, small aperture, weak beams, and sufficiently inexpensive (I then estimated $25K) that every high school physics department could have their own electron synchrotron for training purposes.

Since it now appeared possible to develop proton radiography into a useful and practical medical tool, many people in the ZGS Complex became sufficiently enthusiastic and willing to devote extra effort outside normal working hours, so that we had a team of considerable strength. These people included Steve Kramer, Everette Parker, Russ Klem, Read Moffett, Eugene Colton, Marty Knott, and Ron Timm. Along with Drs. Steward and L. Skaggs from the University of Chicago, we carried out radiographic experiments on many different kinds of phantoms or physical objects for calibration and understanding and on many human tissue specimens supplied by Dr. Steward.

Most of the work was done on the 200-MeV proton beam from Booster-I and we obtained a one-year contract from the National Cancer Institute through the University of Chicago. Continuation of this contract without major modifications, however, was not possible since Booster I·was decommissioned in the spring of 1975 to make way for Booster II installation.

We proposed that we had proven the point that proton radiography was more sensitive at detecting density anomalies in soft human tissue than X-rays and at lower dose, and that the next step in the development would require the PDA for safe proton radiographic studies on live human subjects. The attitude of NCI was that they could not support construction of the PDA until it had been demonstrated that protons were superior to X-rays in detecting cancer in human beings. Our beams were collimated so·strongly that the neutron background made them unsafe for living things including animals. So it appeared like the chicken and egg problem. At about this time the concept of inertial fusion with protons and alpha particles was emerging and the development of proton radiography was set aside.

More recently Steve Kramer has been trying to revive the interest in proton radiography. He has done an excellent job of reworking the previous data, adding corrections to improve the quality of the presentations, and publishing the work. The data includes some work on computerized tomography that had not been analyzed previously and Steve's analysis is quite impressive. We have submitted a new proposal to the National Cancer Institute to carry out more work on tissue slices with a 50-MeV proton beam in a scanning mode. (Note added in proof: The proposal was approved but not funded by the NCI.)

Perhaps proton radiography is before its time. The idea is sound, however, and will refuse to die. I am convinved that it will one day be a most useful medical tool, whether we manage to carry out the development or someone else does - later.

HEAVY ION FUSION (INERTIAL)

Our initial stimulation for heavy ion fusion came directly from the success with H$^-$ charge injection. Over cocktails in San Francisco in the spring of 1974 Glenn Kuswa of Sandia was telling me about their program of inertial fusion with electrons. It occurred to me that protons might be better and that with charge exchange injection of 50-MeV protons into small very high field storage rings (less than one foot in diameter) one could accumulate 300 amperes of circulating beam. One hundred such storage rings could provide 30 kA of protons with a total energy of 1500 joules. The 100 beams could be focussed simultaneously onto a small pellet with nanosecond time durations. At that time, such proton currents were unheard of and sounded very interesting. More recently proton currents of a half million amperes at one MeV have been achieved from pulsed diodes and this process now forms the major inertial fusion thrust of Sandia Laboratories.

In the summer of 1974, however, as hard as I tried, I could not convince myself that 30 kA of protons were adequate. It was clear that α particles would be better since they could have four times the energy for the same range and one could still do charge exchange injection by stripping from α^+ to α^{++} with $\sim$ 100% efficiency. Its six kilojoules was more than the energy of the most powerful glass laser at the time, the ARGUS system at Livermore.

The concept was reported at the accelerator conference in Washington in the spring of 1975. At the same conference Al Maschke of Brookhaven gave a post-deadline paper on an idea using uranium ions for linear implosions. Al's ideas had originated in early 1974 but had been initially classified. They were not declassified early enough to submit an abstract for the accelerator conference before the deadline. Al discussed the effects of very high dE/dx rather than how such beams might be produced.

As far as Argonne s ideas on the subject were concerned the bubble was burst in the spring of 1975 by a computer calculation by Milt Clauser of Sandia. He showed that a beam of 6 megamperes of protons at 10 MeV (and the same current at 50 MeV) were required for scientific breakeven (fusion energy out equals beam energy in) on small deuterium-tritium pellets. This was a total energy of 360 kJ required, far in excess of the 6 kJ of alpha particles I had been talking about. It was immediately clear that heavier ions and much higher ion energies would allow considerably higher beam energies. However, charge exchange injection could not be used on heavier ions because in repeated traversals through the exchange medium the charge would continually change unless the ions were fully stripped. The latter was unacceptable since the number of ions that one can accumulate, other factors being equal, is proportional to the mass and inversely proportional to the square of the charge. So one would like to store heavy ions in a low charge state.

At about this time, Rick Arnold became very interested. With the weak heavy ion sources that we were aware of ($\leq$ 1 mA), we did not believe that we knew how to accumulate enough energy to satisfy Clauser's requirements. Nevertheless, one might still produce a very interesting number of fusion neutrons (10^{15}/burst) at 1% of breakeven.

These ideas were reported at the International Conference on Radiation Test Facilities for the CTR Surface and Materials Program at Argonne in July 1975.

To do better, we felt that an injection trick such as charge exchange was required. Molecular dissociation was an obvious solution if one could find a molecule with all the right properties to make the technique feasible. Joe Berkowitz of Argonne's Physics Division suggested hydrogen iodide to us. The dissociation to neutral hydrogen and singly-charged iodine was a transition requiring 3 electron volt photons. It, therefore, could be accomplished with intense laser light. The opposite reaction with neutral iodine and charged hydrogen as end products was a 6 electron volt transition and would not occur in the laser light. Removing an electron from the singly charged iodine was a 19 electron volt transition so that even though the circulating I^+ would traverse the laser beam many times nothing further would happen to it.

With such an injection technique to avoid Liouville's theorem one could accumulate singly-charged iodine to the space charge limit of a ring even with a very weak source. The idea that it was possible to reach very high beam energies (even many megajoules) solidified in the fall of 1975. Rick and I wrote up the first detailed description of a high energy accelerator system for inertial fusion (subsequently published in Nuclear Instruments and Methods).

An important meeting was called in Germantown in February 1976 in which we presented our ideas, Al Maschke presented his, and Milt Clauser and John Nuckolls (Lawrence Livermore Laboratory) discussed target requirements. The audience was mainly accelerator physicists from Fermilab and Berkeley as well as Brookhaven and Argonne. There was a consensus that the concepts were very interesting and should be followed up. Berkeley and Livermore Laboratories were requested to organize a workshop at a "neutral" location. Prior to this workshop Denis Keefe of Berkeley proposed that the linear induction accelerator, which previously had only been used with electrons, had some significant advantages as the accelerator for heavy ion fusion. From that time on the three accelerator laboratories, Argonne, Brookhaven, and Berkeley have been involved as participants in the heavy ion fusion program.

Rick Arnold coined the word HEARTHFIRE for our program. It stands for High Energy Accelerator and Reactor for Thermonuclear Fusion with Ions of Relativistic Energies. HEARTHFIRE I, using molecular dissociation of hydrogen iodide, was presented to the first workshop. The name is sufficiently appropriate that we have retained it and are now working on HEARTHFIRE IV.

The first two-week workshop, held at the Claremont Hotel in Berkeley/Oakland in July 1976 was a significant milestone. Not only was the group of accelerator physicists from all over the world generally enthusiastic, but source experts felt that beams of singly-charged heavy ions of 100 mA currents and emittances adequate for the purpose could be produced. That meant that the complex beam brightening technique of molecular dissociation would not be needed (it would have required 2 MW of laser energy at 1/sec operation). Furthermore, one could now imagine using any heavy ion as the source.

Argonne's program shifted to xenon ions with mercury as a heavier ion backup, mostly because of attractive features of these ions for source technology.

Acting on the recommendations from the first workshop, DOE initiated funding for the program in the middle of FY 1977. The support was shared between the Division of High Energy and Nuclear Physics and the Office of Inertial Fusion (then part of the Division of Military Applications).

Argonne began a program to demonstrate as much of the front end of the accelerator system as we could get support for. We obtained a 4-MV Dynamitron on surplus from the Goddard Space Flight Center in Maryland and modified it for our purpose in a rather exciting way. We replaced the vacuum rectifiers with solid state rectifiers, changed to full wave rectification for higher current capability and lower voltage, added a quick disconnect opening into the pressurized vessel, and constructed a high gradient column to transport beam in the opposite direction from normal in a Dynamitron.

Hughes Research Laboratories have built several heavy ion sources for us. The first was a very high brightness source of 2 mA of xenon which we have used for neutralization studies and familiarization with heavy ion beams. It is also used as the intense beam of a crossed beam experiment to measure charge changing cross sections of xenon beams at low relative energies. Hughes has also delivered to us two other sources, scaled up to produce 100 mA of xenon and mercury ions. The xenon source has been tested at Argonne, and has been operated with the Dynamitron to produce 30 mA of xenon at 1.3 MeV. Thus the question of sources for heavy ion fusion seems to have been resolved, although there are still some duty cycle questions to be answered.

We have now had three annual workshops with international participants on heavy ion fusion (with a fourth scheduled in November 1979) and all have reinforced the validity of the ideas involved. No fundamental obstacles to achieving the goals have been uncovered, only technological questions for which practical solutions appear feasible. Recognition of this status by the wider community in ICF has only occurred after a review of the entire ICF program by a second reconstitution of the Foster Committee in early 1979. This was an ad hoc committee of experts appointed by DOE, chaired by John Foster, Vice President of TRW, Inc. Their report to DOE is still classified; however, it is clear they recommended support by DOE to prove the feasibility of heavy ion fusion in the two approaches being considered, that of the conventional rf linac/storage ring approach and the linear induction accelerator approach. Each program was to be funded at about $25M and be completed in three years to allow a decision to be made for further development between these two competing concepts for heavy ion fusion.

The Office of Inertial Fusion of DOE prepared to act on these recommendations in the late spring of 1979 to substantially increase the funding for heavy ion fusion in FY 1980. It was with this in mind that we inquired again about the availability of the magnet of the former Princeton-Penn synchrotron, which would be useful as a stacking ring for Argonne's HIF program. We were informed that unless it were removed immediately it would not be accessible for one and a half to two years because a coil-winding facility for Princeton's TFTR tokamak program was being installed in the entrance to the ring tunnel. A series of telephone calls to Princeton and DOE (we had requested $100K from Gregg Canavan, Director of OIF) succeeded in making arrangements for Argonne to acquire the magnet. The sixteen magnets are in the meson area now, thanks to the cooperation of Paul Reardon, Director of the TFTR program, of Princeton University, and of Gregg Canavan of OIF. The excitement was very high at this time.

The House Armed Services Committee, however, ruled that their appropriations were mainly for the military applications of ICF, that the military laboratories would have responsibility for technical management of the program, and that funding for advanced drivers (which include advanced lasers and heavy ions) would be limited to a lower level than proposed by OIF. Therefore our HIF program in FY 1980 will likely not be a substantial increase over the present level and present indications are that we will be a major subcontractor to Los Alamos to carry it out. Supplementary funds for FY 1980 are being requested by OIF. Our main hope is that the support of Argonne's program will be substantially increased by FY 1981 to demonstrate in three years a full scale front end and all of the accelerator technology required of a heavy ion fusion driver for a power plant.

FUTURE NON-HEP PROGRAMS

I have always felt, and sometimes stated, that one could not predict the direction of technology more than two years in advance. Unpredictable surprises occur, attitudes change, and with the latter come changes in funding patterns that determine the course of programs at institutions like Argonne. Therefore, only a few general remarks are in order, with the more detailed ones again being on heavy ion fusion with which I am more intimately involved.

It is very clear that the needs for superconductivity in the CTR and MHD fields are very substantial and Argonne will continue to play a large role in this development. Possibly equally important is the application of superconductivity for storage rings, and beam transport and focussing systems for heavy ion fusion. This application will only become crucial to the HIF program, however, at the stage that driver efficiency becomes very important, i.e., in demonstrating the feasibility of accelerator systems as drivers for fusion power plants.

The future of the IPNS program will depend upon physics results with IPNS I in the next few years. Anticipating very positive results we are enthusiastic about beginning construction of IPNS II in 1983-1984. The ideas about the optimum way to achieve an order of magnitude increase in neutron flux can be expected to evolve continually.

I believe that proton radiography could be developed into an important medical tool in three to four years. Whether support to carry out this development will be forthcoming, however, seems questionable at the present time. Other applications of proton radiography, such as non-destructive testing of materials, might be

developed to a practical stage first. One possibility that might be
looked into at Argonne is in characterization of the quality of metal
spheres, to be applied later to the inspection of heavy ion fusion
targets. A highly accurate technique for batch inspection of HIF
targets will ultimately be needed.

Nuclear physicists are presently very interested in high current
electron beams for nuclear physics. The design of a 2-GeV electron
accelerator with a cw beam of 100 mA is currently being examined and
a proposal to construct such a machine could well be a new initiative
at Argonne.

I believe the future of heavy ion fusion is very good. There is
little doubt in my mind that we can produce the heavy ion beams required. There are a number of technological questions, however, for
which solutions need to be demonstrated. These result from the differences between heavy ions and high energy protons. It seems clear
that Argonne will eventually be supported to carry out our present
plan for an accelerator demonstration facility, Phase 0, because
the justification is quite sound. Whether we will then be allowed
to upgrade the facility to the 10 kJ beam energy level, Phase I, in
order to do energy deposition and plasma instability experiments may
depend on the program needs at the time. If a demonstration of substantial thermonuclear burn, essentially scientific breakeven with
fusion energy out equal to beam energy in, were required with heavy
ions, we have a design of a 300-kJ facility which would still fit
into the ZGS complex. Such a demonstration might not be required
if significant thermonuclear burn is achieved in other parts of the
ICF program, such as lasers or light ion beams. For power plant
drivers, present estimates are that 3 MJ beams at a repetition rate
of 10-20 Hz will be required. Such an accelerator facility is too
large to fit on the Argonne site, so that a totally new site would
be required in which the development could proceed all the way to
fusion power plants.

Since high energy heavy ion accelerator technology seems to offer, for the first time, a solid conceptual solution to an ICF driver
the next major question is the gain vs. input energy that can be obtained from DT targets. The confidence that one can achieve a stable
compression to high density is considerably enhanced with higher input
energy since the required degree of compression for larger targets is
less (the final value of ρr is the critical parameter). A significant
advantage of the heavy ion accelerator technology is the ability to
scale up to high total beam energies (several megajoules).

I personally believe that target gains of 50-100 will be achievable. Once such gains have been demonstrated, and if such targets
can be made economically, then one could reflect on what the future
might hold. The development of ICF would accelerate very rapidly.
A single accelerator facility could supply beam for the many
facility requirements of the development program: materials testing, engineering testing of reactor components, experimental power
reactors, and demonstration power plants. Consequently, these
developments could proceed at a very rapid pace. However, to impact
the energy production in the U.S. is quite another matter. Consider
25 quads of electricity (perhaps one-fourth of the total U.S. energy

in the year 2000) to be generated by fusion power. This quantity
would require 400 plants of 1000 MW's each! The thought of that
number (or even one-fourth as many) accelerator facilities the size
of Fermilab staggers the imagination. Also, producing them would
severely strain the U.S. capability in many ways: rf power tubes,
vacuum pumps, copper, accelerator physicists and engineers - to name
just a few. If even a small part of this speculation were to come
true the economic value of this one spin-off would dwarf the cost of
the high energy physics program to date.

VISUAL REMINISCENCES OF THE ZGS

D. D. Jovanovic
Fermi National Accelerator Laboratory, Batavia, IL 60510

After hearing two days of talks of the grand and glorious
past I almost have the urge to open my comments like Mark Antony:
"Friends, Midwesterners, fellow physicists, we are here to bury
ZGS, not to praise it...," but that is another speech.

In preparation for these 20 minutes I was given a magnificent collection of slides, pictures and diagrams. More than
3000 entries. With such a collection one could have built a
clever animation much like a bicentennial film: History of the
U.S. condensed in the 3 minutes of almost subliminal flashing
images. But I had neither the skill nor talent to do that, so
you have to bear with me through some 61 slides I have selected.

It always begins with the gathering of dignitaries at the
site representing people and taxpayers.

Fig. 1. The first pow-wow.

If some of these gentlemen are in the audience, they will
please forgive me; I was too unimportant to know them.

ISSN:0094-243X/80/600242-27$1.50 Copyright 1980 American Institute of Physics

Fig. 2.   Bulldozers poised for a race.

Fig. 3.   Magnet stands taking shape.

Figs. 4 and 5. The circle is taking place.
Truly embryonic beginnings.

At this point one has the feeling that one is reminiscing
over baby pictures of one's favorite daughter or niece. Biologi-
cally speaking, one has to liken the ZGS to some living creature
whose life span is of the order of eighteen years, so that one
can speak of the first steps, early youth, full blossoming,
maturity, and old age. I have done much research into the ages
of animals and other living creatures. The closest I could come

was the Dodo bird (now extinct, says my Encyclopedia Britannica).
But the same source said that it was a bird with exquisite plumage
which could not fly and always laid a single large egg.... I therefore abandoned the analogy as being inappropriate.

In parallel to the construction other parts were getting
ready:

Fig. 6 Linac tank.

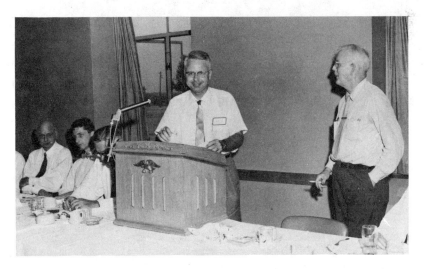

Fig. 7. Some important pow-wow took place.
Observe that Roger Hildebrand looks worried.

So was I. Some of my mentors at the time had warned me:
"If the director of the Laboratory is doing cosmic ray research
flying balloons - one does not have confidence that this machine
will ever work."

Fig. 8.   Flywheel of the motor generator set.

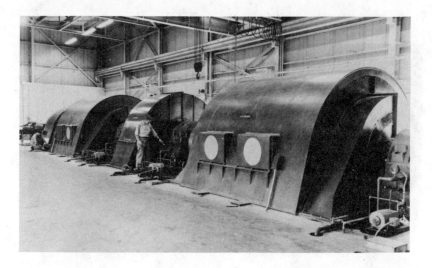

Fig. 9.   Motor generator set complete; heartbeat of the machine.

Fig. 10.　Meson hall full of magnet parts.

　　　This is at the time I joined in, in spite of the previous
admonitions.　This was in the fall of 1962.

Fig. 11.　Coil installation.

Fig. 12.   Some more coil and magnet installation.

As far as I know nobody suggested the true reason why the coils failed. Clearly they were installed upside down. Observe the Westinghouse sign.

Fig. 13.   Testing and mapping of sector of the machine.

These were the good old days; one did not just barely make it. But heroic efforts were invented much later at some other place.

Fig. 14.   The great control room.

I remember Al Crewe telling us students that the machine would be fully computerized, and we were in great awe.   One had just to insert a deck of IBM cards and presto, beam would issue forth instantaneously.   In those days the computers ran with vacuum tubes, no less!

Fig. 15.   Target manipulator.

The target was a magnificent device, driven by a high strength super alloy tape. With an acceleration of 9 g's it would move into a position at the completion of the accelerator cycle. It was computerized. Sometimes, not too often, due to the operator not being able to translate binary numbers into inches, the computer would command the target to move to a coordinate <u>outside</u> the vacuum chamber. With a 9 g acceleration, the target would slam into the wall and break into pieces. Some of us also spent hours in the control room arguing whether the target moved during the slow spill or not. But the machine and the experiments matured. By 1965 things were taking shape and the meson area looked like this.

Fig. 16.   Meson area.

Fig. 17.   Trailers in the proton area.

Fig. 18. L. Teng, R. Sachs, G. Pewitt, and T. Fields.

Yet another important pow-wow. It seems as if they said
to Gale: "Gale, go now and build us a chamber 12 cubits wide
and 12 cubits long." And he did. Perhaps he was slightly confused
between feet and cubits in his measures.

Fig. 19. 12-foot bubble chamber.

Fig. 20. Superconducting coils.

Fig. 21. Omega bellows.

Fig. 22.   Inner bubble chamber vessel.

    All of these illustrate a world first: The largest bubble
chamber in the world, years ahead of its time in technology and
innovation.   It even worked!

Fig. 23.   12-ft bubble chamber tracks
being examined by Gale Pewitt.

Fig. 24. Focusing horn.

     Borrowed from some other neutrino experiment. Was it E-1
or E-1A? Somewhat confusing. But that device worked through
millions of pulses just as the chamber did.

Fig. 25. Another pow-wow.

Fig. 26. Experimental area control room.

Experimenters were using the beams in earnest. Some early experiments were the trend setters.

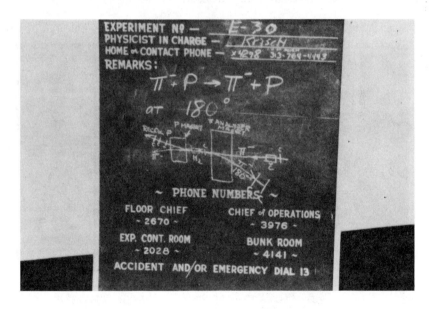

Fig. 27. Alan Krisch set the style.

That experiment discovered a resonance only recently surpassed in mass and by only a factor of three.

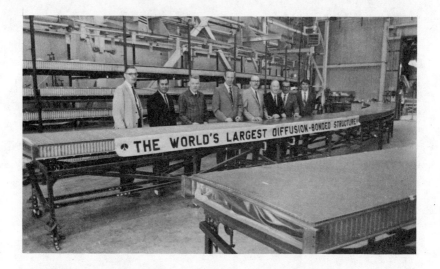

Fig. 28. New titanium vacuum chamber.

Employing space technology and after making a bold decision, the whole ZGS machine was taken apart and the new chambers were installed.

Fig. 29. For successful achievement somebody always gets a plaque.

Here to be distinguished from walking down a plank, which was not practiced - often.

Fig. 30.   Another important pow-wow.

There was no other way to find appropriate pictures of our directors.

Fig. 31.   ZGS Complex in its full bloom.

At this point in the visual reminiscing one would be remiss if not mentioning the users. After all, they were the people for whom the machine worked and who moved the frontiers of our knowledge. Out of hundreds of pictures of users I have selected just a few. Mostly people noted for their youthful appearance then and now. They will be shown alphabetically, fourteen in all. Fourteen is chosen rather than twelve so that some apostolic inference is not made by the uninitiated.

Fig. 32. A. Abashian.

Fig. 33. H. Anderson.

Fig. 34. N. Booth.

Fig. 35. D. Cline.

Fig. 36. J. Fry.

Fig. 37. R. Heinz.

Fig. 38. H. Neal.

Fig. 39. V. Kistiakowsky.

Fig. 40. A. Krisch.

Fig. 41. I. Pless.

Fig. 42. T. Romanowski

Fig. 43. J. Sandweiss

Fig. 44. L. Voyvodic.

Fig. 45. W. Willis.

In the collection of pictures shown to me there was a pre-
ponderance of pictures of some festive occasion or other. Hence
here are a few selected at random, showing people who worked hard
and enjoyed lighter moments.

Fig. 46. Ron Martin trying innovative energy-
conserving idea in flying airplanes.

Fig. 47. Getting a Barbie doll was not innovative.

Fig. 48. Aki Yokosawa pinning a spinor on Alan Krisch.

But most of the pictures showed people in good moods enjoying light spirits.

Fig. 49.   Lundy and Trendler.

Fig. 50.   Some more people in deep conversation
whenever the light spirit would take them.

It would not do if I did not mention important high energy
spin-offs. These same hard-working people, using the latest
technology of superconductivity, built large magnets.

Fig. 51. U-25 superconducting magnet.

Fig. 52. Superconducting magnet ready for export to USSR.

Somewhere at this point in time (to use a slightly worn-out phrase) I relocated some 30 miles to the northwest to a territory known fondly as Buffalo Land.

A new era of the ZGS began, using polarized beams. Unable to obtain any suitable picture showing a clearly recognizable beam of polarized protons, I offer to you these following two figures: The Cockroft-Walton, where the source was located, and a graph showing the result of the polarized beam.

264

Fig. 53. Cockroft–Walton preaccelerator.

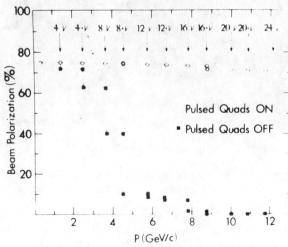

Fig. 54. Beam polarization graph.

That marked an era of unique work in the world using a very high energy polarized proton beam. Who says that old dogs cannot learn new tricks? Alas, visual material from this era was not to be found or I could not, out of my ignorance, readily recognize it.

Fig. 55. Cornell machine rejuvenated to Booster I.

From a vantage point seven years removed I would like to add a comment. The ZGS Complex was uppermost and foremost a veritable haven for bubble chambers. At one point in the history of the ZGS we had five or six bubble chambers active on the floor: the 30-inch, 20-inch, 40-inch, 10-inch, 12-foot, and 500-liter chambers. A veritable harem. Many of my bubble chamber colleagues must lament the passing of such a golden era. Just to show you a few:

Fig. 56. The 40-inch bubble chamber.

Fig. 57. The giant 12-foot bubble chamber.

Fig. 58. A scanner doing her thing.

As if being tired of scanners, some physicist invented a machine replacement. POLLY is shown at work. Probably the most successful automatic scanning device ever built.

Fig. 59.   POLLY-II.

In this haven for bubble chambers, counter people managed to exist, but modestly. Compare this modest trailer with the massive bubble chamber pagodas.

Fig. 60.   Modest and neat trailer of A. Krisch.

268

To conclude this show there should be a picture at the end
showing "riding into a sunset." But as hard as I tried, this
is all I could find.

Fig. 61. The ZGS Complex at high noon.

THE UNIVERSITIES AND THE ZGS IN THE SEVENTIES

Homer A. Neal

Indiana University
Bloomington, Indiana 47405

ABSTRACT

This paper reviews the role of universities in the life of the Argonne Zero Gradient Synchrotron during the 1970s. A chronology of events leading to the closing of the accelerator is presented and the loss to the midwestern universities is assessed.

INTRODUCTION

It was with mixed emotions that I accepted the invitation to speak at the Symposium today on the topic "The Universities and the ZGS in the Seventies". On the one hand, I am fully aware of the need to close certain facilities in the national high energy physics program if our field is to be able to acquire the resources needed for the construction and operation of the new faciltites which we all desire. On the other hand, it is quite disconcerting to note that the level of science support in our country is such that the closing of a productive facility such as the ZGS has to be considered and, indeed, consummated.

This occasion is a particularly somber one for me. I recall becoming interested in the ZGS program over a decade ago and planning one of our group's first experiments in EPB1. Subsequently, I became involved with activities of the vigorous users organization and, in 1970, was elected its chairman. The responsibilities of this position required me to closely interact with colleagues in numerous universities across the country and with the staff of

ISSN:0094-243X/80/600269-21$1.50 Copyright 1980 American Institute of Physics

the Laboratory. Through this link, and through my
service on the the Program Advisory Committee and the
Argonne Universities Association High Energy Physics
Board Committee, I soon realized the enormous value of
the ZGS facility to our universities and to our nation.
I later became a member of the Argonne Universities
Association Board of Trustees and, in this capacity,
gained a perspective of the role the entire Laboratory
played in the nation's R and D efforts. My service on
the Board required me to participate in numerous
meetings with the AEC-ERDA-DOE staff, with the
Commission, with the President's Science Advisors, with
Congressmen and Senators, and to deliver testimony
before joint congressional committees. These efforts,
which often kept me away from my home institution a
significant fraction of the time, were pursued only
because of my high regard for the Argonne program. In
retrospect, I firmly believe that it was truly worth
the commitment. Over the years Argonne National
Laboratory took on a very personal special meaning and
it is for this reason that the occasion today holds for
me a certain degree of sadness. The occasion can,
however, also mark a new beginning for the Laboratory
and our universities, and for this reason there is also
substantial reason for optimism.

The decision-making path associated with the closing of
the ZGS is a very complex and intriguing one. Indeed,
this entire process is one that should be of great
interest to historians. The process involved numerous
individuals, governmental agencies, the Office of
Management and Budget, the Joint Committee on Atomic
Energy, the staff of the Argonne National Laboratory,
the High Energy Physics Advisory Panel, the Argonne
Universities Association, and many universities. From
time to time in my presentation I will refer to the
involvement of these various elements, but in keeping
with the focus of my talk, I will concentrate on the
role of our universities in deliberations regarding the
future of the ZGS.

The outline of my presentation is as follows. Firstly,
I will examine the status of the facility and give a
profile of its users in the late 1960s and early 1970s.
Secondly, I will review the developments which led
policy makers to give serious consideration to the
closing of the ZGS. Thirdly, I will review the posture
of the Argonne Universities Association in the early
1970s and chart its role in the ensuing developments.
Fourthly, I will give a chronology of developments in

the 1970s which have led us to the present Symposium. Finally, I will briefly review what the ZGS meant to our universities and the role that the ZGS staff will continue to play in support of frontier research.

A SNAPSHOT OF THE EARLY SEVENTIES

As viewed approximately ten years ago, the Argonne ZGS represented a capital investment of well over fifty million dollars. The annual operating budget was of the order of 18 million dollars and the manpower level exceeded 350. The synchrotron had become operational in 1963 and the Laboratory had developed the capability to operate several beam lines simultaneously with an intensity that permitted a wide variety of experimental studies. In the context of the other national facilities for high energy physics research, the ZGS provided service to perhaps the highest number of university users and was, in particular, viewed as the regional facility for midwestern high energy physics research. In the early seventies, for example, approximately twelve experiments were being executed at any given time; attendees at the annual meeting of the Users Group numbered several hundred physicists. The Laboratory had already established itself as an excellent facility for examining the strong force through detailed determination of the energy and angular dependence of relatively simple processes. There were three major detector facilities available or under development at the Laboratory for this purpose: namely, the effective mass spectrometer, the 12 foot bubble chamber, and the streamer chamber. In addition, there were numerous smaller scale experiments underway.

Thus, in a snapshot of the early 1970s, the ZGS represented a facility comprising a major capital investment by our government and had embarked on a vigorous research program which involved many of our university faculty, students and staff.

THE PROBLEM

In my view, there were three major elements involved in the decision to close the ZGS. One was the desire on the part of the Office of Management and Budget to put a ceiling on the funding level for the national high energy physics program. The second was the continuing desire of the high energy physics community for major new facilities. Thirdly, there was the expectation by many that the close proximity of Fermilab would reduce the need for the continuation of the ZGS program. These concerns and views, coupled with a continuing constraint on the national R and D budget, have led us to the present juncture.

In reviewing a copy of the minutes of the April 13, 1973 meeting of the AUA Board Committee, I find the following notes. "The AEC, in its five year plan, projected the ZGS and the Bevatron at a funding level of approximately $15.5M, which was defined as viable support. However, the FY74 budget was to be $14.4M, a barely viable level. FY75 was projected to be even worse, with only $12.5M being projected, a level which was not judged by the Laboratory to be viable. The recommendation to the AEC from HEPAP to keep the budget not much below $200M was not accepted by OMB; instead a constant operating budget of $124.4M for FY73 and $128.5M for FY74 was approved. Thus, to keep the Bevatron and the ZGS at approximately $15.0M in FY75, and to meet the demands of Fermilab, an increase of $20M for the AEC operating budget is needed."

AUA IN THE EARLY SEVENTIES

Argonne Universities Association is a collection of thirty universities charged with the policy and review responsibility for Argonne National Laboratory. The geographical distribution of the universities in the Association is quite diverse, extending to Arizona in the West and Pennsylvania in the East. These universities are, however, concentrated in the Midwest in states near Illinois. It is thus quite reasonable to expect that AUA would have become intimately involved in the issues affecting the future of the ZGS program. Indeed, the bulk of the interaction between Argonne National Laboratory and our universities has been through the high energy physics component.

The involvement of AUA with the ZGS has not always been what it should be, however. When I joined the Argonne Universities Association Board of Trustees in 1971, I was one of only two high energy physicists on the Board of 19 members. I detected a feeling on the Board that the high energy physicists had been extremely successful in developing a users group and that this group was capable of handling all negotiations between the Laboratory, the universities, and agencies that might be pertinent to the ZGS program. This view was basically correct, but it unfortunately deprived our faculty of having spokesmen at the level required to

Argonne Universities Association

-Incorporated on July 22, 1965

-Purpose:
...to foster scientific research by formulating, approving and reviewing policies and programs of the Argonne National Laboratory...

-Current membership (30 universities)
University of Arizona
Carnegie-Mellon University
Case Western Reserve University
University of Chicago
University of Cincinnati
Illinois Institute of Technology
University of Illinois
Indiana University
University of Iowa
Iowa State University
University of Kansas
Kansas State University
Loyola University of Chicago
Marquette University
University of Michigan
Michigan State University
University of Minnesota
University of Missouri
Northwestern University
University of Notre Dame
Ohio State University
Ohio University
Pennsylvania State University
Purdue University
Saint Louis University
Southern Illinois University
University of Texas/Austin
Washington University
Wayne State University
University of Wisconsin

Figure 1

Chronology of Decision Making on ZGS

1974	OMB requested AEC and President's science advisor to develop "a plan for shutting down the ZGS accelerator at the earliest reasonable time."
September 1974	Report of the AEC-NSF Study of the Future Role of the ZGS (R. C. Drew, J. M. Teem, M. Bardon, W. A. Wallenmeyer).
	Physics Subpanel - R. L. Walker, Chairman
	Management Subpanel - K. Strauch, Chairman
March 1976	Letter from Kane to Laboratory saying that it is current expectation of ERDA that the ZGS will cease operation by end of 1978 and ERDA is planning review.
June 1976	Laboratory proposes 6 months per year dedicated polarized proton operation for FY79, 80, 81 to carry out a programatic coverage of the obvious experiments.
June 1976	AUA Special Committee for High Energy Physics (F. Loeffler, Chairman) in response to request of AUA Executive Committee of January 1976 recommended that ZGS be kept operating exclusively as a polarized proton facility from 1979 to 1982 and that a continuing program of high energy physics research and development be supported in the post-ZGS period.
August 1976	ERDA Review Panel (R. L. Walker, Chairman) recommends continued operation of the ZGS for polarized proton operation at a rate of 9 months per year of beam operation to the end of CY 1979.
April 1977	AUA Board of Trustees passes resolution supporting continued high energy physics research and accelerator and detector research and development in the post-ZGS period.
October 1978	HEPAP Panel on the Future of the HEP Program at ANL (F. Low, Chairman) recommends that the strong in-house Argonne experimental program should operate in a user mode at other national laboratories.
October 1979	The ZGS machine ceases operation for HEP.

Figure 2

adequately enlist the attention of agency officials, congressmen, and the United States President. In correspondence to the President of Argonne Universities Association in the fall of 1972, I went to great lengths to point out the extreme concern that the Board Committee on High Energy Physics had about the low level of involvement of AUA in issues directly affecting the health of the ZGS program. I have been privileged to observe that in the ensuing years AUA has become much more involved in the Argonne High Energy Physics Program and has been of great assistance to the Laboratory and to the university faculty in seeing that the interests of our universities were well represented. The list of institutions comprising Argonne Universities Association is given in figure 1.

CHRONOLOGY OF EVENTS

I have attempted to analyze the chronology of developments pertaining to the decision to close the ZGS. Figure 2 lists some of the milestones in the process and figure 3 presents a representative distribution of the origin of correspondence having a bearing on the situation during the 1970s. The participants have been separated into several categories: these include the OMB, AEC-ERDA-DOE, national panels, Argonne Universities Association, universities, HEPAP, Congress, Argonne National Laboratory, and the national media. The interactions between the various elements over the years were terribly complex. Figure 4 gives some insight on the principal paths of communication. Virtually every participant in the process had, on at least one occasion, a direct interaction with each of the other participants.

During the year 1972 there was a significant amount of correspondence activity involving AUA, the universities and the Laboratory. This was caused primarily by the continuing budget cuts being levied against the ZGS program and an increasing realization on the part of Argonne Universities Association and the high energy physicists in our universities that some steps should be taken to bolster the ZGS program, since a continuing reduction at the current rate would have indicated a non-viable program within just a few years.

A critical path began to develop in 1974, when the Office of Management and Budget requested that the Atomic Energy Commission and the Science Advisor

276

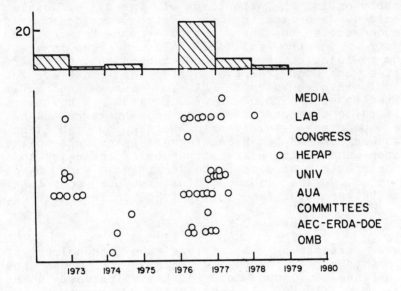

Figure 3.

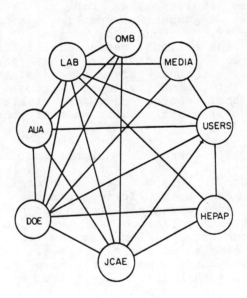

Figure 4.

develop a plan for shutting down the ZGS accelerator at
the earliest reasonable time. A ZGS study committee
and two subpanels were then set up to carry out this
task. This study committee consisted of
representatives of the Science and Technology Policy
Office, the Physics Section of the NSF Research
Directorate, the Division of Physical Research of the
Atomic Energy Commission, and others listed in figure
5. The two subpanels were the Physics subpanel, whose
membership consisted of knowledgeable practitioners of
high energy physics, and the Management and Procedures
subpanel, which consisted of experienced accelerator
laboratory administrators. The final report of the
committee was issued on September 9, 1974. The
committee recommended that the ZGS program should be
continued at least through FY1978. Mid to late FY1979
was projected as the earliest reasonable time to close
the ZGS. Furthermore, the committee recommended that a
review of the situation on a program-wide basis be made
in approximately two years. My participation on this
committee was a most sobering experience, even when the
interruption of my sabbatical at the Niels Bohr
Institute is discounted. Scores of hours of sincere
effort by many persons were invested in the preparation
of the document and, when the process was completed, I
fully believed that the result accurately reflected the
general consensus of the community under the
constraints given. As one can observe from figure 3,
during calender year 1975 there was very little
correspondence activity. As a result of the above
study, the funding level for the ZGS was stabilized, in
keeping with the recommendation of the committee, and
the Laboratory was able to initiate plans for a high
level of operation of the facility at least
through FY1978.

One also notes in the graph in figure 3 an increase in
activity starting near the beginning of 1976. It is
interesting to examine the cause of this increase. The
1974 report of the ZGS Study Committee indicated that a
new study should be made to determine which segment of
the national program might be eliminated, if necessary,
by the middle or end of FY1979 with the least adverse
impact. The timing of this review was to be such as to
permit "consideration of the latest scientific and
programmatic information prior to giving the
recommended advance notice before a shutdown at the
middle or end of FY1979." There was some indication
near the beginning of 1976 that the agency had intended
to forego this proposed review and to proceed with

AEC–NSF Study of the ZGS
September 9, 1974

Participants:

R. Drew
M. Bardon
V. Fitch
H. Neal

J. Teem
W. Wallenmeyer
J. D. Jackson
W. D. Wales

Physics Subpanel

R. Walker
J. Cronin
M. Derrick
G. Fox
A. Krisch
T. O'Halloran
N. Samios
H. Ticho
S. Treiman

Management Subpanel

K. Strauch
R. Cool
T. Fields
W. Hartsough
R. Neal
L. Rosen
W. Wales

Charge:

...develop a plan for shutting down the ZGS accelerator at the earliest reasonable time.

Findings/Recommendations:

-...the ZGS program is vigorous and innovative.
-...the ZGS is not approaching technical obsolescence.
-...operation of the ZGS should be continued, possibly at a more intensive level, through FY1978. Mid-to-late FY1979 is projected as the earliest reasonable time to shut down the ZGS.
-...a policy of advanced notice should be established and announced relative to accelerator shutdown.
-...if funding for high energy physics remains approximately level in constant value dollars, it appears unlikely that all existing HEP facilities can continue to operate effectively when new facilities begin to require operating funds. There should be a global program review within 2 years to identify which segment should be eliminated.

Figure 5.

The 1976 Zero Gradient Synchrotron Review
8/31/76

Panel Membership:

R. Walker   T. A. O'Halloran
R. Diebold   D. Reeder
G. Fox   N. Samios
J. D. Jackson   H. Ticho
A. D. Krisch

Charge:

You are requested to review the status and role of the ZGS in the
U.S. High Energy Physics Program and to update the 1974 report of
the Physics Subpanel of the AEC-NSF ZGS study. We are especially
interested in obtaining your advice on what approach should be
taken prior to shutdown to gain the optimum scientific benefit from
the ZGS and its special and unique facilities during its remaining
lifetime. Present ERDA plans contemplate that the ZGS will be shut
down by the end of calendar year 1978. (J. Kane, June 28, 1976).

Principal Recommendations:

- Continued operation...for the polarized proton program for approximately one year beyond CY1978.
- A level of funding which permits nine months of operation per year
  ...through CY1979.

Figure 6.

Statement by the Executive Committee
of AUA for the Atomic Energy Commission
12/18/72

-research in high energy physics has become an outstanding example
 of Laboratory-university collaboration.
-during the years before AUA came into existence, the Laboratory
 and the universities were largely divided in their high energy
 physics interests.
-many of the universities had been tied to the concept of a separate
 facility to be built by MURA in Wisconsin.
-as a result of planning by a strong user group, the ZGS research
 has become an integral part of the activities of many of the university departments of physics.
-this has helped to give the program its high quality, and at the
 same time, these departments have become strongly dependent upon
 the ZGS.
-the promise for future research with the ZGS remains high, and we
 are very hopeful that funds can continue to be available to
 operate it effectively.

Figure 7.

plans for closing the ZGS during FY1979. This prospect triggered a substantial level of activity from Argonne Universities Association, from the Laboratory and from several users. Indeed, this partly accounts for the substantial amount of correspondence in the first half of 1976. As a result of these efforts and other considerations a new panel was established by the Director of the Division of Physical Research, U.S.E.R.D.A., on June 28, 1976(figure 6). The charge to the new panel was to review the status and role of the ZGS in the U. S. high energy physics program and to update the 1974 report of the Physics subpanel of the AEC-NSF ZGS study. This new study group recommended that the ZGS continue to operate as a polarized proton beam facility for approximately one year beyond calendar year 1978. As one can infer, the recommendation of the study panels directly contributed to the extension of the life of the ZGS accelerator to the present. These reviews and the resulting extension were in large part precipitated by the high level of activity of the Argonne Universities Association and by faculty in our universities. Through the various actions taken to represent the interest of our practicing high energy physicists, what would have likely been a turnoff for the ZGS program as early as 1975, has now been translated into a late 1979 closure. During this time a fantastic number of fundamental contributions have been made to our understanding of the strong interaction through experiments conducted at the ZGS. This accomplishment is one that I believe that Argonne Universities Association, the Users Group, and our community at large can justly be proud of. Statements contained in figures 7, 8, and 9, as well as in the agency reviews, clearly demonstrate the critical role played by the ZGS in the national program. The confidence which we have invested in the ZGS program in prior years has been proven to have been well placed.

WHAT THE ZGS MEANT TO A TYPICAL UNIVERSITY

I wish now to address the issue of the value of the ZGS program to our midwestern universities. As a specific example, I wish to focus on the role that the ZGS has played in the development of the high energy physics program at Indiana University. The points which I shall make here are applicable to many other universities that have utilized the ZGS facility over the years.

The period of the 1960s was one where several

Long Range Planning Study of the
High Energy Physics Program at ANL
June 15, 1976

AUA Special Committee on High Energy Physics:

V. P. Kenney	D. H. Miller
A. D. Krisch	H. A. Neal
F. J. Loeffler	J. B. Roberts

Conclusions/Recommendations:

-...a decision to close down the ZGS...is a step of enormous consequence for the national involvement in...basic research.
-...it is vital that the research at the ZGS be evaluated on an absolute basis and compared with other HEP programs.
-...recommend keeping the ZGS operating...as a polarized proton
facility...at least until 1982. There are compelling reasons to
continue the current healthy, vigorous program in HEP at ANL for
the indefinite future.
-...it is desirable and possible for ANL to maintain an outstanding
staff of scientists, engineers and technicians in the post-ZGS
period to initiate and develop specialized instrumentation and hardware for the support of HEP and other ERDA missions.

Figure 8.

Review of the Future of the HEP Program at ANL
October, 1978: HEPAP

- Participants:

F. Low	E. Courant
R. Diebold	D. Meyer
R. Neal	T. O'Halloran
J. Peoples	G. Trilling

- Recommendations:

...the strong in-house Argonne experimental and theoretical high
energy research program should be continued. The experimental
effort will henceforth operate in the user mode at accelerators
at other national facilities.
...ANL should continue on a trial basis to make available its
support facilities for university users.
...the excellent accelerator group at ANL should continue to receive support and encouragement...so that the national program not lose their singular talents....

Figure 9.

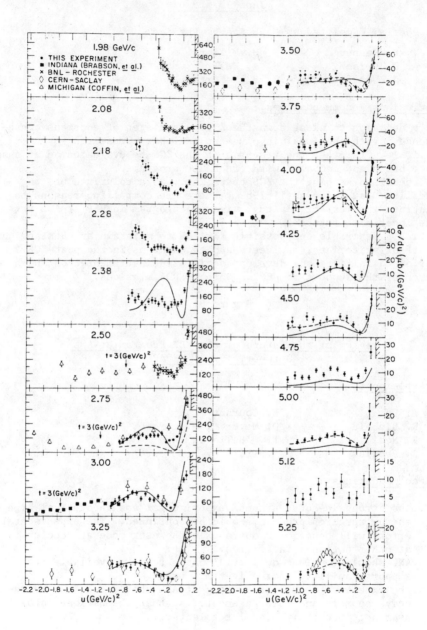

Figure 10a

universities that had previously not had active high energy physics groups made the bold step of hiring young faculty members with interests in this very expensive field. This was done in direct recognition of the fact that no first class university could hope to discharge its obligations to its graduate students, and to maintain the high level of intelluctual challenge appropriate for a university without having a vigorous program in high energy physics. These new groups were faced with the usual unfavorable picture of acquiring running time at national laboratories whose schedules were already quite crowded, and in addition, with the significant difficulties that one encounters while trying to teach regular courses while commuting thousands of miles to participate in an ongoing experiment. Thus, the availability of research facilities at the Argonne ZGS was particularly attractive to the midwestern groups. The ZGS facility had the capability of accomodating a sizeable number of users at any given time, had developed a good users support program, and had demonstrated a high level of responsiveness to the need of outside users.

It was in this context that in the late 1960s the Indiana University group mounted a series of experiments at the Argonne ZGS that spanned a period of a decade. It was through these experiments that our group had its first opportunity to utilize an on-line computer, to develop and utilize magnetostrictive wire spark chambers, multiwire proportional chambers, specialized Cherenkov counters, polarized targets, polarized beams, to design complex beams, and to develop broad expertise in many aspects of counter physics.

The activities at the ZGS were not, however, significant only because of the enhancement of the expertise of faculty and students: fundamental contributions were made to the knowledge of the behavior of the strong interaction. I will provide just a brief overview of some of our experimental efforts carried out at the ZGS:

-we discovered unanticipated structure in the backward and intermediate angle pion-proton elastic scattering distribution, which has been confirmed in later experiments and which still represents a challenge to pertinent models(figure 10).
-we conducted extensive measurements of the pp elastic differential cross section at intermediate energies.

284

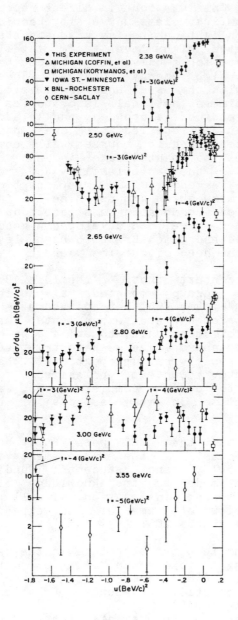

Figure 10b

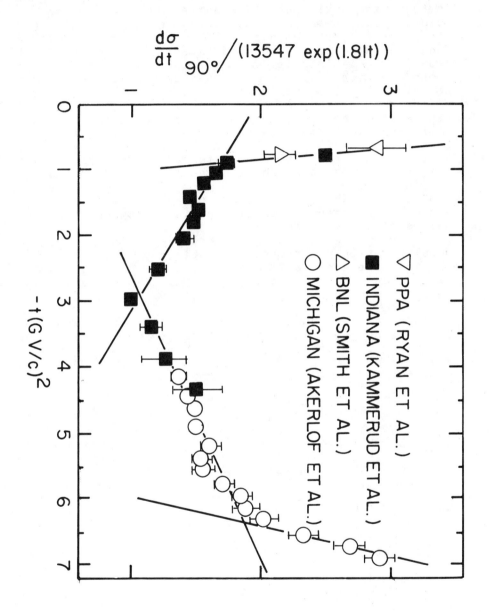

Figure 11

These data led to the discovery of an additional region in the fixed angle p-p differential cross section, a point that has been recently confirmed in other independent measurements. This discovery is of a fundamental nature and indicates that there are at least four distinct regions in the fixed angle p-p cross section (figure 11), an observation which most certainly has its origin in the constituent structure of the proton.

-an experiment conducted by our group on the A2 meson still provides the best evidence yet for the nonexistence of a split in the A2 distribution(figure 12). These results are inferred from the high resolution mass distributions and from studies made on the inclusive polarization (figure 13) of the recoil protons in the production process, both of which indicate no generic dissimilarities between the two halves of the A2 meson.

-our group measured the first t distribution of the depolarization parameter in elastic p-p scattering at intermediate energies(figure 14).

-we made some of the first high t polarization measurements in p-p elastic scattering(figure 15).

It is true that, during the period for which many of these measurements were being conducted at the ZGS, our group was involved in an expanding array of programs at SLAC, Fermilab and CERN; nevertheless, the ZGS always remained a very special component of our overall program.

THE FUTURE

The ZGS has done much to enhance the expertise and interest in the field of high energy physics in the Midwest. This legacy will certainly transcend the existence of the physical machine. The fine staff of physicists, engineers, and technicians currently associated with the Argonne high energy physics program will continue to make contributions to the field in the coming years. Many contributions will likely be in collaboration with groups located at our universities. Some examples of this can already be seen on the horizon. The potential development of a polarized beam at Brookhaven, the proposed polarized beam development at Fermilab and the collaborative efforts on experiment PEP-12 are all examples. In the latter undertaking, a project in which Indiana University is directly involved, the availability of talented ZGS personnel for full scale participation in this project will be

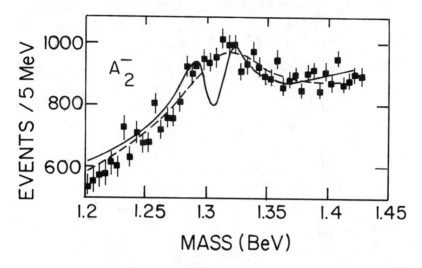

Figure 12.

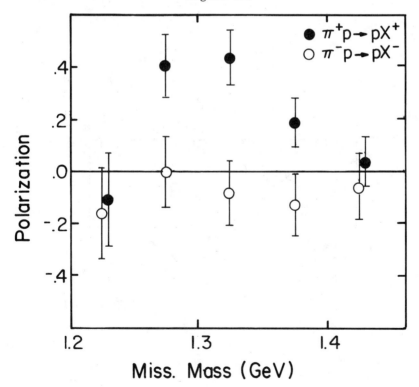

Figure 13.

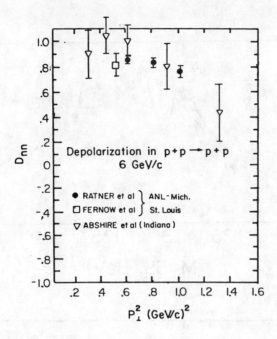

Figure 14.

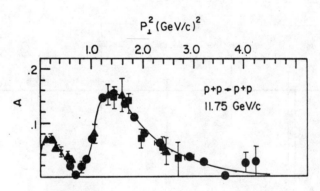

Figure 15.

critical to the success of the experimental program.

CONCLUDING REMARKS

In concluding, I wish to make the following points. We are about to witness the closing of a facility that has had a tremendous impact on the intellectual development of many faculty and students in our universities and on the entire field of high energy physics. The agencies, the universities, and individual users have invested an enormous amount of effort in seeing that the process through which the decision was made to terminate operation of the ZGS was done in a professional and rational manner. The level of cooperation that has existed amongst the various parties in this unpleasant endeavor has been admirable. With the various new programs planned by the staff of the ZGS, and the exciting new areas that are demanding exploration in high energy physics, we can all look forward to the future with great enthusiam.

To the staff of the Laboratory, I wish to say that over the years we have built a very strong bond of university-Laboratory cooperation. This bond has contributed greatly to the strength of the U.S. scientific program and I trust that the coming years will see this bond strengthened as we jointly undertake the resolution of even more difficult problems.

THE ZGS HISTORY AS SEEN FROM WASHINGTON

William A. Wallenmeyer
U.S. Department of Energy
Washington, DC

Many of us here today have had a long-standing connection with
the ZGS. I appreciate this opportunity to make a few comments from
my own substantial and long-standing relationship to the ZGS and the
people of the ZGS, both laboratory and university.

My personal recollections of the ZGS start from the perspective
of MURA in Madison, Wisconsin and my attendance at the early and
intellectually very exciting users meetings held in the ANL Chemistry
Auditorium. The ZGS construction was still in its early stages. As
I recall, these meetings--organized and instigated by Ned Goldwasser--
began in 1959, some four years before the first beam was accelerated
in the ZGS. I remember well the breadth of the presentations on the
physics discoveries and progress that were being made elsewhere,
especially at Berkeley and Brookhaven, the extensive planning for
use of the ZGS, the organizing and preparation of the users and the
vigor of the participants, including the strength with which Art
Rosenfeld and Marty Block would debate points across the stage at
the Chemistry Building auditorium!

I would like to start my comments by listing some recollections
of the ZGS program from my Washington viewpoint which are most
strongly highlighted in my memory.

User Emphasis. The strong involvement and effective participation of the university users from the very beginning, in nearly all
aspects of planning and use of the facility, has been a powerful and
positive feature of the ZGS operations and research efforts. This
effective university/laboratory relationship of cooperative effort
has been mutually beneficial to each and also has added strength to
the overall national effort.

Superconducting Magnets. ANL was an early leader in the
development of superconducting magnets. Today, the U.S. high energy
physics program is planning much of its future around the promises,
efficiencies and successes of superconducting magnets. The superconducting magnet work at ANL, although focused in high energy physics,
has been done for high energy physics and for others. The significant early magnets at ANL included the 44 kg magnet for the 10 inch
Helium bubble chamber and the 18 kg magnet for the 12 foot bubble
chamber. More recently there have been the polarized target magnets,
the Fermilab bubble chamber magnet, and very large magnets for the
MHD program.

The ZGS 12 foot Bubble Chamber. Another recollection is the 12
foot bubble chamber, authorized in FY 1966 along with the Second
External Proton Area for some $17 million and completed on schedule
in 1969 for $17.8 million. This was the first very large bubble
chamber to be built in the world and still retains the reputation
of being the best operating. I've always admired Gale Pewitt, not

only for his outstanding success in building that large device, but also for his courage and vision in specifying a superconducting magnet for its field which was an order of magnitude larger than any superconducting magnet ever attempted before. Later, innovative modifications to the chamber to permit its operation with an internal Track Sensitive Target and to operate with a hydrogen-neon fill for increased gamma detection efficiency, were also quite successful. As we all know, the superconducting magnet with its very large volume of magnetic field has been moved west where it will start a new life as the principal element of a major new detector for PEP.

POLLY: These devices, developed through interaction of the ANL Applied Math Division and the High Energy Physics Division, were very successful and demonstrated one of the most effective approaches in the world to pattern recognition for data analysis of bubble chamber interactions. This approach was widely adapted and used across the country.

Polarized Beams and Targets: The ZGS program has been a leader in polarization physics including both polarized particle beams and targets. Argonne has certainly been in the forefront of polarized target developments and, as we all know, the high energy polarized proton beams from the ZGS are an absolutely unique experimental tool. The ZGS perservered with developing and exploring polarized beams when many theoretical experts were proclaiming that polarization effects would be small and uninteresting, or at least trivial to understand. We now know that the experimentally observed effects--especially those of Alan Krisch and associates--are large, growing, and not understood.

Interactions with Other Disciplines. Argonne to me has always been a place where high energy physics has had a large and mutually beneficial interaction with other disciplines. Part of this has been due to Ron Martin and his crew of accelerator physicists who have been enormously inventive in applications of accelerator technology. Here one thinks immediately of proton radiography, the IPNS, and HIF programs. Fruitful interactions outside of ARF include the ANL artificial kidney and several aspects of the solar program including the Winston collector. Part of this success also has been due, I'm sure, to the multidisciplinary nature of the laboratory and another part to encouragement by laboratory management. On the latter point, Bob Sachs in particular comes to mind, as he frequently reminded us during our visits to ANL and the ZGS of the multidisciplinary nature of the laboratory and the mutual advantages to high energy physics as well as to the rest of the laboratory of having the ZGS as one of the laboratory's programs. Relative to his point, I've already mentioned the POLLY as a success and benefit to the U.S. High Energy Physics Program which resulted from high energy physics interaction with the Applied Math Division.

Innovation and Inventiveness. I've continually been impressed with the overall, widespread high level of innovation and inventiveness of the ZGS staff. This I have alluded to several times earlier, relative to interactions with other disciplines, data analysis developments, the large bubble chamber, superconducting magnets, etc.

Much also relates to the accelerator R&D people and especially to
Ron Martin himself. Some things that quickly come to mind are the
early efforts at computer control of the accelerator operations, H⁻
injection--both in concept and in demonstration--and the development
of the fast-cycling synchrotron.

External Beams and Simultaneous Experiments. Another highlighted recollection of the ZGS has been the large number of
experiments which it has been able to pursue simultaneously. This
pertains more to the program as run a couple of years ago, of course,
than to the more limited polarized beam program the ZGS has been
restricted to recently. The ability to carry out such a large program
of simultaneous experiments is due, of course, in large part to the
idea of extracting the primary beam from the synchrotron and using
external targets. I believe the ZGS may have been one of the very
first to use this approach.

High Productivity and Durability of 30" Bubble Chamber. A word
also needs to be said on the 30 inch bubble chamber, one result of
that strong university/laboratory joint effort I spoke of earlier,
and which was under construction at MURA and the University of
Wisconsin when I left there in the spring of 1962. This chamber
had a very long and productive life at the ZGS and subsequently also
at Fermilab. It promises to continue a productive life at Fermilab
for some time in the future in a hybrid mode of operation.

Project Scoping and Funding. The scope and funding of a
construction project are important matters to us at DOE. Phil McGee,
our sage and expert on all matters of construction and engineering,
reminded me earlier this week of the way the scope of the ZGS kept
changing during construction and an important lesson we learned in
Washington from this. As Phil said, "You know, high energy physics
has the enviable reputation of always building their accelerators on
time, on schedule, and within the cost estimates." "But," he said,
"that was after the ZGS!" "The several scope changes with the ZGS
were very costly, and we learned a tremendous amount from the ZGS on
what to do and what not to do--which is one of the reasons we know
how to have a project properly scoped, and why the program now has
a good reputation."

THE EARLY HISTORY

Although the genesis of the thoughts relative to construction
of the ZGS began well before my arrival in Washington in mid 1962,
I have done some reading of Commission Staff Papers which are in the
DOE archives and of Congressional Hearing Records in the last week
or two, and would like to share with you some of what I learned
from that reading about the early days and the prehistory of the ZGS.

About 1952, or perhaps somewhat earlier, there was strong
midwest agreement on the need for a multi-GeV accelerator in the
midwest. At that time, the Bevatron was under construction on the
west coast, and the Cosmotron was about to come into operation
at Brookhaven. There was not, however, as I gather from my reading,
all that much agreement as to exactly where to put this midwest
accelerator!

A January 27, 1954, letter from the Argonne Director to the AEC Division of Research requested permission to start an accelerator studies group. He indicated that he had on board a physicist by the name of Jack Livingood whom he would put in charge on a half time basis, and that Morton Hamermesh and a couple of other theorists, Ferentz, Crosbie, and Ray Weeks would join; another experimentalist from the Argonne staff would be named and that probably four to five university participants would also join with the group. There is a reply dated March saying, "Well, no, not just now, but don't tell anybody, maybe we will get approval for the accelerator studies in a few months." Apparently the approval did come, because the next step I would like to mention was a July 1955 proposal from Argonne for construction of a "Tandem Accelerator." The proposal requested $30 million for construction of a 2 GeV FFAG synchrocyclotron inject-ing into a 25 GeV FFAG synchrotron, which, with authorization in FY 1957, would be completed in 1963.

August 1955 was the occasion for the first Atoms for Peace Conference, in Geneva, Switzerland, and one of the items that came out at that conference was that the Soviets were building the 10 GeV Synchrophasotron at Dubna, with an estimated completion date of 1957! This came as quite a surprise, and perhaps even a bit of a shock, to the U.S. scientists and AEC officials at the Conference. The Soviets were also planning at that time an even larger accelerator which eventually resulted in the 76 GeV Serpukhov Proton Synchrotron. At a subsequent AEC Commission meeting on November 23, 1955, where the ANL proposal for the "Tandem" accelerator was being discussed, the Commission "authorized the immediate commencement on an urgent basis of the design and construction by the Argonne National Laboratory of a high energy accelerator which would be fully competitive with the Russian 10 BeV accelerator." Following up on this was a memorandum from the General Manager of the AEC to the Chicago Operations Office. Interestingly enough, the first paragraph of this memo talks about MURA (MURA was also under the Chicago Operations Office) and indicated that at some later date there would be a high energy accelerator of advanced design built at a MURA-selected site, with Commission agreement, and this MURA machine would be the world's finest for several years. This implied that it wasn't known at that time how to build such a machine, so MURA should make a study and subsequently design and seek authorization for its construction.

The memorandum also stated to the Chicago Operations Office that on an urgent basis and for early completion there was need for a 10 to 15 BeV accelerator for immediate competition with the Soviet Union; that it was believed that such an accelerator was needed at Argonne anyway to balance the program, and that the Commission was requesting Congressional authorization. If Argonne was able to undertake such a project, the Commission needed the preliminary estimates of cost and the design parameters by February 1, 1956. And, the memo indicated, the Commission was not requesting authorization for the "Tandem" accelerator which had been proposed by Argonne in July 1955 for $30 million. There followed an exchange of communications between the Lab Director and the Operations Office

Manager which I would characterize as being somewhat "testy." The final communication in that exchange was a letter which came from ANL on the 26th of January 1956--very close to the February 1 deadline in the General Manager's request--which said that there was no possible reasonable design which could be readied for 1957 construction. It also stated that a doubling of the Bevatron was something that could be done, but it was not anything that any reasonable person would attempt to even think of because of the cost and other factors and therefore Argonne could not meet the conditions for getting specifications by February 1956 since they could not come up with such a construction design. The Lab indicated, however, that it was prepared to explore designs, "which might meet the Commission's specification of providing immediate competition."

Subsequently in 1956, in the budget authorization request for FY 1957, the AEC did request funding for a midwest accelerator. It was stated that the Argonne site was preferred, that the accelerator would provide competition with the Soviets, and that the estimated cost was $15 million. My understanding is that the $15 million estimate probably came from AEC staff; the implication is that $15 million was a quickly arrived at estimate made under the assumption that a 10 to 15 GeV accelerator would be about half the cost of the 30 GeV AGS. (The AGS was authorized in FY 1953 with an estimate cost of about $30 million.) The budget request for the midwest accelerator was authorized for FY 1957 at $15 million.

In the meantime, in March 1956, the AEC received preliminary specifications from ANL for a 12.5 GeV accelerator estimated to cost $22 million. The proposal which was subsequently received on June 27, 1956, raised that estimate to $25 million. An ad hoc advisory panel on high energy accelerators to the NSF, chaired by Leland Haworth, in their report of October 1956 pointed out the need for a multi-BeV accelerator in the midwest. In November of 1956 there was a Division of Research request for an amendment to the FY 1958 budget to put in an Argonne weak-focussing machine for $27 million. Thus, a 12.5 BeV accelerator was included in the FY 1958 authorization budget, estimated to cost $27 million, to be located at Argonne and with an estimated completion date of 1962 and "using new technology where possible." In December of 1957 ANL was authorized to proceed with machine studies and the architect-engineer selection. About a year later, in November of 1958, the $27 million total estimated cost was increased by $2 million to $29 million to permit flattopping the beam pulse. The ground-breaking was June 27, 1959--the end of FY 1959. I was impressed that here was a machine where the original decision was made for rapid initiation and construction in November of 1955 and the eventual groundbreaking was not until June 1959, a time which must have been near the operating time first hoped for when the initial decision was made!

In 1961 the scope of the project was again increased, accompanied by an increase in cost to $42 million. Also in 1961 a request for FY 1962 authorization for the High Energy Physics Building at $6.9 million was granted. In 1963 the cost for completion was increased to $47.3 million. The first beam was obtained in August of 1963; the construction cost increased to $52 million in 1964 and the first experiment started about June 1964.

SHUTDOWN STUDIES

I would like next to move briefly to the other end of the ZGS lifetime and say a few words about the long series of shutdown considerations made relative to the ZGS. The first formal study was one that we did in May 1973 with some assistance from Argonne. The report was entitled, "Impact of Shutdown of the Zero Gradient Synchrotron and the Bevatron." This report examined the effects of several scenarios in which the ZGS and/or the Bevatron would be shut down. The next major effort was the 1974 study. The directions for this study basically came from the OMB in February 1974, and said that the AEC and the "Science Advisor" should "develop a plan for shutting down the ZGS at the earliest reasonable time." The AEC-President's Science Advisor-NSF study [Report of the AEC-NSF Study of the Future Role of the Zero Gradient Synchrotron (ZGS); September 9, 1974] was a detailed, indepth study which included, in addition to the AEC/PSA/NSF executive committee, the participation of two outstanding and hard working subpanels. A result of this study was the recommendation that the earliest reasonable time to shut down the ZGS would be mid-to-late 1979, and a recommendation that the shutdown consideration should be reexamined in about two years.

The OMB in February of 1976 said that, with the understanding that the ZGS would cease operations in calendar year 1978, PEP would be started. DOE replied, that although the DOE was considering shutting down the accelerator at the end of calendar year 1978, it was appropriate to first make the two year study recommended in the 1974 report.

This study was made in 1976 (Report of the 1976 ZGS Review Panel; August 31, 1976) by a panel which included many of the people who had participated two years earlier. Indeed, Bob Walker from Cal Tech chaired both the 1976 panel and the 1974 Physics Subpanel. The 1976 report reaffirmed that the entire ZGS program was sound and further affirmed the special importance of the polarized proton physics. The report recommended operation of the ZGS "for approximately one year beyond CY 1979." HEPAP at its meetings in September and December of 1976 agreed with the Walker Panel. The final decision was made between the DOE and OMB in November 1977, about two years ago, that the shutdown would commence at the end of Fiscal Year 1979, September 30 - October 1, 1979.

FUNDING HISTORY

A ZGS history as seen from Washington would be incomplete without some discussion of the funding history. Figure 1 is a graph showing a record of the total ZGS funding from FY 1960, close to the beginning of the project, to and including FY 1980, which begins next month. The top line of the graph represents an envelope of the total annual funding to the ANL HEP program over this 21 year period, and represents the sum of the operating money,

296

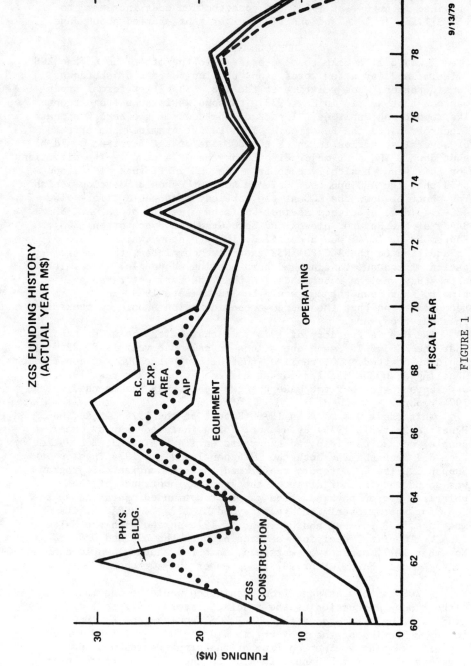

ZGS FUNDING HISTORY
(ACTUAL YEAR M$)

FIGURE 1

9/13/79

used to pay salaries, buy materials and services and keep the
facilities operating; plus the capital equipment used for
secondary beam lines, most of the detectors and detector facilities,
power supplies, etc.; plus line item construction projects, such as
the ZGS itself, the 12 foot bubble chamber and the physics building.
The two equipment peaks in FY 1968 and FY 1974 represent two major
ANL central computers which, although requested and purchased in the
HEP budget, were only fractionally used by the ANL HEP program. The
area between the dashed line and the solid line for the operating
budget in FY 1978, FY 1979, and FY 1980 represents the part of the
ANL HEP funding related to ZGS shutdown costs. The FY 1980 operating
funding below the dashed line represents ANL funding for the physics
research program and accelerator and facility R&D programs which
are continuing at a combined level of about $5 million to $6 million
after the shutdown.

Figure 2 indicates the operating funding level of the ANL HEP
Program from FY 1966 through FY 1980 in terms of constant value
FY 1980 dollars. Also indicated on this somewhat intricate and
perhaps confusing graph are similar funding curves for the other
DOE supported High Energy Physics laboratories.

Figure 3 is a graph which sums the operating funding for each
part of the DOE supported program, with the top line then representing the envelope of total operating support for the Program. The
structure underneath indicates the changing nature of the program
with time, and the birth and death of various high energy
accelerators and laboratories are rather clearly indicated. The
particular time interval represented by this graph, for example,
shows the demise of the Cosmotron, PPA, CEA, Bevatron and ZGS and
the coming into operation of SLAC, Fermilab and PEP.

Figure 4 is a similar summation graph for the total DOE
support to high energy physics including the operating, capital
equipment and line item construction funding. Comparison of
Figure 4 with Figure 5 compares the actual funding for the program
with the funding projected by long range plan which we have been
proceeding under since the fall of 1977. FY 1979 was the first
year included under this plan.

SUMMARY

A few words in summary: the ZGS construction which was to be
urgently initiated and rapidly pursued, had a very slow start; a
productive physics program with the ZGS also got off to a slow
start, well after the first beam was attained in August of 1963.
Indeed the ZGS program was really only coming into its own in the
mid to late sixties, when the funding pressure on the overall
national program started building up strongly. Not too long
thereafter, in the early to mid seventies, the pressure for
shutdown of the ZGS and the series of shutdown studies began!

Nevertheless, as indicated by the recollection highlights I
discussed earlier, and, of course, much more by the content and
substance of the other talks given here yesterday and today, the
ZGS has accumulated an outstanding list of significant achievements
and major contributions. The overall productivity of the ZGS

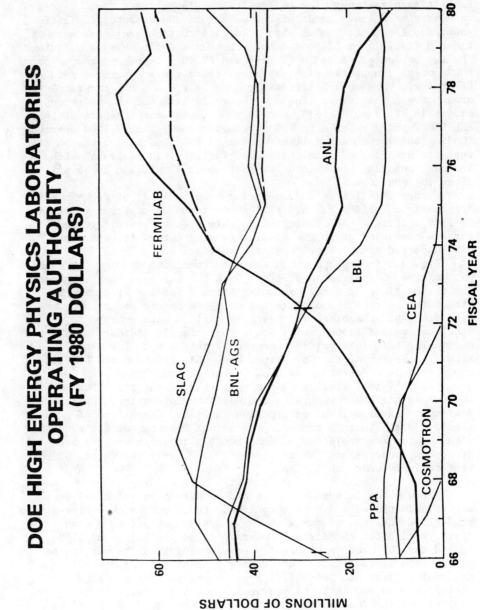

DOE HIGH ENERGY PHYSICS LABORATORIES OPERATING AUTHORITY (FY 1980 DOLLARS)

Figure 2.

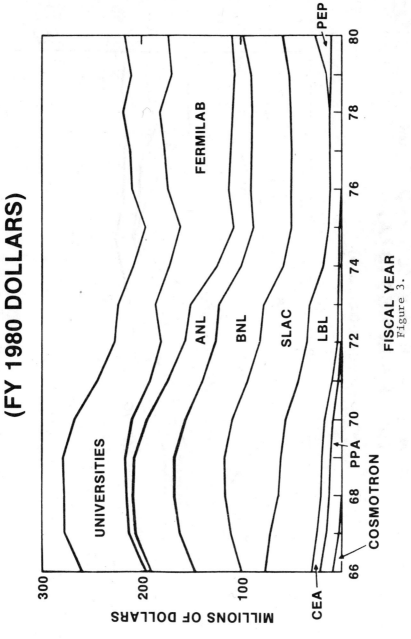

DOE HIGH ENERGY PHYSICS PROGRAM
OPERATING (B/A)
(FY 1980 DOLLARS)

FISCAL YEAR
Figure 3.

300

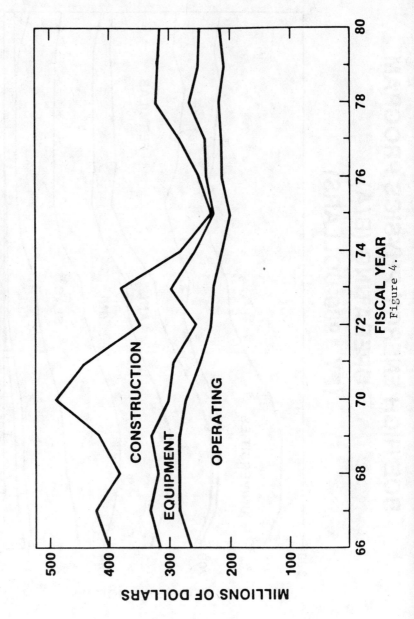

DOE HIGH ENERGY PHYSICS PROGRAM
TOTAL FUNDING (B/A)

(FY 1980 DOLLARS)

CONSTRUCTION

EQUIPMENT

OPERATING

MILLIONS OF DOLLARS

500
400
300
200
100

FISCAL YEAR

66 68 70 72 74 76 78 80

Figure 4.

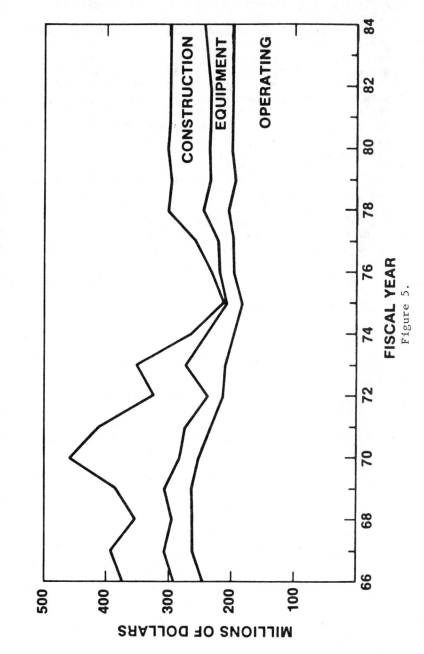

301

DOE HIGH ENERGY PHYSICS PROGRAM
DOE/OMB LONG RANGE FUNDING AGREEMENT
(B/A IN MILLIONS OF FY 1979 DOLLARS)

Figure 5.

program has been large. The ZGS has continued to the very end with
style and strength and with an effective program of first rank
scientific importance.

The future plan for ANL high energy physics is to maintain a
physics research program and an accelerator R&D effort at Argonne
in competition with the rest of the national program. Thus, the
degree to which the ANL program is supported, and indeed even its
continuance in high energy physics depends upon the quality of the
work and how it compares with the rest of the national program.
The people of the ZGS have already intensified their contributions
to other parts of the national high energy physics effort. Their
special insight, knowhow, and innovative approaches are expected to
have major beneficial impact.

The entire ZGS staff and management, as well as the users of
the facility, deserve the highest praise and commendation for both
the excellent operation of the facility, and the strong scientific
importance and productivity of the program that has been outstand-
ingly maintained even through this most difficult final year. The
machine and its people have done exceedingly well.

HIGHLIGHTS AND SPECULATIONS*

T. H. Fields
High Energy Physics Division
Argonne National Laboratory, Argonne, Illinois   60439

ABSTRACT

Several examples of unfinished business of the ZGS program are described. These examples cover physics, apparatus, and institutional subjects. Speculations are given about the evolution of these subjects during the 1980's.

INTRODUCTION

I would like to begin with a few general remarks about the talks which we have heard during the past two days. I think that these talks have summed up the history of the ZGS in fascinating and incisive ways and that they add up to a rather complete picture. Previously, I had harbored some strong doubts as to whether a retrospective symposium like this is a good way to bring forth or to record history, since the customary groundrules for public speaking are not necessarily compatible with telling the whole story. It seems to me that the speakers have met this challenge and have set forth fascinating unvarnished descriptions of the history of the ZGS program from many different points of view.

Since these speakers have done their jobs so well, I shall not attempt here to further summarize the history of the ZGS. Such additional condensation might easily lead to an over-simplification of what scientific research is and how and why it gets done. That is, the history of a large scale scientific research enterprise such as the ZGS can't be accurately described by a few breakthroughs or a few equations. That is the way one usually describes physics results, but it's not the way physics research actually happens. Real physics research, as we all know, involves many long periods of all kinds of work: planning, constructing, testing, running, and analyzing data; doing some of these things very well, some of them not so well; backing up for a second try, getting new kinds of insights which are occasionally breakthroughs but which more often are not. There are also many organizational tasks, fund-raising, and other practical activities which are essential for achieving continuity in a research enterprise. The previous speakers have covered these diverse kinds of activities from many different points of view, and I believe that it all adds up to history and reality. If one should desire to search for the lessons of this history, one should take a broad look at all of the history and reality described at this symposium.

* Work supported by the U.S. Department of Energy.

Rather than further summarizing the ZGS history, then, I shall take the main goal of this talk to be the description of some important kinds of <u>unfinished business</u> of the ZGS program. I hope that it will be interesting and useful to consider some aspects of high energy physics research in whose development the work at the ZGS has played a prominent role, but whose further evolution or impact upon the field remains to be seen. That is, the history of the particular subjects which I have chosen is far from complete, so that foreseeing their future evolution involves some speculation. Of course, it is an important responsibility for future work in our field to further illuminate these subjects and thereby to replace speculation with new knowledge, insights, and capabilities.

Most of these areas of unfinished business have already been described in their respective historical contexts at this symposium, but my goal here is quite different - to emphasize some still-unanswered <u>questions</u>, particularly ones which have been closely associated with work at ZGS and which also may have a large impact on the future evolution of high energy physics research.

Before describing these subjects, I should like to digress in the next section by giving two significant examples of <u>finished</u> business which have not apparently been covered in previous talks and should be briefly mentioned here for completeness' sake.

TWO EXAMPLES OF FINISHED ZGS BUSINESS

These are examples of ZGS work whose consequences have already been well incorporated into the present day mainstream of high energy physics.

The first example is the choice of the window frame dipoles for the bending magnets of the ZGS. This type of magnet design allowed the first use of a 20 kilogauss guide field in a synchrotron, and hence yielded the maximum practical energy for a given bending radius. The choice, a decade later, of window frame dipoles for the Fermilab separated function lattice allowed the Fermilab accelerator to reach 500 GeV in a tunnel of radius 1 kilometer. Figure 1 gives the energy per unit radius for various proton synchrotrons, and shows that 0.5 GeV/meter is still the present limit. This limit will be exceeded only when synchrotrons using superconducting magnets come into operation (more about this later).

A second example of finished ZGS business concerns the first search for direct muons from hadron collisions. This method of searching for the intermediate vector boson was invented, at least at the ZGS, by M. L. Good, then at the University of Wisconsin, and led to the first published report of such an experiment.[1] By now we know that much higher proton energies will be required to produce the intermediate vector boson, but in the meantime the generalization of this method by L. Lederman and his coworkers at the AGS to the study of direct production of muon <u>pairs</u> has led to the opening of whole new areas of particle physics.

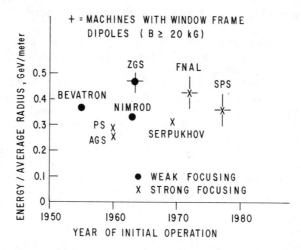

Fig. 1. Energy divided by mean radius for various proton synchrotrons.

SUBJECTS FOR SPECULATION - UNFINISHED BUSINESS

Now I come to the main part of this talk. I shall describe nine examples of important subject areas in which the ZGS program has made basic contributions, but where there is still much room for speculation, for new ideas, and for new approaches. The first four examples are in the area of high energy physics itself, the next three concern apparatus for high energy research, and the last two are institutional matters.

1. SCALAR MESONS

Notable early work at the ZGS involved the first observation of the $\delta_2(980) \rightarrow \eta\pi$ in K^-p collisions by the Northwestern-Argonne groups[2] and the search for the $\epsilon(700) \rightarrow \pi^+\pi^-$ under the ρ^o peak by the Wisconsin-Toronto groups.[3] Important recent work at the ZGS on scalar mesons decaying to $\overline{K}K$ has been carried out by the Notre Dame-Argonne Streamer Chamber group[4] and the Argonne Effective Mass Spectrometer group.[5]

Two kinds of challenging questions concern the 0^+ mesons. The first kind consists of theoretical questions which center on the contrast between, on the one hand, the simple nonet structure of the 0^-, 1^- mesons which seem to be $\overline{q}q$ systems in a relative S wave, as well as the corresonding simple nonet pattern of the 2^+ mesons, and, on the other hand, the observed complexities of the 0^+ mesons. Perhaps the 0^+ mesons contain a substantial admixture of $\overline{qq}qq$ states or even of glueballs (a gluon-gluon bound state).[6] The second kind concerns avenues for making further experimental progress. What types of next-generation spectrometer experiments should be carried out to better understand the 0^+ mesons? Will there be sufficient priority and funding to permit experimental progress in this kind of meson spectroscopy in the 1980's?

2.   HADRON DYNAMICS AT SMALL TRANSVERSE MOMENTUM

As already described at this symposium, several kinds of ZGS experiments have made important contributions to the phenomenological understanding of hadron dynamics at small p_T. Some pioneering examples are elastic scattering at small and medium values of momentum transfer, two-body inelastic hadron reactions, elastic scattering at 180° as a function of beam energy, inclusive production experiments, and polarization measurements of various hadron reactions.

Here the present-day unsolved theoretical questions mainly concern the problem of connecting these low p_T phenomena with the behavior of quarks and gluons, the fundamental strongly interacting quanta of quantum chromodynamics (QCD). (As a related historical note, we recall that an early quark model for the dynamics of small p_T reactions[7] found important support in experimental data from the 30-inch bubble chamber.[8]) It seems compelling to try to understand low p_T data in terms of QCD ideas even though these data involve the quark confinement regime where quantitative QCD calculations cannot yet be made. Perhaps low p_T reactions may eventually yield new insights into QCD phenomena comparable to those which are now being obtained from high p_T phenomena. Of course, a similar challenge exists for hadron spectroscopy. One can reverse the emphasis by observing that quark confinement effects surely need to be studied in depth, both experimentally and theoretically, for both their fundamental interest and their practical importance.

3.   DIBARYON STATES

As described earlier in this symposium, experiments at the ZGS have shown that there is considerable structure in the energy dependence of the polarized beam and target total pp cross section differences $\Delta\sigma_L$ and $\Delta\sigma_T$, as shown in Fig. 2. This structure has been interpreted as evidence for dibaryon resonances.[9] More phase shift analyses and polarized beam polarized target experiments will be necessary to fully determine the properties of these structures - just as was necessary during the past two decades to sort out the baryon resonances. Some of these experiments can be carried out at medium energy acelerators such as LAMPF but others will require higher energy proton polarized beams.

In any case, it is clear that these previously unexpected dibaryon phenomena have now become experimentally accessible by using polarized beams and targets. Further work is needed to determine whether these phenomena can be interpreted as six-quark bound states, and if so, what the implications of this will be for "quark chemistry." A very important related question concerns the role of multiquark structures within nuclei. Again, confinement effects in QCD are what we are speculating about.

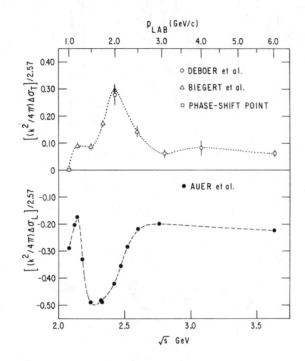

Fig. 2. Proton-proton total cross section differences using polarized beam and polarized target.

Many of the speculations about the physics of multiquark bound states are likely to be clarified by the totally new level of precision in $\bar{p}p$ studies which will be made possible in 1982 by the CERN Low Energy Antiproton Ring (LEAR). If narrow baryonium ($\overline{qq}qq$) states exist, this will be a powerful method for their study. Will comparable progress in methods for observing new properties of dibaryon systems take place during the next few years?

4.   <u>SPIN EFFECTS AT LARGE P_T</u>

Figure 3 shows the very large spin-spin effects which have been observed by the Michigan-AUA-Argonne group for pp elastic scattering at 11.75 GeV/c.[10] Of course, it is very tempting to try to interpret these effects as manifestations of a spin-spin dependence of the scattering of fundamental constituents of the proton: quarks of spin $\frac{1}{2}$ and gluons of spin 1. If this is indeed the case, then this type of experiment will yield a direct measure of the spin dependence of the basic interactions in QCD.

To test these ideas quantitatively requires a better understanding of QCD and its application to elastic hadron scattering at

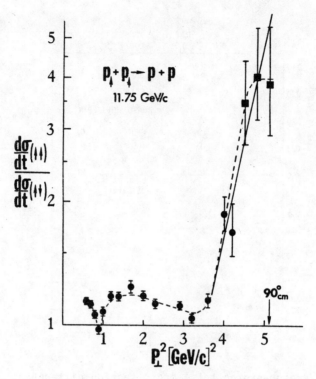

Fig. 3. Spin-spin effect in elastic pp scattering as measured by the Michigan-AUA-Argonne collaboration.

large P_T than has yet been achieved. Perhaps achieving this under-standing will be expedited by measuring single and double spin effects (i.e. spin-orbit and spin-spin forces) at larger values of energy and transverse momentum. Perhaps a better phenomenological understanding of $\bar{q}q$ and qqq potentials for describing the masses of known hadrons can provide a related quantitative measure of the large spin-spin interaction between quarks.

There are also unsolved experimental challenges in finding ways to carry out these large transverse momentum polarization measure-ments at Fermilab and SPS, and someday at very high energy $\bar{p}p$ and pp colliders.

A general observation about the above four physics topics is that they illustrate the progress of the quark concept during the years of the ZGS program. At the time of the first ZGS experiments in 1964, the quark had just been invented and served mainly as a mathematical convenience for remembering SU_3 multiplets. Although quark models have come a long way since then, quantitative tests of QCD, the theory of quarks and gluons, are still few. Of course, there is also a continuing, tantalizing, ultimate question of whether free quarks can exist.

5. APPLICATION OF SUPERCONDUCTING MAGNETS TO HIGH ENERGY PHYSICS

Since the first use of a superconducting magnet in a high energy physics experiment occurred at the ZGS some fourteen years ago, and since the giant superconducting magnet for the 12-ft bubble chamber was first operated ten years ago, it is reasonable to wonder why superconducting magnet technology could still be described as unfinished business. The reasons for this are well known to some of you: the exceptional technical challenges involved in the design and fabrication of accelerator-quality superconducting dipoles, the lack of funding for appropriate development and demonstration projects, and the underestimation of the overall technical obstacles involved in new superconducting accelerator projects. The net effect is that today no superconducting accelerator ring has been built, nor is one close to completion. So we have very little real data on such basic matters as beam heating effects or the operational reliability of large rings of superconducting magnets and their associated cryogenic systems.

The question for the future is not only obvious but also is a key to the future of much of the U.S. high energy physics program: Will the technology of large systems of superconducting accelerator magnets now begin to progress at the hoped-for rate so that Isabelle and the Fermilab Energy Doubler/Collider will achieve their design goals with a reasonable degree of operational reliability and overall practicality? There is also an important present question: Should the U.S. research and development program on superconducting accelerator magnets be strengthened?

6. BEAM BRIGHTENING

Over the years, work at the ZGS has made some important contributions to the development of new methods for brightening the proton beam in an accelerator. The use of H^- stripping injection, in which one can achieve large increases in beam brightness within a synchrotron by injection (and stripping) of a continuous H^- beam through a thin foil, an idea which originated at Novosibirsk, was first achieved in a synchrotron at the ZGS in 1969. Since then, several existing and planned accelerators, including the Fermilab machine, have adopted H^- injection. Important open questions include: whether this will become the standard injection method for proton synchrotrons, whether an intense H^- polarized proton source can be developed, and whether there are major new applications for beam brightening by charge-changing processes in addition to those already in use for proton synchrotrons and for Tokomak-type devices.

The most exciting present-day questions about beam brightening in high energy accelerators concern the application of stochastic cooling (invented at CERN) and electron cooling (invented at Novosibirsk) to achieve $\bar{p}p$ colliding beams in the CERN SPS and at Fermilab. (In fact, a significant early step along the path toward the concept of stochastic cooling was taken in the mid 1960's with the development by MURA and Argonne of feedback systems for damping the coherent transverse instabilities of the ZGS circulating proton beam.) Clearly, the course of high energy physics during the 1980's and beyond will be crucially affected by the overall amount of brightening which can be achieved for antiproton beams. The CERN LEAR project, mentioned above, also depends on such antiproton beam brightening.

7. ACCELERATION OF POLARIZED PROTON BEAMS TO HIGH ENERGY

The pioneering contributions of work at the ZGS in this area are described in the talk of Everett Parker. Here I wish to remark mainly on the possible impact of this ZGS work upon higher energy proton accelerators. (Of course, there are also important physics opportunities connected with achieving polarized beams in e^+e^- machines such as PETRA, PEP, and LEP, as well as with the invention of the Siberian Snake, but I shall not describe these here.)

Two kinds of future uses for accelerated polarized proton beams are clear. One kind is an extension of the large p_T studies which have been begun here at the ZGS to energies where jet production and other "simple" QCD processes can be clearly identified and their spin dependence studied. Although first-generation experiments of this kind may be performed using a secondary polarized proton beam, the high level of precision which seems appropriate for detailed measurements of the spin structure of basic quark-gluon interactions will require intense and thus directly-accelerated beams of polarized protons. Achieving such acceleration with alternating gradient accelerators will be the next challenge, one which may be addressed using the AGS at Brookhaven.

The second kind of possible use for high energy polarized proton beams will be for use in a pp collider (Isabelle) or perhaps in $\overline{p}p$ colliders. Such an achievement would offer the chance to measure strong interaction spin effects in a totally new energy range, where large momentum transfer QCD effects are expected to be essentially background-free and unambiguous. In addition, the polarized colliding beams could offer a practical means for detecting weak interaction effects by their parity violation.

8. END OF THE ERA OF REGIONAL ACCELERATORS

With the shutdown of the ZGS, we are entering a period where each of the three DOE-supported U.S. high energy physics accelerators is designed for completely different kinds of experiments. Thus each one should be expected to serve the entire U.S. high energy physics program. The increasing scale of new accelerators as well as the difficult budget situation (described below) seem to leave no practical alternative to such an arrangement. But a host of deep issues are raised, one of which is a managerial and institutional question: What form of organizational arrangement will allow an accelerator laboratory to best serve the interests of the <u>entire</u> U.S. high energy physics community? Note that the three remaining DOE-HEP accelerator centers each have different contractor arrangements: a single university, a regional consortium of universities, and a national consortium of universities. Perhaps the safest speculation about organization is that historical precedent will prevail and that these three different arrangements will continue beyond the ending of the era of the regional high energy accelerator.

Related problems that may become increasingly evident as the work is confined to fewer accelerator centers are: a loss of diversity in styles and directions of research, decreasing leadership opportunities for younger scientists, and a narrowing of opportunities for the development of new ideas in accelerator science and technology. Moreover, at some later time, the painful issues involved in further decreasing the number of national high energy accelerator centers will no doubt have to be addressed.

9. THE ERA OF SINGLE PURPOSE HIGH ENERGY PHYSICS LABORATORIES

The ZGS has been the only U.S. high energy physics accelerator to be imbedded within a very large multipurpose energy research laboratory. The new era of mainly single purpose high energy accelerator laboratories appears to be inevitable because of the large cost and physical size of new machines and, for international laboratories, to avoid complications concerning commercial and other aspects of international technology development. The decreasing role of high energy physics in multipurpose laboratories may impede our ability to exchange personnel and advances in technology with other areas of R&D work. Such mutual exchange might become a more important goal now that energy problems and their technological challenges and controversies have become critical issues for industrialized nations. This kind of issue may be of particular importance in the U.S. since most federal support for high energy physics research is budgeted through the Department of Energy.

REMARKS

The above examples of issues for further work and thought show clearly that the field of high energy physics offers many fundamental challenges in physics, instrumentation, and organization. Additional examples from other areas of high energy physics could easily be given. We are now entering a decade in which new theoretical ideas, new accelerators, and new experimental techniques offer the real possibility of achieving a very fundamental level of understanding of the "elementary particles" and their interactions. There is also the possibility of finding a total surprise which does not fit into present schemes for the quarks and leptons. The extent to which these possibilities are realized will depend upon the quality of the answers to questions of the types I have outlined, as well as upon the level of funding for the work.

In fact, the budget problems which have led to many difficulties in carrying out the ZGS physics program are still casting a shadow on the future of U.S. high energy physics research. My calculations indicate that for the entire decade of the 1970's, the annual rates of change of various budgets (measured in constant value dollars) were fairly steady, with the following average values:

U.S. Gross National Product	+ 2% per year
U.S. Federal Budget	+ 4% per year
U.S. High Energy Physics Budget	- 3% per year
ZGS Budget	- 9% per year

In such a situation, it is not surprising that the ZGS program was unable to fully utilize its new capabilities during the 1970's. These new capabilities included the 12-ft bubble chamber and its neutrino research program, the polarized proton beam (particularly for 12-GeV running), the polarized deuteron beam, and the development of superconducting accelerator magnets. The Argonne proposal in 1971 for the construction of a Superconducting Stretcher Ring for the ZGS, a demonstration project which could have made major contributions to the development of practical superconducting magnet accelerators, was not seriously considered by the funding agency in view of the overall ZGS budget outlook. Another important opportunity to develop practical superconducting dipole magnets was lost in 1976 when the POPAE high energy colliding beam design project was not supported.

Contraction of support for the U.S. high energy physics program during the 1970's has created several kinds of problems. First, as illustrated by the above specific ZGS examples, some important opportunities have been lost forever. Some of those losses are likely to lead to delays and difficulties in the projects of the 1980's. Second, even after the shutdown of the ZGS, present budget levels will not permit full utilization of U.S. accelerators during the 1980's.

ACKNOWLEDGEMENT

As an ex-administrator, I recall that one of the duties of an administrator is to try to keep things coordinated. But, as shown in Homer's Neal's talk, the question of who is coordinating whom is a complicated one in high energy physics. There are complex webs of relationships among the user research groups, the people who actually make the accelerator and detectors work, the laboratory management, and the funding agency people who have found support for the work even in very difficult times. These relationships involve more than a thousand people who have contributed to the ZGS program over a period of some two decades. To end on a coordinating note, I should like to take the liberty of speaking on behalf of all of the ZGS administrations and offer thanks to all of these people who have worked so well together on the ZGS program and have accomplished so much.

References

1. R. C. Lamb et al., Phys. Rev. Lett. $\underline{15}$, 800 (1965).

2. R. Ammar et al., Phys. Rev. Lett. $\underline{21}$, 1832 (1968).

3. B. Y. Oh et al., Phys. Rev. $\underline{D1}$, 2494 (1970).

4. N. M. Cason et al., Phys. Rev. Lett. $\underline{36}$, 1485 (1976).

5. A. J. Pawlicki et al., Phys. Rev. Lett. $\underline{37}$, 1666 (1976).

6. P. Estabrooks, Phys. Rev. $\underline{D19}$, 2678 (1979).

7. G. Alexander et al., Phys. Rev. Lett. $\underline{17}$, 412 (1966).

8. J. Mott et al., Phys. Rev. Lett. $\underline{18}$, 355 (1967).

9. I. P. Auer et al., Phys. Rev. Lett. $\underline{41}$, 1436 (1978).

10. D. G. Crabb et al., Phys. Rev. Lett. $\underline{41}$, 1257 (1978).

SYMPOSIUM ON THE HISTORY OF THE ZGS

ATTENDEES

Name	Affiliation
Abashian, A.	National Science Foundation
Abolins, M. A.	Michigan State University
Abrams, R. J.	University of Illinois at Chicago Circle
Adams, H. H.	Argonne National Laboratory
Ambats, I.	Argonne National Laboratory
Andrle, C. O.	Fermi National Accelerator Laboratory
Arenton, M. W.	Argonne National Laboratory
Arnold, R. C.	Argonne National Laboratory
Aspel, E. B.	Argonne National Laboratory
Auer, I. P.	Argonne National Laboratory
Ayres, D. S.	Argonne National Laboratory
Bacci, F. F.	Argonne National Laboratory
Baker, W. F.	Fermi National Accelerator Laboratory
Balka, L. J.	Argonne National Laboratory
Barnett, B. A.	Johns Hopkins University
Berger, E. L.	Argonne National Laboratory
Berrill, E. C.	Argonne National Laboratory
Bertermann, R. E.	G. D. Searle and Company
Bertucci, P. E.	Argonne National Laboratory
Biggs, J. A.	Argonne National Laboratory
Biwer, R. L.	Fermi National Accelerator Laboratory
Blackman, R. B.	Argonne National Laboratory
Blaskie, R. T.	Argonne National Laboratory
Blume, L. O.	Fluor Power Services
Bogaty, J. M.	Argonne National Laboratory
Bohm, H. V.	Argonne Universities Association
Bohringer, D. E.	Argonne National Laboratory
Bouie, R.	Argonne National Laboratory
Bourkland, K. R.	Fermi National Accelerator Laboratory
Boye, J. L.	Fermi National Accelerator Laboratory
Brandeberry, F. E.	Argonne National Laboratory
Brescia, A. A.	NCR Corporation
Bresof, I. H.	Argonne National Laboratory
Broholm, R. H.	Argonne National Laboratory
Brown, L. M.	Northwestern University
Brumwell, F. R.	Argonne National Laboratory
Bryan, G. W.	Argonne National Laboratory
Brzegowy, C.	Argonne National Laboratory
Burba, K.	Argonne National Laboratory
Burnstein, R. A.	Illinois Institute of Technology
Bywater, J. A.	Argonne National Laboratory
Calkin, M. M.	Rice University
Cason, N. M.	University of Notre Dame
Clark, C. E.	Argonne National Laboratory

Name	Affiliation
Cho, Y.	Argonne National Laboratory
Colton, E. P.	Argonne National Laboratory
Cork, B.	Lawrence Berkeley Laboratory
Cosgrove, D. F.	Fermi National Accelerator Laboratory
Crabb, D. G.	University of Michigan
Craig, S. L.	Argonne National Laboratory
Creer, A. J.	Argonne National Laboratory
Crewe, A. V.	University of Chicago
Crosbie, E. A.	Argonne National Laboratory
Czyz, W. S.	Argonne National Laboratory
Daniels, R. E.	Princeton University
Day, J. S.	Argonne National Laboratory
Derrick, M.	Argonne National Laboratory
DeWitt, A. D.	Argonne National Laboratory
Diaz, R. J.	Community Unit School District #201
Diebold, R. E.	Argonne National Laboratory
Dinkel, J. A.	Fermi National Accelerator Laboratory
Dittrich, J. E.	Argonne National Laboratory
Ditzler, W. R.	Argonne National Laboratory
Doherty, W. L.	Argonne National Laboratory
Donley, L. I.	Argonne National Laboratory
Dvorak, J. A.	Argonne National Laboratory
Ducar, R. J.	Fermi National Accelerator Laboratory
Duffield, R. B.	Los Alamos Scientific Laboratory (retired)
Ely, R. S.	Argonne National Laboratory
Erickson, M. O. I.	Argonne National Laboratory
Evans, W. J.	Argonne National Laboratory
Faber, M. M.	Argonne National Laboratory
Failla, P.	Argonne National Laboratory
Fields, T. H.	Argonne National Laboratory
Forrestal, J. R.	Argonne National Laboratory
Foss, M. H.	Argonne National Laboratory
Fouts, G. N.	Argonne National Laboratory
Fuja, R. E.	Argonne National Laboratory
Genens, L. E.	Argonne National Laboratory
Getz, D. R.	Stanford Linear Accelerator Center
Glowacki, A. M.	Fermi National Accelerator Laboratory
Goldwasser, E. L.	University of Illinois
Gorka, A. J.	Argonne National Laboratory (retired)
Gottlieb, S. A.	Argonne National Laboratory
Grahn, A.	Argonne Universities Association
Gray, S. W.	Indiana University
Gregorash, D. A.	Argonne National Laboratory
Groves, T. H.	Fermi National Accelerator Laboratory
Gunderson, G. R.	Argonne National Laboratory
Haffner, R. L.	Argonne National Laboratory
Haldeman, M.	Fermi National Accelerator Laboratory
Haldeman, P.	

Name	Affiliation
Hansen, P. H.	University of Michigan
Hanson, W.	Fermi National Accelerator Laboratory
Hardek, T. W.	Argonne National Laboratory
Hemmi, J. L.	Argonne National Laboratory
Heyden, L. G.	Heyden Associates
Heyn, E. E.	Argonne National Laboratory
Higgins, E. F.	Los Alamos Scientific Laboratory
Hildebrand, R. H.	University of Chicago
Hill, N. F.	Fermi National Accelerator Laboratory
Hoffman, J. A.	Argonne National Laboratory
Hogrefe, R. L.	Argonne National Laboratory
Hoover, J. A.	Fermi National Accelerator Laboratory
Hornstra, F.	Argonne National Laboratory
Hothan, A. W.	Retired
Hulet, D. V.	Argonne National Laboratory
Hunckler, R. J.	Fermi National Accelerator Laboratory
Hyman, L. G.	Argonne National Laboratory
Janes, R. J.	Fermi National Accelerator Laboratory
Johns, L. D.	Argonne National Laboratory
Jones, N. F.	Self-employed
Joswick, J. M.	Argonne National Laboratory
Jovanovic, D. D.	Fermi National Accelerator Laboratory
Kalmus, P. I. P.	Queen Mary College
Kaminskas, S.	Argonne National Laboratory
Keig, M. W.	Self-employed
Kelleck, K. T.	Argonne National Laboratory
Kelly, A. L.	Argonne National Laboratory
Kenney, V. P.	University of Notre Dame
Khoe, T. K.	Argonne National Laboratory
Kickert, R. B.	Argonne National Laboratory
Klaisner, L. A.	Eldec Corporation
Klindworth, C. M.	Argonne National Laboratory
Kliss, R. M.	Argonne National Laboratory
Knott, M. J.	Argonne National Laboratory
Konecny, R. S.	Argonne National Laboratory
Kramer, S. L.	Argonne National Laboratory
Krell, R. E.	Argonne National Laboratory
Krieger, C. I.	Argonne National Laboratory
Krisch, A. D.	University of Michigan
Krisciunas, A. B.	Argonne National Laboratory
Krizek, R. J.	Allis Chalmers
Kulovitz, E. C.	Argonne National Laboratory
Kustom, R. L.	Argonne National Laboratory
LaFave, W. E.	Argonne National Laboratory
Lanham, R. N.	Argonne National Laboratory
Lari, R. J.	Argonne National Laboratory
Laurence, G. L.	Argonne National Laboratory

Name	Affiliation
Lawrentz, F. J.	Argonne National Laboratory
Leach, D. J.	Argonne National Laboratory
Leclercq, W. J.	Fermi National Accelerator Laboratory
Leonard, C. F.	Argonne National Laboratory
Lesikar, J. D.	Rice University
Levman, G. M.	Argonne National Laboratory
Lewis, L. G.	Argonne National Laboratory
Lipkin, H. J.	Argonne National Laboratory
Little, D. C.	Argonne National Laboratory
Livdahl, P. V.	Fermi National Accelerator Laboratory
Livingood, J. J.	Argonne National Laboratory (retired)
Loeffler, F. J.	Purdue University
Loos, J. S.	Argonne National Laboratory
Lucks, H.	Augustana Hospital
Ludwig, H. F.	Argonne National Laboratory
Lund, H. R.	Argonne National Laboratory
Madsen, J. J.	Argonne National Laboratory
Manson, D. S.	Argonne National Laboratory (retired)
Marcks, D. F.	Radiation Polymer Company
Marcowitz, S. M.	Private Medical Practice
Marek, L. J.	Argonne National Laboratory
Markley, F. W.	Fermi National Accelerator Laboratory
Marmer, G. J.	Argonne National Laboratory
Martin, R. L.	Argonne National Laboratory
Martin, W. C.	Kinetic Systems Corporation
Marzec, B. A.	Argonne National Laboratory
Mataya, K. F.	Argonne National Laboratory
May, E. N.	Argonne National Laboratory
Mazarakis, M. G.	Argonne National Laboratory
Miettinen, H. E.	Rice University
Millar, E. B.	Argonne National Laboratory
Miller, S. A.	Argonne National Laboratory
Miller, W. C.	University of Notre Dame
Missig, J. R.	Liquid Carbonic Corporation
Missig, M. J.	Liquid Carbonic Corporation
Mizicko, D. M.	Fermi National Accelerator Laboratory
Moenich, J. S.	Argonne National Laboratory
Moffett, D. R.	Argonne National Laboratory
Moyer, D. F.	Northwestern University
Mulera, T. A.	Rice University
Musgrave, B.	Argonne National Laboratory
Muszynski, G. L.	Argonne National Laboratory
McDonald, P. A.	Hines Veteran Administration Hospital
Neal, H. A.	Indiana University
Needell, A.	American Institute of Physics
Nestander, W. W.	Fermi National Accelerator Laboratory
Neubauer, A. W.	Fermi National Accelerator Laboratory

Name	Affiliation
Niemann, R. C.	Argonne National Laboratory
Nodulman, L. J.	Argonne National Laboratory
Norton, R. L.	Fermi National Accelerator Laboratory
Novey, T. B.	T. A. Associates
O'Fallon, J. R.	Argonne Universities Association
O'Hare, E. W.	Argonne National Laboratory
Otavka, J. G.	Fermi National Accelerator Laboratory
Otavka, M. A.	General Atomic
Parker, E. F.	Argonne National Laboratory
Passi, A. R.	Argonne National Laboratory
Paugys, A. A.	Argonne National Laboratory
Peerson, J. J.	Argonne National Laboratory
Pelczarski, W. J.	Argonne National Laboratory
Perry, R.	Brigham Young University
Pewitt, E. G.	Argonne National Laboratory
Phelan, J. J.	Bell Telephone Laboratory
Phillips, G. C.	Rice University
Piatak, D. J.	Argonne National Laboratory
Polychronakos, V. A.	Fermi National Accelerator Laboratory
Pondrom, L. G.	University of Wisconsin
Potts, C. W.	Argonne National Laboratory
Praeg, W. F.	Argonne National Laboratory
Price, L. E.	Argonne National Laboratory
Prien, R. J.	Argonne National Laboratory
Putnam, C. C.	Argonne National Laboratory
Ratner, L. G.	Argonne National Laboratory
Rauchas, A. V.	Argonne National Laboratory
Reay, N. W.	Ohio State University
Reibel, K.	Ohio State University
Reinke, S. A.	Argonne National Laboratory
Richied, D. E.	Fermi National Accelerator Laboratory
Rihel, R. K.	Fermi National Accelerator Laboratory
Rivetna, R. R.	Argonne National Laboratory
Roberts, J. B.	Rice University
Roman, R. D.	Argonne National Laboratory
Romanowski, T. A.	Ohio State University
Romy, M. J.	Argonne National Laboratory
Royston, R. J.	Argonne National Laboratory
Rubin, H. A.	Illinois Institute of Technology
Ruddick, K.	University of Minnesota
Rudnick, S. J.	Argonne National Laboratory
Ryk, J.	Fermi National Accelerator Laboratory
Sachs, R. G.	University of Chicago
Sapp, D. E.	Argonne National Laboratory
Sauer, J. R.	Argonne National Laboratory
Saunders, C. F.	Argonne National Laboratory
Scherr, R. E.	Fermi National Accelerator Laboratory
Schlereth, J. L.	Argonne National Laboratory

Name	Affiliation
Schultz, P. F.	Argonne National Laboratory
Schweingruber, F. L.	G. D. Searle and Company
Sellers, J. F.	Sargent and Lundy
Senders, C. W.	Self-employed
Siljander, W. A.	Argonne National Laboratory
Simanton, J. R.	Fermi National Accelerator Laboratory
Sinclair, D.	University of Michigan
Singer, R. A.	Argonne National Laboratory
Singleterry, M. K.	Argonne National Laboratory
Sivers, D. W.	Argonne National Laboratory
Smith, R. P.	Argonne National Laboratory
Sobocki, L. J.	Fermi National Accelerator Laboratory
Spinka, H. M.	Argonne National Laboratory
Sprau, G. A.	Argonne National Laboratory
Stapay, J. R.	Argonne National Laboratory
Steinberg, E. P.	Argonne National Laboratory
Stejskal, B. A.	Argonne National Laboratory
Stipp, V. F.	Argonne National Laboratory
Stockley, R. L.	Argonne National Laboratory
Strom, L. A.	Kirkland and Ellis
Suddeth, D. E.	Argonne National Laboratory
Sumner, R. L.	Princeton University
Sunde, R. A.	Argonne National Laboratory
Swallow, E. C.	Elmhurst College/University of Chicago
Takeda, H.	Argonne National Laboratory
Tamura, N.	Argonne National Laboratory
Teng, L. C.	Fermi National Accelerator Laboratory
Terandy, J.	Argonne National Laboratory
Terwilliger, K. M.	University of Michigan
Theodosiou, G. E.	Argonne National Laboratory
Thomas, A.	Argonne National Laboratory
Thomas, G. H.	Argonne National Laboratory
Thompson, R. E.	Argonne National Laboratory
Toshioka, K. K.	Argonne National Laboratory
Trcka, R. A.	Standard Oil Company of Indiana
Trendler, R. C.	Fermi National Accelerator Laboratory
Underwood, D. G.	Argonne National Laboratory
Upstrom, A. C.	Gardner, Carton and Douglas
Urban, J. J.	Argonne National Laboratory
Uretsky, J. L.	Ruff and Grotefeld, Ltd.
Vasilopulos, G. F. P.	Argonne National Laboratory
Visser, A. T.	Fermi National Accelerator Laboratory
Volk, G. J.	Argonne National Laboratory
Voyvodic, L.	Fermi National Accelerator Laboratory
Wagner, R. G.	Argonne National Laboratory
Wali, K. C.	Syracuse University
Walker, W. D.	Duke University
Wallace, E. J.	Argonne National Laboratory

Name	Affiliation
Wallenmeyer, W. A.	Department of Energy
Ward, C. E. W.	Bell Telephone Laboratories
Watson, J. M.	Argonne National Laboratory
Wattenberg, A.	University of Illinois
Weberg, K. H.	Cardwell-Westinghouse
Wehrle, R. B.	Argonne National Laboratory
Weis, D. R.	Argonne National Laboratory
Welch, W. A.	Argonne National Laboratory
Wesel, R. H.	Argonne National Laboratory
West, G. C.	Argonne National Laboratory (retired)
Wilslef, D. M.	Fermi National Accelerator Laboratory
Wilson, R. L.	Argonne National Laboratory
Woods, R. M.	Department of Energy
Yokosawa, A.	Argonne National Laboratory
Zanolla, D. J.	Midway Manufacturing
Zonick, C. A.	Fermi National Accelerator Laboratory

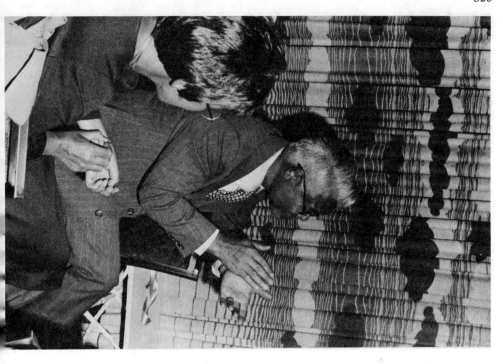

ZGS EXPERIMENTS

PROP. NO. (DATE)	EXPER. NO. (DATE)	SPOKESMAN (USER GROUPS)	TITLE	BEAM	DETECTOR	AMOUNT OF RUN	REMARKS (PUBLICATIONS)
P-5 (3/21/63)	E-1 (4/24/63)	T. B. Novey (ANL)	Neutrino and Antineutrino Interactions	ν Beam	SC, Ctrs	51 shifts	Run completed (12/2/66) PRL22, 1014(1969)
P-11 (4/1/63)	E-2 (4/24/63)	W. D. Walker (U. of Wisconsin)	3.0 and 7.0 GeV/c π^- in H_2	7°	30" HBC	200,000 pictures	Run completed (4/10/65) NC49A, 399(1967);PR166, 1430(1968) PRL20, 1414(1968);PRL18, 630(1967) PRL20, 133(1968);PL24B, 307(1967) PRD4, 133(1971);PL35B, 457(1971) NPB32, 366(1971);PR127, 1025 (1971);NPB33, 1(1971)
P-15 (4/2/63)	E-4 (4/24/63)	A. Abashian (U. of Illinois)	T-Invariance and μ Polarization in $K_2^0 \rightarrow \pi + \mu + \nu$	30°	SC, Ctrs	185 shifts	Run completed (1/25/65) PRL17, 606(1966);PR176, 1603 (1968)
P-16 (4/2/63)	E-5 (4/24/63)	R. A. Schluter (ANL)	Excited State in ΛHe^4	30°	Ctrs	8 shifts	Run completed (11/27/64)
P-17 (4/2/63)	E-6 (4/24/63)	T. H. Fields (Northwestern U./ANL, U. of Illinois, U. of Wisconsin)	5.5 GeV/c K^- in H_2	7°	30" HBC	400,000 pictures	Run completed (8/14/66) PL23, 171(1966);PRL20, 171(1968); PRL58, 1509(1968);PR173, 1330 (1968);PRL18, 266(1967);PRL18, 355(1967);PRL19, 1071(1967);PR 166, 1317(1968);PL27B, 532(1968); PRL74, 2165(1968);PRL21, 1832 (1968);PR177, 1951(1969);PR177, 1966(1969);NPB13, 565(1969);PRD 1960(1970);PRD2, 430(1970);PRL 26, 1502(1971);PRD5, 2162(1972); NPB52, 70(1973)
P-23 (8/23/63)	E-8 (10/24/63)	R. A. Schluter (ANL)	Excited States in ΛLi^7 and K^- X rays	30°	Ctrs	20 shifts	Run completed (11/27/64) PRL15, 70(1965);PR158, 1343(1967)

ISSN:0094-243X/80/600328-43$1.50 Copyright 1980 American Institute of Physics

ZGS EXPERIMENTS

PROP. NO. (DATE)	EXPER.NO. (DATE)	SPOKESMAN (USER GROUPS)	TITLE	BEAM	DETECTOR	AMOUNT OF RUN	REMARKS (PUBLICATIONS)
P-25 (8/28/63)	E-18,E-19 (10/31/63)	E.P. Steinberg (ANL) N. Sugerman (U. of Chicago)	Investigation of (p,π^+) Reactions/ Study of Fission-Spallation Reactions	Internal		29 hours	Run Completed (7/23/65) NP-B5,188(1968); NP-A113,265(1968) NC-29,2859(1967); JINC-32,1775(197... PR-C3,1631(1971);JINC-33,32I(1971) NP-A179,645(1972);PR-C7,1410(1973) PR-C7,1597(1973);PRL-30,1221(1973) PR-C8,1448(1973); PR-C8,775(1973); PR-C10,167(1974);PR-C10,1925(1974) PR-C10,2268(1974);PR-C8,1091(1973) JINC-29,2859(1967);NI&M-59,192(1968) PR-C6,1153(1972);NI&M-59,192(1968) PR-C11,1258(1975);PR-C12,938(1975) NI&M-99,309(1972);PL-42B,141(1972) PR-C7,353(1973);NI&M-115,47(1974); NI&M-C10,2467(1974)
P-28 (8/30/63)	E-10 (10/25/63)	H. J. Martin, Jr. (Indiana U.)	6.0 GeV/c π^+ and π^- in H_2	7°	30" HBC	100,000 pictures	Run Completed (2/19/66) PL-27B,250(1968); PRL-21,1421(1968) NP-B16,102(1970);PR-D1,54(1970); PR-D1,3077(1970); PR-D5,1097(1972)
P-37 (9/3/63)	E-16 (10/25/63)	G. W. Tautfest (Purdue U.)	4.5 GeV/c K^- in D_2	7°	30" HBC	415,000 pictures	Run Completed (2/14/67) PLB-28B,356(1968); PRL-22,963(1969) PL-29B,255(1969);NP-B12,9(1969); PR-188,2011(1969); PRL-24,327(1970) PR-D2,30(1970);NP-B20,565(1970); NP-B21,407(1970);PL-33B,536(1970); PL-34B,533(1971);NP-B35,317(1971) PR-D7,1345(1973);NP-B126,397(1977)
P-43 (12/19/63)	E-3 (2/20/64)	K. M. Terwilliger (U. of Michigan)	$\pi^\pm$ - p Elastic Diff. Cross Sections 2.5 - 6.0 GeV/c	17°	SC, Ctrs. (LH_2 Tgt.)	73 Shifts	Run Completed (3/15/65) PRL-15,838(1965); PRL-17,458(1966) PR-159,1169(1967)

ZGS EXPERIMENTS

PROP. NO. (DATE)	EXPER. NO. (DATE)	SPOKESMAN (USER GROUPS)	TITLE	BEAM	DETECTOR	AMOUNT OF RUN	REMARKS (PUBLICATIONS)
P-45 (3/2/64)	E-22 (5/27/64)	D.T. King (U. of Tennessee, U. of South Carolina)	π° Production by 6.0 GeV/c π^-	7°	Emulsion	1,600 Pulses	Run Completed (1/14/65)
P-46 (3/12/64)	E-21 (5/27/64)	T. A. Romanowski (ANL, U. of Chicago, Ohio State U., Washington)	Stopping K Beam Studies	EPB	Ctrs.	46.5 Shifts	Run Completed (2/1/67) PR-179,1294(1969)
P-49 (6/9/64)	E-24 (7/6/64)	A. Abashian (U. of Illinois)	Anomalous Regeneration and Decays of K°	30°	SC, Crrs.	99 Shifts	Run Completed (8/65) PRL-17,669(1966);PRL-18,138(1967) PR-174,1674(1968)
P-54 (9/8/64)	E-41 (2/11/66)	L. Marshall (U. of Colorado)	6.7 GeV/c π^- in H_2	7°	30" HBC	100,000 pictures	Run Completed (3/10/66) PL-27B,253(1968); PR-D1,771(1970)
P-55 (9/15/64)	E-30 (3/31/65)	A.D. Krisch (U. of Michigan)	π^- - p, 180° Elastic Scattering	17°	Ctrs. (LH_2 Tgt.)	165 Shifts	Run Completed (2/21/66) PRL-16,709(1966);PR-164,166(1967)
P-60 (11/13/64)	E-31 (4/2/65)	E. P. Steinberg (ANL)	p-Induced Nuclear Reactions in Complex Nuclei	EPB	Scattering Chamber	80 Hours	Run Completed (11/11/68)
P-62 (1/20/65)	E-43 (2/21/66)	W. D. Walker (U. of Wisconsin, ANL, U. of Toronto)	7.0 GeV/c π^- and p in H_2	7°	30" HBC	207,529 pictures	Run Completed (12/22/66) PRL-20,1414(1968);PRL-18,630(1967) PRL-20,133(1968);PL-28B,564(1969) PRL-23,331(1969);NP-B18,1(1970); PR-D1,2494(1970);PRL-25,1393(1970) NP-B24,253(1970);NP-B27,605(1971) PR-D4,133(1971); PL-35B,457(1971) NP-B32,366(1971);PRL-27,1025(1971) NP-B33,1(1971); PRL-28,318(1972); NP-B43,77(1972);PR-D7,1977(1973)

Z G S   EXPERIMENTS

PROP. NO. (DATE)	EXPER. NO. (DATE)	SPOKESMAN (USER GROUPS)	TITLE	BEAM	DETECTOR	AMOUNT OF RUN	REMARKS (PUBLICATIONS)
P-63 (1/20/65)	E-34 (12/14/65)	M. M. Block (Northwestern U.)	Stopping K$^-$ in He	28°	20" HeBC	219,943 pictures	Run completed (6/5/67) PRD1,20(1970);NPB67,269(1973); NC24A,294(1974); NC-31A,401(1976)
P-65 (1/18/65)	E-23 (1/22/65)	D. D. Jovanovic (ANL)	$\pi^\pm$ and K$^\pm$ Production Cross Section	EPB	Ctrs (LH$_2$ Tgt)	50 shifts	Run completed (4/65) PRL14,504(1965)
P-66 (1/18/65)	E-25 I-(3/16/65) II-(6/17/65)	D. D. Jovanovic (ANL, UCSD)	Baryon Number 2 Resonances	EPB	Ctrs (LH$_2$ Tgt)	31 shifts 41 shifts	Run completed (1/19/66) PRL17,100(1966)
P-68 (1/29/65)	E-40 (2/11/66)	K. M. Terwilliger (U. of Michigan)	ρ Production and Decay	17°	SC, Ctrs (LH$_2$ Tgt)	976,000 SC pictures	Run completed (12/7/66) PRD7,637(1973)
P-70 (1/29/65)	E-17 (2/22/65)	S. Suwa (ANL, U. of Chicago)	Polarization in π^- - p Elastic Scattering 1.7 - 2.5 GeV/c	17°	Ctrs (Pol. Tgt)	172 shifts	Run completed (3/24/65) PRL15,560(1965);PRL16,714(1966) PRL16,1019(1966);PRL159,1431 (1967);PRD1,729(1970)
P-71 (2/1/65)	E-29 (3/16/65)	R. H. Hildebrand (ANL/U. of Chicago)	4.5 GeV/c K$^+$ in H$_2$	7°	30" HBC	255,000 pictures	Run completed (10/6/66) PRD3,2610(1971);PRD7,621(1973)
P-72 (2/1/65)	E-42 (2/16/66)	R. A. Schluter (ANL/Northwestern U.)	Polarization and Lifetime of Σ^+ and Λ^+	EPB	SC, Ctrs (LH$_2$ Tgt)	55.7 shifts	Run completed (10/31/67)

Z G S   EXPERIMENTS

PROP. NO. (DATE)	EXPER. NO. (DATE)	SPOKESMAN (USER GROUPS)	TITLE	BEAM	DETECTOR	AMOUNT OF RUN	REMARKS (PUBLICATIONS)
P-76 (2/1/65)	E-12 (4/13/65)	M. Derrick (ANL, Carnegie Tech.)	Stopping and In-Flight K^- in He	28°	10" HeBC	250,000 pictures	Run completed (3/30/66) PL25B,376(1967);PRL19,1405(1967) PRL20,819(1968);PRL20,822(1968) PRL20,1215(1968);PRL77,1945 (1969);PRD1,66(1970);NCL2,389 (1969);PRD1,1267(1970);PRD2,98 (1970);PRD2,119(1970);PRA2,7 (1970);PRD2,1803(1970);PRD5, 1063(1972);PRD6,3069(1972); PR-D11,3065(1975);NP-A240,485 (1975)
P-78 (2/1/65)	E-9 (4/13/65)	H. L. Anderson (U. of Chicago, ANL, U. of London, Nat'l Res. Council of Canada, Queen Mary College)	p-p Isobar Production	EPB	SC, Ctrs (TV Scan)	104 shifts	Run completed (5/9/66) PRL18,89(1967)
P-83 (4/19/65)	E-83 (6/9/66)	D. Cline (U. of Wisconsin)	2.1 GeV/c π^+ in D_2 (Search for X^0)	7°	30" DBC w/plates	151,158 pictures	Run completed (12/15/66) PRL21,1275(1968);NPB18,77(1970) PRD8,3794(1973)
P-84 (4/26/65)	E-84 (3/15/66)	W. F. Fry (U. of Wisconsin)	700 - 800 MeV/c K^+ in $C_3 H_8$	28°	40" PBC	275,444 pictures	Run completed (1/5/70) PR-D12, 2570(1975)
P-85 (5/10/65)	E-38 (2/11/66)	T. A. Romanowski (ANL/Ohio State U., U. of Chicago, Washington U.)	$\Lambda \to p + e^- + \nu$	EPB	SC, Ctrs (LH_2 Tgt)	327.5 shifts	Run completed (10/13/69) 1,074,854 SC Pictures PRL27,612(1971);JPG-2,L211(1976) PR-D10,2104(1977)
P-87 (6/2/65)	E-35 (1/5/66)	T. F. Hoang (ANL, U. of Notre Dame, Purdue U.)	Production of $K_1^0 - K_1^0$ Pairs	17°	SC, Ctrs (LH_2 Tgt)	31 shifts	Run completed (5/9/66) PL25B,615(1967);PRL21,317(1968) PRT84,1363(1967)

ZGS   EXPERIMENTS

PROP. NO. (DATE)	EXPER. NO. (DATE)	SPOKESMAN (USER GROUPS)	TITLE	BEAM	DETECTOR	AMOUNT OF RUN	REMARKS (PUBLICATIONS)
P-89 (6/29/65)	E-32 (6/29/65)	R. A. Lundy (ANL, U. of Wisconsin)	Search for μ Excess (W)	EPB	SC	77 shifts	Run completed (9/4/65) PRL15,800(1965)
P-93 (9/28/65)	E-36 (2/15/66)	A. Wattenberg (U. of Illinois)	Partial Decay Rates of $K^+ \rightarrow \pi^+ + \pi^+ + \pi^+$ and $K^- \rightarrow \pi^- + \pi^- + \pi^+$	EPB	SC, Ctrs	76.7 shifts	Run completed (1/16/67) PRL19,98(1967)
P-94 (10/4/65)	E-94 (3/7/66)	M. Derrick (ANL, Carnegie Tech.)	Stopping $\bar{p}$ in C_3H_8 and $C_3H_8 - CF_3Br$	28°	40" PFBC	56,821 pictures	Run completed (11/18/68) PRD2,2538(1970);PR-D14,1182(1976)
P-95 (10/13/65)	E-95 (3/17/66) (5/5/70)	J. G. Fetkovich (Carnegie Tech.)	Interactions of Slow K^- in Propane	28°	40" PBC	401,433 pictures	Run completed (3/26/69) PRD9,1248(1974);PR-D7,3295(1973)
P-96 (10/18/65)	E-96 (3/7/66)	M. Derrick (ANL)	730 MeV/c K^- in C_3H_8	28°	40" PBC	484,851 pictures	Run completed (11/14/69) PRD7,1331(1973)
P-97 (10/25/65)	E-95 (3/17/66) (5/5/70)	J. G. Fetkovich (Carnegie Tech.)	Stopping K^- in C_3H_8 (Test of $\Delta S = \Delta Q$, T-Invariance)	28°	40" PBC	401,433 pictures	Run completed (3/26/69) Joint experiment for P-95 and P-97
P-98 (11/2/65)	E-98 (6/10/66)	J. E. A. Lys (U. of Michigan)	1.63 - 2.20 GeV/c $\bar{p}$ in H_2	7°	30" HBC	151,253 pictures	Run completed (3/1/67) PRL21,116(1968);PRL21,1718(1968);NPB24,445(1970);PRD2,2525(1970);PRD4,1275(1971);NPB42,1(1972);PRD7,610(1973);PRD9,77(1974);PRD9,90(1974)
P-101 (12/2/65)	E-101 (5/5/66)	N. E. Booth (U. of Chicago, ANL)	Polarization in $\pi^{\pm}p$ Elastic Scattering 3.10 - 5.50 GeV/c	EPB	Ctrs (Pol. Tgt)	327.8 shifts	Run completed (11/11/68) PRL21,651(1968);PRL21,1410(1968);PRL23,192(1969);PRL25,898(1970);PRD8,451(1973);PRD8,1263(1973)

Z G S EXPERIMENTS

PROP. NO. (DATE)	EXPER. NO. (DATE)	SPOKESMAN (USER GROUPS)	TITLE	BEAM	DETECTOR	AMOUNT OF RUN	REMARKS (PUBLICATIONS)
P-102 (12/29/65)	E-102 (4/7/66)	A. D. Krisch (U. of Michigan, ANL)	p-p Elastic Scattering at 90°	EPB	Ctrs	90 shifts	Run completed (9/26/66) PRL17,1105(1966);PRL159,1138(1967);PRL165,1442(1968);PRL19,1149(1967)
P-105 (2/4/66)	E-39 (2/15/66)	A. Wattenberg (U. of Illinois)	Feasibility of K_2^0 Experiments in 7° Branch Beam	7° Branch	SC, Ctrs	30 shifts	Run completed (9/2/66)
P-107 (2/14/66)	E-44 (5/5/66)	A. Cooper (U. of Wisconsin, ANL, U. of Toronto)	5.5 GeV/c π^+ in H_2	7°	30" HBC	250,000 pictures	Run completed (9/18/66) PRL20,472(1968);NPB23,605(1970);NPB35,79(1971);NPB35,133(1971);NPB37,203(1972);NPB39,525(1972);NPB46,402(1972);NPB59,365(1973);NPB81,231(1974)
P-108 (2/15/66)	E-108 (5/5/66)	W. D. Walker (U. of Wisconsin, ANL, U. of Toronto)	π^+ in D_2 7 GeV/c	7°	30" DBC w/plates	180,000 pictures	Run completed (12/5/66) PRL23,331(1969);PRL25,1393(1970)NC1,361(1971);PRD3,2561(1971);PRD4,133(1971);PL35B,457(1971);NPB32,366(1971);PRL27,1025(1971)NPB33,1(1971);PRD5,2730(1972)PRD10,1430(1974)
P-109 (2/24/66)	E-109 (3/17/66)	D. Cline (U. of Wisconsin)	Stopping K^+ in CF_3 Br (Decay Modes of K^+ and τ'/τ Branching Ratio)	28°	40" FBC	123,292 pictures	Run completed (5/24/68) PRL23,326(1969)
P-111 (3/14/66)	E-111 (5/5/66)	A. Yokosawa (ANL)	Polarization in πN Charge Exchange Scattering 2 - 5 GeV/c	17°	Ctrs (Pol. Tgt)	197.5 shifts	Run completed (8/31/67) PRL20,275(1968);PRL20,353(1968);PRL20,566(1968)
P-113 (4/11/66)	E-113 (4/13/66)	M. M. Block (Northwestern U.)	1.5 - 5.0 GeV/c π^- in He	17°	20" HeBC	126,000 pictures	Run completed (8/31/66)

ZGS EXPERIMENTS

PROP. NO. (DATE)	EXPER. NO. (DATE)	SPOKESMAN (USER GROUPS)	TITLE	BEAM	DETECTOR	AMOUNT OF RUN	REMARKS (PUBLICATIONS)
P-114 (4/29/66)	E-114 (5/5/66)	B. P. Roe (U. of Michigan)	900 MeV/c π^+ or π^- in CF_3Br (η Decay)	28°	40" FBC	120,000 pictures	Run completed (11/21/66)
P-116 (5/6/66)	E-116 (5/24/66)	T. F. Hoang (ANL, U. of Illinois, U. of Notre Dame, Purdue U)	Production of $K_1^\circ K_1^\circ$ Pairs by $\pi^- p$ at 5.0 GeV/c	17°	SC, Ctrs (LH_2 Tgt)	36 shifts	Run completed (6/3/66) PL25B,615(1967);PRL21,317(1968)
P-119 (5/1/66)	E-119 (6/9/66)	V. P. Kenney (U. of Notre Dame)	5.5 GeV/c π^+ in D_2	7°	30" DBC	137,600 pictures	Run completed (11/20/66) PL31B,82(1970);PL32B,391(1970); PRD3,365(1971);NPB29,237(1971)
P-121 (6/6/66)	E-39 (6/9/66)	A. Wattenberg (U. of Illinois)	Presence of Neutral Current $K_2^\circ \to \mu^+ + \mu^-$	7° Branch	SC, Ctrs	Small	Run completed (9/2/66) Part of E-39
P-124 (6/20/66)	E-124 (11/15/66)	R. G. Ammar (Northwestern U., ANL)	5.5 GeV/c K^- in D_2	7°	30" DBC	426,481 pictures	Run completed (2/27/68) PR188,2023(1969);PL32B,409(1970); NPB23,37(1970);PL34B,428(1971); PRD3,1561(1971);NPB44,45(1972); PRD7,585(1973);PRD7,2537(1973); NP-B9?,426(1975)
P-125 (6/27/66)	E-125 (10/11/66)	R. M. Heinz (Indiana U.)	Large Angle $\pi^\pm p$ Elastic Scattering 2 - 4 GeV/c	EPB	SC, Ctrs (LH_2 Tgt)	99.8 shifts	Run completed (8/6/68) PRL23,186(1969);PRD1,3050(1970); PRD3,1523(1971)
P-126 (7/7/66)	E-126 (7/19/66)	K. M. Terwilliger (U. of Michigan)	π - p Elastic and Inelastic Angular Distribution Around 180°	17°	SC, Ctrs (LH_2 Tgt)	71.4 shifts	Run completed (3/6/67) PRL21,1605(1968);PRL26,38(1971) PRL27,687(1971);PRD6,740(1972)
P-128 (7/14/66)	E-128 (12/13/66)	D. D. Reeder (U. of Wisconsin)	5.5 GeV/c K^+ in D_2	7°	30" DBC w/plates	189,953 pictures	Run completed (6/27/68) NPB22,247(1970)

ZGS EXPERIMENTS

PROP. NO. (DATE)	EXPER. NO. (DATE)	SPOKESMAN (USER GROUPS)	TITLE	BEAM	DETECTOR	AMOUNT OF RUN	REMARKS (PUBLICATIONS)
P-129 (9/13/66)	E-129 (11/1/66)	A. Abashian (U. of Illinois, Northeastern U.)	Study $K_S^0 \to \mu^+ \mu^-$	$7°$	SC, Ctrs	457,518 pulses	Run completed (1/16/67) PR177,2009(1969)
P-132 (10/17/66)	E-132 (12/13/66)	A. Shapiro (Brown U.)	π^+ p Interactions at 4.1 GeV/c	$7°$	30" HBC	70,000 pictures	Run completed (11/20/67) PRD8,3765(1973);PRD9,3015(1974)
P-133 (10/17/66)	E-133 (12/13/66)	L. Guerriero (U. of Padova)	π^+ p Reactions at 3.7 GeV/c	$7°$	30" HBC	100,000 pictures	Run completed (11/23/67)
P-134 (10/17/66)	E-134 (12/13/66)	Y. Eisenberg (Weizmann Institute)	π^+ p Reactions at 3.5 GeV/c	$7°$	30" HBC	100,000 pictures	Run completed (11/25/67) NPB38,20(1972)
P-135 (10/17/66)	E-135 (12/13/66)	I. A. Pless (MIT)	π^+ p Interactions at 3.9 GeV/c	$7°$	30" HBC	100,000 pictures	Run completed (11/22/67) PRD3,2047(1971);PRD10,1387(1974) NPB75,93(1974)
P-136 (10/25/66)	E-136 (4/11/67)	R. A. Schluter (ANL, Northwestern U.)	K-Mesic X rays	EPB (2nd focus)	SC, Ctrs	60 shifts	Run completed (9/24/68) NPB16,389(1970);PRA2,1630(1970)
P-137 (10/28/66)		D. Cline (U. of Wisconsin, U. of Padova)	p̄p Interactions at 0.960 GeV/c and 1.35 - 1.40 GeV/c	$28°$	40"PHLBC	300,000 pictures	Withdrawn (3/26/71)
P-138 (11/1/66)	E-138 (12/13/66)	E. Marquit (U. of Minnesota)	π^+d at 3.65 GeV/c	$7°$	30" DBC w/plates	69,350 pictures	Run completed (7/12/68)
P-139 (11/1/66)	E-139 (11/15/66)	H. Courant (U. of Minnesota)	p̄ Interactions in Freon	$28°$	40"FHLBC	40,000 pictures	Run completed (10/27/67)

337

ZGS EXPERIMENTS

PROP. NO. (DATE)	EXPER. NO (DATE)	SPOKESMAN (USER GROUPS)	TITLE	BEAM	DETECTOR	AMOUNT OF RUN	REMARKS (PUBLICATIONS)
P-140 (11/7/66)	E-140 (7/31/67)	K. Ruddick (U. of Minnesota)	Large Angle Charge Exchange Scattering and Neutral Boson Resonance Production	17°	SC, Ctrs	90.4 shifts	Run completed (4/1/68) PRL22, 1137(1969);PL30B, 659(1969 PRD2, 2588(1970)
P-142 (11/14/66)	E-142 (11/15/66)	A. Brody (SLAC, LRL)	$\pi^- p$, 0.5 – 0.9 GeV/c π^- in H_2	7°	30" HBC	185,099 pictures	Run completed (3/22/67) PRL22, 1401(1969);PRD3, 2619(1971
P-144 (11/15/66)	E-144 (12/13/66)	V. L. Telegdi (U. of Chicago, U. of Ill. Chicago Circle)	Measurement of Relative Phase of the $K_S \rightarrow \pi^+\pi^-$ and $K_L \rightarrow \pi^+\pi^-$ Amplitudes Without Regenerator	EPB	SC, Ctrs	146.3 shifts	Run completed (8/12/68) PRL23, 615(1969)
P-145 (11/22/66)	E-145 (1/10/67)	A. D. Krisch (U. of Michigan, ANL U. of Iowa)	π^-, K^-, and $\bar{p}$ Production in p-p Collisions and Structure within the Proton	EPB	Ctrs	72.0 shifts	Run completed (4/27/67) PRL18, 1218(1967);PR166, 1353 (1968)
P-146 (12/2/66)	E-146 (12/13/66)	B. Musgrave (ANL)	A Measurement of the Natural Width of the ω^0 1.5 GeV/c π^+	7° Sep.	30" DBC w/plates	103,663 pictures	Run completed (7/28/68)
P-147 (12/2/66)	E-147 (2/14/67)	D. Meyer (U. of Michigan, ANL)	Measurement of $\pi^+ + p \rightarrow \Sigma^+ + K^+$ Angular Distribution and the Σ^+ Polarization at 4.0 – 7.0 GeV/c	EPB	SC, Ctrs	194.6 shifts	Run completed (12/16/68) PRL23, 189(1969);PL30B, 289(1969)
P-148 (12/9/66)	E-148 (6/14/67)	D. D. Carmony (Purdue U.)	4.5 GeV/c π^+	7°	30" HBC w/plates	750,000 pictures	Withdrawn (6/23/69)
P-151 (1/10/67)	E-151 (3/7/67)	E. L. Goldwasser (U. of Illinois)	$K^+ p$ at 2.5, 2.75, and 3.0 GeV/c	7°	30" HBC	448,356 pictures	Run completed (9/30/68) PRL21, 1407(1968);PRD1, 2433(1970 PRD3, 1092(1971);NPB8T, 189(1974) NPB8T, 365(1975)
P-152 (1/11/67)	E-152 (3/7/67)	V. Hagopian (U. of Pennsylvania)	$\pi^- p$ at 3.5 GeV/c	7°	30" HBC	101,260 pictures	Run completed (11/29/67)

ZGS EXPERIMENTS

PROP. NO. (DATE)	EXPER. NO. (DATE)	SPOKESMAN (USER GROUPS)	TITLE	BEAM	DETECTOR	AMOUNT OF RUN	REMARKS (PUBLICATIONS)
P-154 (2/1/67)	E-154 (3/7/67)	H. L. Anderson (U. of Chicago, ANL, Carleton U., Los Alamos, Nagoya U., Nat'l Res. Council of Canada)	Study of I = 1 Bosons with a Deuteron Missing Mass Spectrometer	EPB	SC, Ctrs	63.3 shifts	Run completed (4/1/68) PRL21,853(1968);PRL26,108(1971) PRD3,1536(1971)
P-155 (2/3/67)	E-155 (3/7/67)	A. Abashian (U. of Illinois Northeastern U.)	Study K_{e3} and $K_{\pi 3}$ Decay Modes of Neutral K Mesons	17°	SC, Ctrs	194.1 shifts	Run completed (7/23/68) NC9A,151(1972);NC9A,160(1972); NC9A,166(1972)
P-156 (2/6/67)	E-156 (2/14/67)	E. Beier (U. of Illinois)	Partial Decay Rates of $K^+ \to \pi^+ + \pi^\circ + \pi^\circ$ and $K^- \to \pi^- + \pi^\circ + \pi^\circ$	EPB	SC, Ctrs	101.3 shifts	Run completed (6/30/67) PR186,1403(1969);PRD5,2720(1972)
P-157 (2/22/67)	E-157 (3/7/67)	W. A. Cooper (ANL)	1.2 - 1.6 GeV/c $\bar{p}$ in H_2	7°	30" HBC	109,469 pictures	Run completed (5/23/67) PRL20,1059(1968);NPB16,155(1970)
P-159 (3/1/67)	E-159 (3/7/67)	D. Cline (U. of Wisconsin)	1.2, 1.3, 1.4 GeV/c $\bar{p}$ in D_2	7°	30" DBC	108,506 pictures	Run completed (4/3/67) PRL21,770(1968);PL27B,573(1968) NPB33,505(1971)
P-160 (5/11/67)	E-160 (7/31/67)	A. Fridman (Nuclear Research Center - France)	$\bar{p}$d Expt. at 5.5 GeV/c $\bar{p}$ Incident Momentum	7°	30" DBC	154,210 pictures	Run Completed (11/4/67) NPB 14,699 (1969);PRD2, 488(1970); PRD2,1212 (1970);PRD3, 2572(1971);PRD6,767 (1972);PRD6, 2311(1972);NPB54,61 (1973);PRD8,2765(1973);PRD10,3573 (1974);NP-B95,503(1975);NP-B89,120 (1975);PR-D12,3414(1975);PRD15, 1293(1977);PRD20,587(1979)
P-161 (5/9/67)	E-161 (5/9/67)	A. Brody (SLAC, LRL)	π + p at 1.06 and 1.50 GeV/c	7°	30" HBC	304,000 pictures	Run completed (12/4/67) PL34B,665(1971);PRD3,2619(1971)

Z G S EXPERIMENTS

PROP. NO. (DATE)	EXPER.NO. (DATE)	SPOKESMAN (USER GROUPS)	TITLE	BEAM	DETECTOR	AMOUNT OF RUN	REMARKS (PUBLICATIONS)
P-163 (5/19/67)	E-163 (6/14/67)	N. Gelfand (U. of Chicago)	4.3 GeV/c K^+ in H_2	7°	30" HBC	222,674 pictures	Run completed (3/30/68) PRD3,2610(1971);PRD7,621(1973); NPB1$\underline{79}$,28(1977)
P-164 (6/9/67)	E-164 (7/31/67)	W. Frisken (Case Western Reserve U.)	Radiative $K^+_{\pi 2}$	EPB	SC, Ctrs	60 shifts	Withdrawn (4/19/68)
P-166 (6/12/67)	E-166 (12/5/67)	G. A. Smith (Michigan State U., LRL)	1.6 - 2.0 GeV/c p in D_2	7°	30" DBC	134,680 pictures	Run completed (2/19/68) PRL24,1257(1970);PRL27,344(1971) PRL28,779(1972);NPB40,151(1972) NPB5$\underline{1}$,29(1973);NPB51,57(1973); NPB5$\underline{1}$,77(1973);PL47B, 177(1973); NPB63,1(1973);NP-B103,537(1976);
P-168 (6/14/67)	E-168 (6/14/67)	D. Cline (U. of Wisconsin)	1.23, 1.65 GeV/c p̄ in D_2	7°	30" DBC	61,105 pictures	Run completed (7/16/68)
P-169 (6/23/67)	E-169 (4/16/68)	D. D. Reeder (U. of Wisconsin, State U. of New York)	5.5 GeV/c Polarized p in D_2	7°	30" DBC	101,216 pictures	Run completed (4/4/69)
P-171 (7/6/67)	E-171 (9/18/67)	W. F. Fry (U. of Wisconsin)	850 MeV/c K^+ in C_3F_8	28°	40" HLBC	780,950 pictures	Run completed (12/13/68) NC25A, 688(1974)
P-172 (7/11/67)	E-172 (10/10/67)	P. H. Steinberg (U. of Maryland, ANL)	Measurement of K^+p Polarization Parameter from 1 to 2 GeV/c K^+	EPB I (6)	Ctrs	230.9 shifts	Run completed (12/15/69) PRL23,194(1969); PRD8,2751(1973)
P-174 (7/20/67)		T. A. Romanowski (ANL/Ohio State U., U. of Chicago, Washington U.)	Study of Lambda Muon Decay with Spark Chambers	EPB I (3)	SC, Ctrs	100 shifts	Withdrawn (10/15/70)

ZGS EXPERIMENTS

PROP. NO. (DATE)	EXPER. NO. (DATE)	SPOKESMAN (USER GROUPS)	TITLE	BEAM	DETECTOR	AMOUNT OF RUN	REMARKS (PUBLICATIONS)
P-176 (7/27/67)		D. D. Carmony (Purdue U.)	4.6 GeV/c K^+ in D_2	7°	30" DBC	250,000 pictures	Withdrawn (6/4/68)
P-178 (7/28/67)	E-178 (9/18/67)	E. W. Jenkins (U. of Arizona)	$K^\pm$ Nucleon Total Cross Sections Measurement 0.4 - 1.2 GeV/c	EPB	Ctrs	205 shifts	Withdrawn (8/12/68)
P-179 (7/31/67)	E-179 (9/18/67)	D. Sinclair (U. of Michigan)	Measurement of K_s Branching Ratio	28°	40" HLBC with CF_3Br	278,343 pictures	Run completed (5/21/68) PRL23,660(1969)
P-180 (8/11/67)	E-180 (4/16/68)	I. A. Pless (ANL, MIT, St. Louis U.)	180° Pion Charge Exchange from 1.0 5.0 GeV/c	17°	Ctrs (LH_2 Tgt)	190.9 shifts	Run completed (4/8/69) PRL22,618(1969);PRL26,1498(1971 PRD6,1882(1972)
P-181 (8/18/67)	E-181 (12/5/67)	J. I. Rhode (Iowa State U.)	4.2 GeV/c K^+ in D_2	7°	30" DBC	250,979 pictures	Run completed (10/28/68)
P-182 (8/21/67)	E-182 (11/7/67)	A. D. Krisch (U. of Michigan, ANL, NAL)	Particle Production at High Transverse Momentum	EPB	Ctrs (LH_2 Tgt)	34.4 shifts	Run completed (5/27/68) PRL21,830(1968);PRL21,1097(1968 PR178,2086(1969)
P-183 (9/14/67)	E-183 (12/5/67)	W. D. Walker/J. D. Prentice (U. of Wisconsin/U. of Toronto)	π^+ + d at 7.0 - 7.5 GeV/c	7°	30" DBC w/plates	505,429 pictures	Run completed (7/12/68) PRL25,1393(1970);PRL26,400(1971 PRD3,256(1971);PRD4,133(1971); PL35B,457(1971);NPB32,366(1971) PRL27,1025(1971);NPB33,1(1971); PRL28,318(1972);PRD5,2730(1972 PRD7,22(1973),PRD7,1977(1973) PRD10,1430(1974)
P-184 (9/14/67)	E-184 (9/14/67)	I. A. Pless (MIT)	Double Charge Exchange Investigation	17°	Ctrs	37 shifts	Run completed (9/19/67) PRL20,283(1968)

Z G S EXPERIMENTS

PROP. NO. (DATE)	EXPER. NO. (DATE)	SPOKESMAN (USER GROUPS)	TITLE	BEAM	DETECTOR	AMOUNT OF RUN	REMARKS (PUBLICATIONS)
P-185 (10/9/67)	E-185 (12/5/67)	W. A. Cooper (ANL)	$\bar{p}$ Interactions	7°	30" HBC	78,841 pictures	Run completed (12/13/68) PRL24,618(1970)
P-186 (10/27/67)		T. H. Fields (ANL)	Bubble Chamber Study K_L Decay	28°	40" FHLBC	1 million pictures	Withdrawn (2/20/68)
P-188 (11/29/67)	E-188 (12/5/67)	E. Marquit (U. of Minnesota, U. of Colorado)	π^- at 2.3 GeV/c	7°	30" DBC	63,946 pictures	Run completed (2/21/68)
P-189 (12/26/67)	E-189 (10/9/73)	H. A. Neal (Indiana U.)	Depolarization Parameter in Elastic Proton-Proton Scattering at 1.0 to 1.7 GeV/c	EPB I (1)	SC, Ctrs	32.5 shifts	Run completed (8/6/74) Additional Run 3/27-28/75 PR-D12, 3393(1975)
P-190 (1/15/68)		L. J. Koester, Jr. (U. of Illinois)	$K^+ K^- \pi^+ \pi^-$ Cross Sections on Complex Nuclei in the Energy Region from 1.0 - 2.0 GeV/c	EPB I (6)	WSC	60 shifts	Withdrawn (6/5/72)
P-191 (1/15/68)	E-191 (5/14/68)	R. C. Lamb (Iowa State U., U. of Minnesota)	Elastic Scattering of Protons and Positive Pions from Deuterons at Large Angles	17°	Ctrs	145.9 shifts	Run completed (10/22/68) PRL22,1265(1969);PRL23,600(1969) PRL23,811(1969);PRD2,1777(1970)
P-193 (1/22/68)	E-193 (6/4/68)	B. Eisenstein T. O'Halloran (U. of Illinois)	Study of K^+ Interactions in Deuterium at 3.8 GeV/c	7°	30" DBC	298,882 pictures	Run completed (12/3/68) PRD8,38(1973)
P-194 (2/21/68)	E-194 (5/27/68)	D. Cline (U. of Wisconsin)	Study of Isospin Dependence of $\bar{p}N$ Annihilation as a Function of Energy	7°	30" DBC	7,271 pictures	Run completed (2/28/68)
P-197 (2/28/68)		D. McLeod (U. of Illinois - Chicago Circle)	Study $\pi^- + p \to X^- + p$ by Proton Angle and Mass Correlation	17° EPB II	WSC Ctrs.	90 shifts	Withdrawn (1/4/69)

ZGS EXPERIMENTS

PROP. NO. (DATE)	EXPER. NO. (DATE)	SPOKESMAN (USER GROUPS)	TITLE	BEAM	DETECTOR	AMOUNT OF RUN	REMARKS (PUBLICATIONS)
P-198 (2/28/68)		A. Yokosawa (ANL)	Polarization Measurement in $\pi^- p \to \eta\, n$ Process from 1.8 to 3.8 GeV/c	17°	Gamma Detector/Neutron Counters	140 shifts	Withdrawn (6/4/68)
P-199 (3/1/68)	E-199 (8/13/68)	H. A. Neal (Indiana U.)	Differential Cross Section in Large Angle Elastic Proton-Proton Scattering from 1.5 to 5.0 GeV/c	EPB	Scint. Ctrs WSC	100.4 shifts	Run completed (4/2/69) Extension for π-p scattering has been designated as E-235 PRL23,1306(1969);PRD4,1309(1971)
P-201 (4/3/68)	E-201 (8/13/68)	L. Holloway (U. of Illinois, ANL)	Study of $\pi^- p \to \omega\, n$ at 3.65, 4.5, and 5.5 GeV/c	17°	SC, Ctrs	197.5 shifts	Run completed (11/16/69) PRL27,171(1971);PL42B,117(1972 PRD8,2814(1973)
P-202 (4/9/68)		T. Alväger (Indiana State U.)	π^+ Decay Rate in Flight from 1.0 to 7.0 GeV/c	EPB II (46)	Ctrs	10 shifts	Withdrawn (1/12/70)
P-203 (4/10/68)	E-203 (10/1/68)	N. W. Reay (Ohio State U., ANL, Michigan State U.)	A Study of Neutron Proton Scattering in the Energy Range 3.0 to 11.0 GeV/c	7°N (14)	WSC Ctrs	193.9 shifts	Run completed (3/17/70) PRL26,984(1971)
P-204 (4/18/68)	E-204 (6/4/68)	I. A. Pless (MIT)	3.9 and 5.8 GeV/c π^+ and π^- Exposures in the ANL-MURA 30" Bubble Chamber	7°	30" HBC	1,194,954 pictures	Run completed (3/8/70) NPB78,29(1974);PL51B,187(1974); PRD10,1387(1974);NPB75,93(1974) PRD11,495(1975);PL56B,297(1975) NP-B103,12(1976)
P-205 (5/1/68)	E-205 (10/1/68)	N. Kwak (U. of Kansas)	$K^- d$ Interactions at 3.3 GeV/c Beam Momentum	7°	30" DBC	113,300 pictures	Run completed (6/9/69) NC7A,605(1972);NC8A,173(1972); NC14A,363(1973)
P-206 (5/10/68)	E-206 (10/1/68)	R. M. Edelstein (Carnegie-Mellon U.)	K^0_2 Momentum Spectrum and Charge Exchange in Copper	EPB I (15)	WSC Ctrs	130.0 shifts	Run completed (2/9/70) PRD9,1242(1974)

Z G S   EXPERIMENTS

PROP. NO. (DATE)	EXPER. NO. (DATE)	SPOKESMAN (USER GROUPS)	TITLE	BEAM	DETECTOR	AMOUNT OF RUN	REMARKS (PUBLICATIONS)
P-207 (5/10/68)	E-207 (6/4/68)	T. L. Jenkins (Case Western Reserve U., Oberlin College)	High Momentum Transfer Studies of $\pi^- p \to \pi^0 n$	$17°$	SC, Ctrs	143.5 shifts	Run completed (6/24/69) PRL26,527(1971)
P-208 (5/13/68)	E-208 (6/4/68)	B. A. Munir (Ohio U.)	$\bar{p}$ - d Experiment at 3.5 GeV	$7°$	30" DBC	91,791 pictures	Run completed (8/5/68) PRL24,1251(1970);PRD2,1167(1970 NPB42,85(1972);PRD8,2004(1973)
P-209 (5/22/68)	E-209 (8/13/68)	R. K. Adair (BNL, Yale U.)	Measure the Energy Spectra and the Decay $K^+ \to \mu^+ + \nu + \pi^0$	EPB II (42)	SC, Ctrs	136.1 shifts	Run completed (9/30/70) PRL28,1472(1972)
P-210 (5/31/68)		R. Orr (U. of Illinois)	Look for Decay $K^0_s \to \mu^+ \mu^-$	$17°$	SC, Ctrs	60 shifts	Withdrawn (10/25/68)
P-211 (6/3/68)	E-211 (10/1/68)	L. Pondrom (U. of Wisconsin)	$K^+_{\ell3}$ Branching Ratio and Spectra	EPB	SC, Ctrs	83 shifts	Run completed (5/27/69) PRD6,1254(1972)
P-212 (6/4/68)	E-212 (6/4/68)	W. A. Cooper (ANL)	Study of Mesons Produced in $\bar{p}p$ Annihilations	$7°$	30" HBC	562,805 pictures	First run completed (12/10/68) Extension granted and completed 5/27/69. PRA2,1834(1970);PRL 27,1681(1971);PRL27,1749(1971); NPB38,62(1972);PL40B,503(1972); NPB53,269(1973);NPB56,1(1973); NPB71,189(1974);NP-B130,269(1977) PRD17,42(1978)
P-213 (6/17/68)	E-213 (10/1/68)	G. B. Yodh (U. of Maryland, ANL)	K^- Interactions in Neon-Hydrogen Mixture	$7°$	30"HNeBC	61,320 pictures	Run completed (7/27/69)
P-214 (6/26/68)	E-214 (10/1/68)	A. Subramanian (Tata Institute)	$\pi^\pm$ Interactions in Neon-Hydrogen Mixture at 3.5 GeV/c	$7°$	30"H_2NeBC	87,699 pictures	Run completed (7/21/69) NPB67,333(1973)

ZGS EXPERIMENTS

PROP. NO. (DATE)	EXPER. NO. (DATE)	SPOKESMAN (USER GROUPS)	TITLE	BEAM	DETECTOR	AMOUNT OF RUN	REMARKS (PUBLICATIONS)
P-215 (7/22/68)	E-215 (8/13/68)	W. J. Willis (Yale U., ANL)	Time Reversal Invariance in $K^0_{\mu3}$ Decay	EPB II (43)	Ctrs	319.6 shifts 172.0 tapes	Run completed (3/9/70) PRL30,1002(1973)
P-216 (7/29/68)	E-216 (11/12/68)	A. D. Krisch (U. of Michigan, ANL, NAL)	Study of Breaks in Inelastic Proton - Proton Scattering	EPB II	Ctrs	135.6 shifts	Run completed (6/20/69) PRL23,1469(1969);PRD2,1808(1970) PRD3,645(1971);NPB67,333(1973)
P-217 (8/6/68)	E-217 (8/13/68)	V. L. Telegdi (U. of Chicago, U. of Ill., Stanford U.)	A Precision Measurement of the K_L - K_S Mass Difference	EPB I (2)	SC	230.4 shifts	Run completed (8/26/69) Extension completed (1/22/70) PRL25,1057(1970)
P-218 (8/16/68)	E-218 (10/1/68)	H. J. Martin (Indiana U.)	$\pi^+ p$ at 6.5 GeV/c	7°	30" HBC	317,531 pictures	Run completed (2/23/70)
P-219 (9/9/68)	E-219 (11/12/68)	K. Ruddick (U. of Minnesota)	Investigate the Radiative Decay Modes of the Vector Mesons	EPB II (8)	SC, Ctrs	184.0 shifts	Run completed (4/6/70) PRL27,885(1971)
P-220 (9/16/68)	E-220 (10/1/68)	L. Marshall Libby (U. of Colorado)	$\bar{p}$ at 4.6 GeV/c	7°	30" DBC	100,303 pictures	Run completed (6/14/69)
P-221 (9/30/68)	E-221 (11/12/68)	R. A. Schluter (ANL, McGill U., U. of Pennsylvania)	Investigation of Meson Resonances of Isospin I = 0,3 and 1 in the Mass Region M_X = 1600 to 3700 MeV produced in $p + d \rightarrow He + X^0$	EPB II	Ctrs	173.4 shifts	Run completed (9/30/69) NPB60,396(1973)
P-222 (10/1/68)	E-222 (1/28/69)	J. A. Poirier (U. of Notre Dame)	S-Wave Pi-Pi Phase Shifts at Low Dipion Mass	EPB II (8)	SC, Ctrs	151.3 shifts	Run completed (9/30/70) PR-D12,681(1975)
P-223 (10/1/68)	E-223 (11/12/68)	M. Derrick (ANL, U. of Michigan NAL)	Measurement of Pion Production Cross Sections in Proton-Be Collisions	EPB	Ctrs	11.5 shifts	Run completed (6/16/69) PRD3,1089(1971);PRD4,1967(1971)

ZGS EXPERIMENTS

PROP. NO. (DATE)	EXPER. NO. (DATE)	SPOKESMAN (USER GROUPS)	TITLE	BEAM	DETECTOR	AMOUNT OF RUN	REMARKS (PUBLICATIONS)
P-224 (10/17/68)		A. Roberts (NAL, U. of Chicago, Northwestern U., U. of Notre Dame)	Search for X rays from Antiprotonic Atoms	EPB	SC, Ctrs	21 shifts	Withdrawn (11/12/68)
P-225 (11/4/68)	E-225 (3/18/69)	D. C. Miller (MIT)	Antiproton-Proton Interactions in the 4.0 - 5.0 GeV/c Momentum Region	7°	30" HBC	439,548 pictures	Run completed (8/24/69)
P-226 (11/11/68)	E-226 (3/18/69)	D. Rust (ANL, U. of Michigan)	Study K^+p Elastic Scattering and $\pi^+ p \to K^+ \Sigma^+$ at Large Angles	EPB I (1)	SC, Ctrs	168.6 shifts	Run completed (1/24/70) PRL 24,1353(1970);PRL 24, 1361 (1970);PRL 26,1278(1971)
P-227 (12/9/68)		M. Nussbaum (U. of Pennsylvania)	Search for Strangeness one Boson Resonances in the Mass Region 1.0 to 2.0 GeV	17° or EPB	SC, Ctrs	100 shifts	Withdrawn (7/22/69)
P-228 (12/30/68)	E-228 (1/28/69)	R. Hildebrand (U. of Chicago, LRL)	Search for $K^+ \to \pi^+ + \nu + \bar{\nu}$	EPB	SC, Ctrs	80 shifts	Withdrawn (2/19/69)
P-229 (1/13/69)	E-229 (1/28/69)	G. A. Smith (Michigan State U.)	$\bar{p}p$ and pn Interactions from 2.0 to 3.0 GeV/c	7°	30" HDBC	254,848 pictures	Run completed (5/13/69) NPB32,29(1971);NPB40,151(1972); NPB51,29(1973);NPB51,57(1973); NPB51, 77(1973);PL47B,177(1973); NPB63, ¼(1973);NP-B103,537(1976);
P-230 (1/24/69)		J. M. Daniels (U. of Toronto, ANL)	Polarization Measurements in p - n Elastic Scatt., in Range 8 to 12 GeV/c Using a Polarized Neutron (He³) Target	EPB I (1)	SC, Ctrs	90 shifts	Withdrawn (2/10/70)
P-231 (1/24/69)		R. C. Lamb (Iowa State U., ANL, Northwestern U.)	Study of Wide Angle Elastic Scattering of Positive and Negative Pions from Protons	EPB I (1)	Ctrs (LH_2 Tgt)	215 shifts	Withdrawn (9/15/69)
P-232 (1/28/69)	E-232 (6/17/69)	C. A. Mistretta (U. of Wisconsin)	Triggering the 30" Chamber on Neutral Particles, 7.0 GeV/c $\pi^- p$	7°	30" HBC	49,348 pictures	Run completed (4/21/70)

346

ZGS EXPERIMENTS

PROP. NO. (DATE)	EXPER. NO. (DATE)	SPOKESMAN (USER GROUPS)	TITLE	BEAM	DETECTOR	AMOUNT OF RUN	REMARKS (PUBLICATIONS)
P-233 (3/3/69)	E-233 (5/27/69)	T. S. Yoon (U. of Toronto)	7.0 GeV/c $\pi^- p$	7°	30" HBC	204,274 pictures	Run Completed (4/21/70) PRL25,1393(1970);NPB24,253(1970) NPB37,621(1972)
P-234 (3/26/69)	E-234H (6/17/69)	M. Derrick (ANL)	Neutrino Experiment in Hydrogen	EPB II (9)	12' HBC	423,877 pictures	Hydrogen Phase Completed(9/30/71) Deuterium Phase Complete(12/14/74) PRL30,335(1973);PRL30,339(1973); PRL31,844(1973);NPB58,333(1973); PRL33,448(1974);PRL33,1446(1974); PRL36,179(1976);PL65B,174(1976); PL66B,291(1977);PRD15,1(1977);PRL 38,1049(1977);PRD16,3103(1977);
	E-234D (6/17/69)	M. Derrick (ANL, Purdue U.)	Neutrino Experiment in Deuterium		12' DBC	1,007,559 pictures	PRD19,2521(1979)
P-235 (4/24/69)	E-235 (3/18/69)	H. A. Neal (Indiana U.)	Differential Cross Section in Large Angle Elastic π^--Proton Scattering 3.0 - 5.0 GeV/c	EPB I	Scint. Ctrs WSC	100.1 shifts	Run Completed (8/24/69) PRL25,553(1970)
P-236 (4/11/69)		C. A. Mistretta (U. of Wisconsin)	Triggering the 30" Chamber on Neutral Particles - 7.0 GeV/c π^- d	7°	30" DBC SC, Ctrs	1x10⁶ BC expansions	Withdrawn (7/14/70)
P-237 (4/14/69)	E-237 (10/14/69)	B. Huld (U. of Illinois)	Study Possible Structure in the f^o (1260) and A_2^o (1300) Mesons Produced in $\pi^- p$ Interactions at 4.5 GeV/c	17°	SC, Ctrs	204 shifts	Run Completed (5/24/71) PRD9,1161(1974)
P-238 (5/8/69)	E-238 (7/22/69)	R. C. Lamb (Iowa State U., ANL)	Survey of the Missing Mass Spectra from $\pi^- p \to X^0 n$ and $\pi^+ p \to X^{++} n$	EPB II (8)	Ctrs	138 shifts	Run Completed (9/21/71) PRL28,520(1972);PRL29,671(1972) PRL29,1477(1972)
P-239 (5/15/69)		H. O. Cohn (Oak Ridge Nat'l Lab., ANL, Computing Tech. Center, U. of Tennessee)	Interaction of 7.0 GeV/c π^+	7°	20" HeBC	400,000 pictures	Withdrawn (5/13/71)
P-240 (5/16/69)	E-240 (7/22/69)	T. A. Romanowski (ANL, Ohio State U., U. of Chicago)	Study of the Associated Production Reaction $\pi^- p \to \Lambda K$ Near Σ Threshold	EPB I (3)	SC, Ctrs	62.8 shifts	Run Completed (11/14/69) PRL31,901(1973); PRD11,1(1975)
P-241 (5/19/69)		J. W. Chapman (U. of Michigan)	Further Study of S-Channel Resonance Effects in $\bar{p}p$ Interactions	7°	30" HBC	300,000 pictures	Withdrawn (10/9/70)

Z G S EXPERIMENTS

PROP. NO. (DATE)	EXPER. NO. (DATE)	SPOKESMAN (USER GROUPS)	TITLE	BEAM	DETECTOR	AMOUNT OF RUN	REMARKS (PUBLICATIONS)
P-242 (5/19/69)	E-242 (6/14/69)	J. C. VanderVelde (U. of Michigan)	Study of Energy Dependence of $\bar{p}n$ Channels Near 1.8 GeV/c	7°	30" DBC	156,092 pictures	Run completed (6/18/69)
P-243 (5/20/69)	E-243 (7/22/69)	C. Akerlof (U. of Michigan, NAL)	Differential Cross Sections for $\pi p \to \Sigma K$	EPB I (1)	Ctrs, WSC	119.8 shifts	Run completed (9/30/70) PRL27,74(1971);PRL27,539(1971)
P-244 (5/21/69)	E-244 (6/17/69)	Y. Cho (ANL, Carnegie-Mellon U.)	K_L Experiment	EPB II (46)	12' HBC	317,999 pictures	Run completed (9/17/73) Pl-60B,293(1976);PL63B,231(1976) PRD15,587(1977);PRD18,623(1978); PRD18,3061(1978)
P-245 (5/22/69)		B. G. Reynolds (Ohio U.)	$K^+ p$ Interactions at 4.0 GeV/c	7°	30" HBC	300,000 pictures	Withdrawn (6/17/69)
P-246 (5/23/69)	E-246 (6/17/69)	B. P. Roe (U. of Michigan)	$K^+_{\mu 3}$ Decay Study	28°	40" HLBC	247,599 pictures	Run completed (12/11/69) PRD9,1221(1974)
P-247 (5/23/69)	E-247 (6/17/69)	W. Fry (U. of Wisconsin)	Stopping K^+ in CF_3Br Extension of E-109	28°	40" HLBC	589,346 pictures	Run completed (8/10/70) PRL28,523(1972);PRL28,1287(1972 PRD8,1307(1973)
P-248 (5/26/69)		W. F. Fry (U. of Wisconsin, U. of Padova)	Test of the $\Delta S = \Delta Q$ Rule Extension of E-171	28°	40" HLBC	500,000 pictures	Withdrawn (6/17/69)
P-249 (6/6/69)		W. F. Baker (NAL)	Development of a Tagged Neutron Beam and Neutron Total Cross Sections	EPB II	Ctrs	≃100 shifts	Withdrawn (7/22/69)
P-252 (7/3/69)		I. A. Pless (MIT, ANL, U. of Bari, St. Louis U.)	Momentum Dependence of the 180° Pion Charge Exchange from 5.0 to 8.0 GeV/c	EPB IA (22)	Ctrs	195 shifts	Withdrawn (2/4/70)

ZGS EXPERIMENTS

PROP. NO. (DATE)	EXPER. NO. (DATE)	SPOKESMAN (USER GROUPS)	TITLE	BEAM	DETECTOR	AMOUNT OF RUN	REMARKS (PUBLICATIONS)
P-253 (7/17/69)	E-253 (7/22/69)	A. Yokosawa (ANL, U. of Maryland, Northwestern U.)	Polarization Measurements in Kp and πp Scattering up to 2.2 GeV/c	EPB I (6)	Ctrs	198.7 shifts	Run completed (4/7/70) PRL24,615(1970);PRL26,338(1971); PL34B,655(1971);NPB37,401(1972)
P-254 (7/15/69)	E-254 (7/22/69)	D. K. Robinson (Case Western Reserve U. Carnegie Mellon)	π^+ d Interactions at 6.0 GeV/c	7°	30" DBC	501,872 pictures	Run completed (12/27/70) PL45B,165(1973);PRL31,562(1973); PL45B,521(1973);PRD10, 2070(1974 PR-D12,1272(1975);NPB119,1(1977)
P-255 (7/17/69)	E-255 (10/14/69)	A. M. Shapiro (Brown U.)	π^+ p Bubble Chamber Exposure at 4.1 GeV/c	7°	30" HBC	414,406 pictures	Run completed (10/29/70) PRD8, 3765(1973);PRD9,3015(1974)
P-256 (7/17/69)	E-256 (7/22/69)	A. Wattenberg (U. of Illinois)	Tests of CP Violation and CPT Invariance in the 2π Decay of K° and $\bar{K}^\circ$	EPB II (42)	WSC	153.9 shifts	Run completed (9/8/71) PRD7,1989(1973)
P-257 (8/25/69)		R. R. Crittenden (Indiana U.)	Differential Cross Section for π^-p Scattering at Backward Angles from 3.0 to 6.0 GeV/c	EPB I (1)	WSC, Ctrs	95 shifts	Withdrawn (6/5/72)
P-258 (10/1/69)		J. R. O'Fallon (St. Louis U., Iowa State U.)	π^-d Elastic Scattering Near 180°	EPB I	Ctrs	100 shifts	Withdrawn (10/14/69)
P-259 (10/3/69)	E-259 (10/30/69)	G. Conforto/R. Prepost (U. of Chicago, U. of Wisc., NAL, U. of Toronto	$\eta \pi$ Mass Spectrum from Threshold to 1500 MeV	EPB I (1)	Optical SC Ctrs	200.1 shifts	Run completed (2/28/72) PRL30,503(1973);PL45B,154(1973); PRDII, 2345(1975)
P-260 (10/7/69)		R. A. Carrigan, Jr. (NAL)	Phi Production in Pion-Nucleon Interactions Near Threshold	EPB II	WSC Ctrs	100 shifts	Withdrawn (10/14/69)
P-261 (10/8/69)	E-261 (10/14/69)	T. L. Jenkins (Case Western Reserve U.)	Study of the Reaction π^- p → π° n at 5.9 GeV/c	17°	SC, Ctrs	225.5 shifts	Run completed (4/7/70) PL51B,390(1974)

ZGS EXPERIMENTS

PROP. NO. (DATE)	EXPER. NO. (DATE)	SPOKESMAN (USER GROUPS)	TITLE	BEAM	DETECTOR	AMOUNT OF RUN	REMARKS (PUBLICATIONS)
P-262 (10/31/69)	E-262 (10/14/69)	W. F. Baker (NAL, Notre Dame)	Energy Dependence of the Backward Peak in π^+ p Elastic Scattering, 2.4 - 6.0 GeV/c	EPB II (8)	Ctrs	189.9 shifts	Run completed (4/26/72) PRL32,251(1974);PRL32,908(1974) PRD11,1777(1975);PRD15,59(1977);
P-263 (10/31/69)	E-263 (2/10/70)	R. B. Sutton (Carnegie-Mellon U., Virginia Polytechnic Institute, College of William and Mary, ANL)	Kaonic (and Antiprotonic) X rays	EPB II (42)	Ctrs	157.4 shifts	Run completed (4/26/72) PRL29,230(1972);PRL29,1132 (1972); PR-C11,1056(1975)
P-264 (12/1/69)		L. Libby (U. of Colorado)	Λ - p Interaction at High Energy, K^- at 5.0 GeV/c	7°	30" HBC	200,000 pictures	Withdrawn (2/10/70)
P-265 (12/8/69)		R. J. Plano (Rutgers U., Stevens Inst. of Technology)	$\bar{p}p$ Interactions in the T (2190) Meson Region	7°	30" HBC	500,000 pictures	Withdrawn (7/14/70)
P-266 (12/26/69)	E-266 (3/12/70)	R. M. Heinz (Indiana U.)	Spin and Parity of the A_2 Meson, 4 GeV/c π^-	EPB I (1)	WSC, Ctrs (LH_2 Tgt)	174.5 shifts	Run complete (11/10/74) PRL29,1688(1972);PRD8,2785(1973)
P-267 (1/12/70)	E-267 (1/12/70)	G. Charlton (ANL, U. of Notre Dame)	pp Interactions, 12 GeV/c	EPB II (9)	12' HBC	98,500 pictures	Run complete (10/6/71) PRD8,3824(1973);PL48B,479(1974) PRD1,1756(1975)
P-268 (2/10/70)	E-268 (1/23/70)	R. E. Diebold (ANL)	Study Meson Production with a Wire Spark Chamber Spectrometer	EPB IA (21)	WSC, Ctrs	155.4 shifts	Run complete (1/14/72) PRL32,904(1974);PRL33,505(1974); PR-D17,1197(1978)
P-269 (1/26/70		D. O. Huwe (Ohio U., U. of Calif., Florida State U., U. of Wisc.)	Energy-Dependent Phenomena in $\bar{p}$ - n Scattering from 3.05 to 3.35 GeV/c	7°	30" DBC	150,000 pictures	Withdrawn (2/10/70)

Z G S EXPERIMENTS

PROP. NO. (DATE)	EXPER. NO. (DATE)	SPOKESMAN (USER GROUPS)	TITLE	BEAM	DETECTOR	AMOUNT OF RUN	REMARKS (PUBLICATIONS)
P-270 (2/3/70)	E-270 (2/10/70)	R. M. Edelstein (Carnegie-Mellon U., ANL, Iowa State U.)	Measurement of Differential Cross Sections for $K_L^0 + p \rightarrow K^+ + n$ in the Momentum Interval 500 - 1500 MeV/c	EPB I (15)	WSC	213.2 shifts	Run completed (6/30/71) PR-D14, 702 (1976)
P-271 (2/4/70)	E-271 (2/10/70)	C. E. W. Ward (ANL)	Study Reactions $\pi^- p \rightarrow K^0 \Lambda^0$ and $\pi^- p \rightarrow K^0 \Sigma^0$ in the range 3.0 to 6.0 GeV/c	EPB IA (21)	WSC	88.6 shifts	Run completed (12/6/71) PRL31,1149(1973);PL-48B,471(1974) NPB77,269(1974);PRD11,1802(1975); PR-D16,2041 (1977)
P-272 (2/6/70)	E-272 (7/14/70)	A. Yokosawa (ANL, Northwestern)	Polarization Measurements in Backward πp Elastic Scattering and in Forward πp Charge-Exchange Scattering from 2.75 to 5.15 GeV/c	EPB I (22B)	Ctrs	150 shifts	Run Completed (1/16/75) PRL-35,1738(1975);PRL-37,83(1976) NP-B113,279 (1976)
P-273 (2/11/70)	E-273 (5/5/70)	B. Musgrave (ANL, U. of Chicago)	$K^- p$ Interactions at 5.0 GeV/c K^- Momentum	7°	30" HBC	250,000 pictures	Withdrawn (10/13/70)
P-274 (2/12/70)	E-274 (2/13/70)	B. A. Munir (Ohio U.)	$\bar{p}$ - d Interactions at 3.5 GeV/c	7°	30" DBC	110,000 pictures	Withdrawn (9/14/71)
P-275 (3/2/70)	E-275 (5/5/70)	V. L. Telegdi (U. of Chicago, U. of Ill. Northeastern Ill. State College, Stanford U., Swiss Government Fellow)	Study of $K^0 \rightarrow \pi^\mp \ell^\pm \nu$	EPB I (2)	WSC	111.1 shifts	Run completed (10/21/71)
P-276 (3/18/70)		R. Ott (McGill U., Carleton U.)	A Measurement of the Spin - Parity of the A_2 and R Mesons	EPB I	WSC (LH$_2$ Tgt)	90 shifts	Withdrawn (1/29/71)
P-277 (3/27/70)	E-277 (6/26/70)	P. H. Steinberg (U. of Maryland, ANL)	Differential Cross Sections in Elastic $K^+ p$ Scattering from 1.0 to 1.5 GeV/c	EPB I (6)	WSC (LH$_2$ Tgt)	337.6 shifts	Run completed (1/1/72) PRD11,495(1975);PR-D12,1(1975)

ZGS EXPERIMENTS

PROP. NO. (DATE)	EXPER. NO. (DATE)	SPOKESMAN (USER GROUPS)	TITLE	BEAM	DETECTOR	AMOUNT OF RUN	REMARKS (PUBLICATIONS)
P-278 (3/31/70)	E-278 (5/5/70)	T. B. Novey (ANL, NAL, Northwestern U.)	Measurements of Polarization in $\pi^- p$ Backward Elastic Scattering Between 1.80 and 2.30 GeV/c	EPB I (6)	Ctrs	201.7 shifts	Run completed (1/9/71) PRL27, 1241(1971)
P-279 (4/13/70)	E-279 (5/5/70)	D. I. Meyer (U. of Michigan, ANL)	Study the Dip in Differential Cross Sections at $-t$ = 2.85 $(GeV/c)^2$	EPB I (1)	SC	84.3 shifts	Run completed (1/9/71) PRL27, 219(1971)
P-280 (4/13/70)	E-280 (5/5/70)	H. A. Neal (Indiana U., Los Alamos Scientific Lab.)	Polarization in Large Angle p - p Elastic Scattering from 3.0 to 12.5 GeV/c	EPB I (1)	WSC	186.4 shifts	Run completed (9/3/73) PRL32,1261(1974);PRD9,555(1974)
P-281 (4/30/70)	E-281 (5/5/70)	B. Y. Oh (Michigan State U.)	Direct Channel Investigations of $\bar{p} p$ Interactions from 2 - 4 GeV/c	$7°$	30" HBC	252,370 pictures	Run completed (11/14/70) NPB51,57(1973);NPB51,77(1973); PL47B,177(1973);NPB63,1(1973)
P-282 (5/18/70)	E-282 (10/20/70)	T. L. Jenkins (Case Western Reserve U., Northwestern U.)	Study the Backward Peak of the Reaction $\pi^- p \rightarrow \Lambda^0 K^0$	$17°$, 1, or 8	Ctrs, SC (LH_2 Tgt)	150 shifts	Withdrawn (3/21/72)
P-283 (6/4/70)	E-283 (12/8/70)	J. Sacton (U. of Belgrade, U. of Brussels, University College-Dublin, University College - London, U. of Warszaw)	$K^+ \rightarrow \pi^+ + \pi^0 + \pi^0$ Decay Using Nuclear Emulsion	$28°$	Emulsion Stack	50 shifts	Withdrawn (3/18/71)
P-284 (6/18/70)		W. J. Kernan (Iowa State U.)	Investigation of p p Interactions at $\sim$ 10.8 GeV/c (Revised Proposal - 4.25 GeV/c)	$7°$	30" HBC	500,000 pictures	Withdrawn (10/20/70)
P-285 (7/7/70)	E-285 (7/14/70)	M. A. Abolins (Michigan State U., Carleton U., Ohio State)	n-p Elastic Scattering Near 180° Using A Polarized Target	7° N (14)	WSC Ctrs	270 shifts 187,100 events	Run completed (12/28/71) PRL30,1183(1973)

ZGS EXPERIMENTS

PROP. NO. (DATE)	EXPER. NO. (DATE)	SPOKESMAN (USER GROUPS)	TITLE	BEAM	DETECTOR	AMOUNT OF RUN	REMARKS (PUBLICATIONS)
P-286 (7/10/70)	E-286 (7/14/70)	C. A. Mistretta (U. of Wisconsin)	π^- + n Interactions at 5 GeV/c in 30-Inch Bubble Chamber with Neutral Tagging	7°	30" HBC	617,714 expansions	Run completed (12/15/70)
P-287 (7/24/70)		J. Chapman (U. of Michigan)	A Search for Exotic States Produced with Fast Forward Neutrons and Kaons	7°	30" HBC	30,000 pictures	Withdrawn (12/8/70)
P-288 (8/18/70)		B. G. Reynolds (Ohio University)	Energy-Dependent Phenomena in $\bar{p}$ - n Scattering from 3 to 4 GeV/c	7°	30" DBC	200,000 pictures	Withdrawn (12/8/70)
P-289 (9/16/70)	E-289 (10/20/70)	G. Smith (Michigan State U., ANL)	Study Ξ States, K^- at 6.5 GeV/c	EPB II (10)	12' HBC	1.114 million pictures	Joint experiment for E-289 and E-292 Run Completed (3/29/76)
P-290 (9/23/70)	E-290 (9/25/70)	J. A. Poirier (Notre Dame, ANL)	I=2 $\pi\pi$ Phase Shifts at Low Dipion Mass	EPB II (8)	SC, Ctrs	37.3 shifts	Run completed (10/30/70) PRD10,2055(1975)
P-291 (10/5/70)		H. J. Martin (Indiana U.)	π^+ p Interactions at 6.5 GeV/c	7°	30" HBC	400,000 pictures	Withdrawn (12/8/70)
P-292 (10/12/70)	E-292 (10/20/70)	B. Musgrave (ANL, Michigan State U.)	Study K^-p Interactions	RF	12' HBC	1.114 million pictures	Joint experiment for E-289 and E-292;PRD19,3197(1979) Run Completed (3/30/76)
P-293 (10/12/70)	E-293 (10/20/70)	B. Musgrave (ANL)	Study K^+d Interactions	7°	30" DBC	250,000 pictures	Withdrawn (4/13/71)
P-294 (10/9/70)	E-294 (12/8/70)	J. Whitmore (ANL, C.N.R.S./ Strasbourg, Iowa State U., Ohio U.)	$\bar{p}$d Exposures	7°	30" DBC	600,000 pictures	Withdrawn (4/13/71)

352

ZGS EXPERIMENTS

PROP. NO. (DATE)	EXPER. NO. (DATE)	SPOKESMAN (USER GROUPS)	TITLE	BEAM	DETECTOR	AMOUNT OF RUN	REMARKS (PUBLICATIONS)
P-295 (10/8/70)	E-295 (10/20/70)	A. Engler (Carnegie-Mellon U.)	π^+d Interactions at 6.0 GeV/c	7°	30" DBC	419,514 pictures	Run completed (1/9/71) PL45B,165(1973);PRL31,562(1973); PL45B,521(1973);PRL32,260(1974) NPB88,202(1975);PL-63B,461(1976)
P-296 (10/15/70)		T. A. Romanowski (ANL/Ohio State U., U. of Chicago, Washington U.)	Study of Lambda Muon Decay with Spark Chambers	EPB I (19)	SC, Ctrs	215 shifts	Withdrawn (9/19/72)
P-297 (10/19/70)	E-297 (4/13/71)	T. O'Halloran (U. of Illinois, NAL, ANL)	Backward Meson Production in π^-p Interactions Between 5 and 8 GeV/c	EPB I Annex (22)	Streamer Chamber Ctrs	98.1 shifts 355,164 pictures	Run completed (4/11/73) PRL34,691(1975);PR-D13,5(1976); PRD18,1370(1978)
P-298 (10/20/70)		W. D. Shephard (U. of Notre Dame)	Backward Resonance Production via Baryon Exchange in 4 GeV/c π^-p Interactions	7°	30" HBC	1 million expansions	Withdrawn (10/20/70)
P-299 (10/20/70)		J. D. Prentice (U. of Toronto, State U. of New York at Buffalo)	π^+p Interactions at 5.45 GeV/c	7°	30" HBC	600,000 pictures	Withdrawn (12/8/70)
P-300 (10/30/70)	E-300 (11/3/70)	J. A. Poirier/G. A. Smith (U. of Notre Dame, ANL, Michigan State U.)	$p + p \rightarrow p + \pi^+ + n$, 6.0 GeV/c	EPB II (8)	SC, Ctrs	28.3 shifts	Run completed (12/14/70) PR-D12,1211(1975)
P-301 (11/11/70)		C. Akerlof (U. of Michigan)	Total Cross Section for Λ^0 Collisions	EPB I (Annex or Beam 2)	SC, Ctrs LH$_2$ Tgt	50 shifts	Withdrawn (6/5/72)
P-302 (11/11/70)		C. Akerlof (U. of Michigan)	Measuring Diff. Cross Section for the Elastic Scattering Reaction $\Lambda^0 + p \rightarrow \Lambda^0 + p$	EPB I (Annex or Beam 2)	SC, Ctrs LH$_2$ Tgt	≈80 shifts	Withdrawn (6/5/72)
P-303 (12/3/70)	E-303 (11/9/71)	D. Cline (U. of Wisconsin)	Study Very Low Energy Antineutron Proton Scattering, np Interactions	EPB II (10)	12' HBC	281,200 pictures	Run Completed (6/25/75)

ZGS EXPERIMENTS

PROP. NO. (DATE)	EXPER. NO. (DATE)	SPOKESMAN (USER GROUPS)	TITLE	BEAM	DETECTOR	AMOUNT OF RUN	REMARKS (PUBLICATIONS)
P-304 (12/7/70)	E-304 (12/8/70)	J. D. Prentice (U. of Toronto, U. of Indiana, U. of Michigan, State Univ. of New York)	6.0 GeV/c π^+p Exposure	7°	30" HBC	1,000,000 pictures	Withdrawn (4/13/71)
P-305 (12/29/70)	E-305 (2/9/71)	K. Ruddick (U. of Minnesota)	A Study of the Process π^- p → ϕ n	EPB (8)	Ctrs WSC	251.1 shifts	Run Completed (9/17/73) PRD16,1 (1977)
P-306 (2/25/71)	E-306 (3/22/71)	P. Koehler (ANL, NAL, Northwestern Univ.)	Measurement of the Polarization Parameter in π - p Charge-Exchange Scattering at 4.5, 3.7, and 3.2 GeV/c	17°	Ctrs	264.3 shifts	Run completed (3/18/72) PRD30,239(1973);PRD30,242(1973)
P-307 (3/24/71)	E-307 (1/19/72)	H. Neal (Indiana U.)	Recoil Proton Polarization in the Reaction $\pi^{\pm}$ p → $A_2^{\pm}$ p	EPB I (1)	WSC, Ctrs (LH$_2$ Tgt)	112.7 shifts	Run completed (4/26/72) PRD9,603(1974)
P-308 (3/29/71)	E-308 (4/13/71)	T. Kitagaki (Tohoku U.)	Proton-Proton Interactions at 7 and 11 GeV/c	EPB II (9)	12' HBC	36,700 pictures	Run completed (2/10/72)
P-309 (8/24/71)	E-309 (4/26/73)	G. Wolsky (Tufts U.)	Study K$^+$ Interactions in the 12-Foot Bubble Chamber	EPB II (10)	12' HBC	410,000 pictures	Withdrawn (11/30/73)
P-310 (8/30/71)	E-310 (9/14/71)	M. F. Gormley (U. of Illinois)	Measurement of Λ-Nucleus Cross Sections	EPB II (42)	WSC, Ctrs	17.4 shifts	Run completed (9/18/71)
P-311 (9/7/71)	E-311 (9/14/71)	V. L. Telegdi (U. of Chicago, U. of Ill. Chicago Circle)	A Search for and Possible Measurement of, Decay K_S → $\mu^+ \mu^-$	EPB I (2)	WSC, Ctrs	291.6 shifts	Run completed (9/28/73) NC-32A,235(1976)

ZGS EXPERIMENTS

PROP. NO. (DATE)	EXPER. NO. (DATE)	SPOKESMAN (USER GROUPS)	TITLE	BEAM	DETECTOR	AMOUNT OF RUN	REMARKS (PUBLICATIONS)
P-312 (9/9/71)	E-312 (9/14/71)	D. R. Rust (ANL)	Measure Elastic Scattering at Small Angles	EPB I A (21)	WSC, Ctrs	22.6 shifts	Run completed (10/13/71) PRL 29, 1415(1972); PR-D9, 1179(1974
P-313 (9/13/71)		N. Gelfand (U. of Chicago, Northwestern U.)	A Study of K Pair Production by a 5.3 BeV/c π^+ Beam in the 12-Foot Bubble Chamber	EPB II (10)	12' HBC	350,000 pictures	Withdrawn (11/9/71)
P-314 (9/30/71)	E-314 (10/1/71)	L. J. Koester (U. of Illinois)	Excitation of the 4.4 MeV/c State in C^{12} by Pion Scattering and Related Experiments	17°	WSC, Ctrs	48.8 shifts	Run completed (12/15/71) PRD15, 47(1977)
P-315 (10/11/71)	E-315 (11/9/71)	C. Akerlof (U. of Michigan)	Study the Double Charge Exchange Reactions, $\pi^- p \rightarrow K^+ \Sigma^-$, $K^- p \rightarrow \pi^+ \Sigma^-$, $K^- p \rightarrow K^+ \Xi^-$	17°	WSC, Ctrs	193.3 shifts	Run completed (3/19/73) PRL33, 119(1974)
P-316 (10/7/71)	E-316 (10/7/71)	G. Burleson (Northwestern, ANL)	Preliminary Investigation of the Inclusive Reaction $\pi^- + p \rightarrow \pi^0 +$ anything	17°	Ctrs	56.4 shifts	Run completed (12/27/71) PR-D12, 2557 (1975)
P-317 (10/25/71)	E-317 (11/9/71)	M. J. Longo (U. of Michigan)	Large Angle n-p Elastic Scattering from 5 to 12 GeV	7° N (14)	WSC, Ctrs	232.2 shifts	Run completed (9/3/73) PRL38, 1315(1977); PRL38, 1317(1977) NPB143, 1(1978)
P-318 (12/6/71)	E-318 (1/19/72)	P. H. Steinberg (U. of Maryland, ANL)	Differential Cross Sections in Elastic $\pi^+ p$ Scattering from 1.0 to 2.3 GeV/c	EPB I (6)	WSC, Ctrs	85.5 shifts	Run completed (3/20/72) PRD10, 3556(1974); PR-D12, 6(1975)
P-319 (12/20/71)	E-319 (12/20/72)	E. P. Steinberg (ANL)	Investigate the Mechanism of "Fragmentation"	EPB I (12)	Scatter Chamber, Ctrs	45.8 shifts	Run completed (11/29/74)
P-320 (11/23/71)	E-320 (1/26/72)	E. P. Steinberg (ANL)	Determine the Absolute Cross Section for the Reaction C^{12} (p, pn) C^{11} at the ZGS	EPB II (8)	Ctrs	2 shifts	Run completed (3/25/72) PRC6, 1153(1972)

ZGS EXPERIMENTS

PROP. NO. (DATE)	EXPER. NO. (DATE)	SPOKESMAN (USER GROUPS)	TITLE	BEAM	DETECTOR	AMOUNT OF RUN	REMARKS (PUBLICATIONS)
P-321 (1/10/72)	E-321 (1/19/72)	L. Holloway (U. of Illinois)	Study the Reaction $\pi^- p \to \rho^0 n$ at High Momentum Transfer	17°	WSC, Ctrs (LH_2 Tgt.)	75.4 shifts	Run completed (4/25/72)
P-322 (1/11/72)	E-322 (1/19/72)	T. M. Knasel (U. of Chicago, ANL, Ohio State U.)	Study of the Λ° Production in 6 GeV/c Proton-Proton Interactions	EPB IA (21)	EMS (LH_2 Tgt.)	34.8 shifts	Run Completed (2/22/72)
P-323 (2/15/72)	E-323 (9/6/72)	U. E. Kruse (U. of Illinois)	A Measurement of the Three Pion Mass Spectrum in the Reaction $\pi^- C^{12} \to \pi^+ \pi^- \pi^- C^{12*}$ (4.45 MeV)	EPB I (21)	EMS	73.1 shifts	Run Completed (1/30/73) PRL31,795(1973)
P-324 (2/16/72)	E-324 (5/9/72)	A. D. Krisch (U. of Michigan, ANL, St. Louis U.)	Total p-p Cross Sections Using a Polarized Target and a Polarized Beam	EPB I (1)	Ctrs (PPT) (LH_2 Tgt.)	56.1 shifts	PRL31,783(1973) Run completed (10/29/74) PRL34, 558(1975)
P-325 (3/3/72)	E-325 (3/6/72)	D. S. Ayres (ANL)	Study $f^\circ - A_2^\circ$ Interference in $\pi^- p \to K^+ K^- n$	EPB I (21)	EMS	30.5 shifts	Run Completed (4/16/72) PRL31,665(1973);PRL32, 1463(1974) PR-D17,631 (1975)
P-326 (3/3/72)		B. A. Barnett (U. of Maryland, ANL)	Differential Cross Section in Elastic $K^- p$ Scattering from 1.0 to 2.3 GeV/c	EPB I (6)	WSC, Ctrs (LH_2 Tgt.)	140 shifts	Withdrawn (5/25/73)
P-327 (3/6/72)	E-327 (3/6/72)	L. Pondrom (U. of Wisconsin, U. of Michigan, Rutgers U.)	$K^+ \to \mu^+ \pi^0 \nu$ in Flight and a Measurement of $Re\xi(q^2)$ from the μ^+ Polarization	EPB I (6)	WSC, Ctrs	216.9 shifts	Run completed (9/28/73)
P-328 (3/10/72)		J. Mapp (U. of Wisconsin)	Study Low Energy $K_2^0 p$ Interactions in the 12-foot Chamber	EPB II (46)	12' HBC	300,000 pictures	Withdrawn (11/7/72)
P-329 (3/10/72)		J. Mapp (U. of Wisconsin)	Study Low Energy $K_2^0 d$ Interactions in the 12-foot Chamber	EPB II (46)	12' HBC	300,000 pictures	Withdrawn (11/7/72)

ZGS EXPERIMENTS

PROP. NO. (DATE) / EXPER. NO. (DATE)	SPOKESMAN (USER GROUPS)	TITLE	BEAM	DETECTOR	AMOUNT OF RUN	REMARKS (PUBLICATIONS)
P-330 (3/10/72)	M. Robinson (U. of Wisconsin)	Attempt a Definitive Search for a Z_0^* at $M_0 \simeq 1780$ MeV using the Reaction $K_L^0 + p \to K_S^0 + p$	EPB II (46)	12' HBC	400,000 pictures	Withdrawn (9/1/72)
P-331 (3/13/72) E-331 (7/18/72)	B. Musgrave (ANL, California Inst. of Tech.)	Study the Reactions $K^+ p \to K^0 \Delta^{++}$ and $K^- n \to \bar{K}^0 \Delta^-$ Using the Effective Mass Spectrometer	EPB IA (21)	EMS (LH_2 Tgt.)	148.4 shifts	Run Completed (12/18/73) PL-61B,483(1976)
P-332 (3/14/72)	O. Piccioni (U. of California)	Study of Multiple Pion Productions Associated with Nuclear Transitions	EPB IA (21)	EMS	120 shifts	Withdrawn (9/6/72)
P-333 (3/15/72) E-333 (3/22/72)	J. W. Cronin (U. of Chicago, ANL)	Measure the Decay Rate $K_L \to \mu^+ \mu^-$	EPB II (43)	WSC, Ctrs	413.8 Shifts	Run Completed (1/2/76) PRL39,59(1977); PRD16,565(1977); PRD19,1565(1979)
P-334 (3/16/72) E-334 (4/11/72)	J. Watson (ANL, U. of Illinois)	A Search for Backward Production of Exotic Mesons in 8 GeV/c π^- Interactions in Deuterium	EPB IA (22)	Streamer Chamber Ctrs	7.5 shifts 61,800 pictures	Run Completed (4/16/72)
P-335 (3/21/72) E-335 (3/22/72)	J. A. Poirier (U. of Notre Dame)	Small Angle Elastic Scattering and Total Cross Section Measurements for πp, Kp, and pp Interactions	EPB II (8)	Ctrs (LH_2 Tgt.)	156 shifts	Withdrawn (8/1/77)
P-336 (5/1/72) E-336 (7/18/72)	R. Winston (U. of Chicago, ANL, Ohio State U.)	Measurement of Λ's Produced by a Polarized Proton Beam	EPB IA (21) Pol. Protons	EMS (LH_2 Tgt.)	59 shifts	Run Completed (5/31/74) PRL-35,770(1975);PRD15,1826(1977)
P-337 (5/2/72) E-337 (5/9/72)	N. R. Stanton (Ohio State U., Carleton U., Michigan State U.)	A High Statistics Study of ω^0 Production	EPB II (8)	Ctrs (LH_2 Tgt.) WSC, PC	186.1 shifts	Run Completed (1/22/75) PRL-36,5(1976); PRL-36,8(1976)
P-338 (5/4/72) E-338 (7/18/72)	E. W. Beier (U. of Pennsylvania, U. of Michigan)	A Measurement of the Asymmetry in Polarized Proton + Proton → { $\pi^\pm$, $K^\pm$, p, $\bar{p}$ } + Anything	EPB I (1) Pol. Protons	Ctrs (LH_2 Tgt.) PC	70 shifts	Withdrawn (6/7/74)

Z G S   EXPERIMENTS

PROP. NO. (DATE)	EXPER. NO. (DATE)	SPOKESMAN (USER GROUPS)	TITLE	BEAM	DETECTOR	AMOUNT OF RUN	REMARKS (PUBLICATIONS)
P-339 (5/8/72)	E-339 (7/18/72)	A. B. Wicklund (ANL)	Study of Resonance Production with a Polarized Proton Beam Using the Effective Mass Spectrometer	EPB IA (21) Pol. Protons	EMS (LH$_2$ Tgt.) (LD$_2$ Tgt.)	180 shifts	Run Completed (4/29/75) PRD15,1826(1977)
P-340 (6/5/72)	E-340 (7/18/72)	M. J. Longo (U. of Michigan)	Neutron-Deuteron Elastic Scattering Measurements	7° N (14)	WSC, Ctrs (LD$_2$ Tgt.)	118.6 shifts	Run completed (9/17/73) PL-61B,93(1976)
P-341 (6/5/72)	E-341 (7/18/72)	M. J. Longo (U. of Michigan)	Antineutron-Proton Backward Scattering Study	7° N (14)	WSC, Ctrs (LH$_2$ Tgt.)	20.6 shifts	Run completed (9/28/73)
P-342 (7/3/72)	E-342 (7/18/72)	D. D. Reeder (U. of Wisconsin)	A Measurement of the $\bar{n}$p Annihilation Cross Section at Low Momentum	EPB II (42)	SC, Ctrs	148.4 shifts	Run Completed (1/30/75)
P-343 (7/3/72)		J. R. Albright (Florida State U.)	Study Leptonic Decays of Hyperons Using the 12-Foot Bubble Chamber	EPB II (47)	12' HBC	1 million pictures	Withdrawn (8/29/72)
P-344 (7/3/72)	E-344 (10/18/72)	F. C. Peterson (Iowa State U.)	Neutral Meson Spectrum Near 1000 MeV	EPB IA (21)	EMS (LH$_2$ Tgt)	115.6 shifts	Run Completed (4/26/74) PRL-35,970(1975);PR-D13,1153(1976) PR-D16,2054 (1977)
P-345 (7/13/72)	E-345 (1/8/73)	B. Eisenstein (U. of Illinois, ANL, Iowa State U.)	Study Backward Meson Production in the Reaction π^- p → n X° at 8 GeV/c	EPB IA (22A)	St. Ch. Neut. Ctrs (LH$_2$ Tgt)	86.4 shifts 475,000 pictures	Run completed (1/30/75)
P-346 (7/14/72)	E-346 (10/3/72)	C. Ward (ANL)	Measure Λ° Polarization with High Statistics in π^- p K° Λ° at 5 GeV/c	EPB IA (21)	EMS	71.2 shifts	Run completed (12/4/73) PR-D16,2041 (1977)
P-347 (7/17/72)	E-347 (8/29/72)	E. C. Swallow (U. of Chicago, ANL, Ohio State U.)	Measurement of the Electron Asymmetry in Polarized Σ⁻→n e⁻ ν Decay	EPB II (3)	WSC, Ctrs (LH$_2$ Tgt)	268 shifts	NIM-138, 61 (1976) Run Completed (7/26/77)

ZGS EXPERIMENTS

PROP. NO. (DATE)	EXPER.NO. (DATE)	SPOKESMAN (USER GROUPS)	TITLE	BEAM	DETECTOR	AMOUNT OF RUN	REMARKS (PUBLICATIONS)
P-348 (9/21/72)	E-348 (10/23/72)	K. C. Stanfield (Purdue, ANL)	Measure the Differential Cross Sections for Several Δ^{++} Exchange Reactions	17°	WSC, Ctrs (LH$_2$ Tgt)	197.1 shifts	Run completed (12/2/74) PRL35,138(1975)
P-349 (9/21/72)	E-349 (10/3/72)	T. Kitagaki (Tohoku U.)	Proton-Proton Interactions at 7, 8.7, and 11 GeV/c	EPB II (10)	12' HBC	303,301 pictures	Run completed (2/12/73) . Japanese Phys. Soc.28,336(1973)
P-350 (10/27/72)	E-350 (11/3/72)	R. E. Diebold (ANL)	Further Study of Forward Elastic Scattering	EPB IA (21)	WSC, Ctrs EMS	47.4 shifts	Run completed (2/27/73)
P-351 (11/13/72)	E-351 (6/5/74)	A. Fridman (Institut de Physique Nucleaire de Strasbourg)	Study the Coherent Reactions: $p^4 He \rightarrow p^4 He \pi^+ \pi^- (\pi^0)$ $\rightarrow p^4 He 2\pi^+ 2\pi^- (\pi^0)$	EPB IA (22)	Streamer Chamber PC	80.5 Shifts	Run Completed (5/17/77)
P-352 (11/2/72)	E-352 (1/8/73)	R. Segel (ANL)	Fast Pion Induced Nuclear Reactions	EPB I (6)	Ctrs	30.7 shifts	Run completed (12/10/73) PRL31,1353(1973)
P-353 (12/21/72)	E-353 (4/16/73)	W. Venus (RHEL, ANL)	Study 6 GeV/c π p Interactions in a Track Sensitive Target in the 12 Foot Chamber	EPB II (10)	12' HBC with TST	100,000 pictures	Withdrawn (11/29/74)
P-354 (12/29/72)	E-354 (4/11/73)	D. E. Nagle (LASL, U. of Chicago, U. of Illinois)	Parity Violation in Proton Scattering Processes	EPB I (2) Pol. Protons	Ctrs	103.7 shifts	Run Completed (4/29/75) PRL34,1184(1975)
P-355 (1/2/73)	E-355 (1/8/73)	N. M. Cason (U. of Notre Dame)	Study of the Reaction $\pi^- p \rightarrow K_1^0 K_1^0$ at 6.0 and 7.0 GeV/c	EPB IA (22A)	Streamer Chamber PC, Crs	101 shifts 409,198 pictures	Run completed (11/29/74) PRD11,2400(1975);PRL-36,1485 (1976); PRL-41,271(1978);PRD9,1317 (1979)
P-356 (3/2/73)	E-356 (3/30/73)	C. Akerlof (U. of Michigan)	$\pi^- p \rightarrow K^+ \Sigma^-$ Near 2.7 GeV/c	17°	WSC, Ctrs	36.7 shifts	Run completed (4/16/73) PRL33,119(1974)

ZGS EXPERIMENTS

PROP. NO. (DATE)	EXPER. NO. (DATE)	SPOKESMAN (USER GROUPS)	TITLE	BEAM	DETECTOR	AMOUNT OF RUN	REMARKS (PUBLICATIONS)
P-357 (3/29/73)		R. W. Kraemer (Carnegie-Mellon U., Weizmann Inst. of Science)	p-p Interactions at 6 GeV/c Using the Polarized Proton Beam and the 12-Foot Bubble Chamber	EPB II (9) Pol. Prot.	12' HBC	200,000 pictures	Withdrawn (6/6/73)
P-358 (4/16/73)	E-358 (6/6/73)	A. B. Wicklund (ANL)	$\bar{K}K$ System Using the Effective Mass Spectrometer and a Deuterium Target	EPB I (21)	EMS	139.8 shifts	Run Completed (1/30/75) Additional Run 5/2-16/75 PRL-37,971 (1976); PRL-37,1666(1976) PRL-38,269(1977); PRL-5,3196(1977)
P-359 (4/16/73)		E. P. Colton (ANL)	Investigate the Interactions of Polarized Protons in the 12-Foot Liquid-Hydrogen Bubble Chamber	EPB II (9) Pol. Prot.	12' HBC	60,000 pictures	Withdrawn (6/6/73)
P-360 (5/17/73)	E-360 (6/6/73)	D. Rhines (U. of Illinois, ANL, Iowa State U.)	Study Backward Meson Production in π^+p Interactions at 8 GeV/c	EPB IA (22A)	Streamer Chamber	8.5 Shifts	Run Completed (12/31/73)
P-361 (5/18/73)		Z. Ming Ma (Michigan State U.)	A Precision Measurement of η_{+-0}	EPB IA (22A)	Streamer Chamber	200 shifts 468,000 pictures	Withdrawn (6/6/73)
P-362 (5/24/73)		T. Wangler (ANL, Concordia Teachers College, Utah State U.)	Measurement of Charge Exchange of Positive and Negative Pions in Nuclei	EPB IA (22B)	Lead Glass Hodoscope	50 shifts	Withdrawn (5/28/74)
P-363 (6/29/73)	E-363 (10/9/73)	Z. Ming Ma (Michigan State U., ANL)	A Precision Measurement of η_{+-0}	EPB IA (22A)	Streamer Chamber	79.8 shifts	Run Completed (10/28/75)
P-364 (7/28/73)	E-364 (10/17/73)	D. R. Rust (Indiana U., U. of Chicago, Ohio State U.)	Measure the Polarization Parameter at Small Angles in pp Elastic Scattering	EPB IA (21) Pol. Protons	EMS (LH_2)	5.5 shifts	Run Completed (10/24/73) PL-58B,114(1975)
P-365 (11/29/73)	E-365 (6/5/74)	E. Peterson (U. of Minnesota, ANL, Columbia U.)	Elastic Scattering Near 90° CM	EPB I (5)	MWPC $(LH_2$ Tgt)	207 shifts	Run Completed (4/14/76) PRL-40,425(1978); PRL-40,429(1978)

ZGS EXPERIMENTS

PROP. NO. (DATE)	EXPER. NO. (DATE)	SPOKESMAN (USER GROUPS)	TITLE	BEAM	DETECTOR	AMOUNT OF RUN	REMARKS (PUBLICATIONS)
P-366 (11/29/73)	E-366 (1/8/74)	A. D. Krisch (U. of Michigan, ANL, St. Louis U.)	Feasibility Study for Measurement of the Recoil Spin in p-p Elastic Scattering with a Polarized Beam and Polarized Target	EPB I Pol.(1) Protons	Ctrs (PPT) LH_2 Tgt.	108.9 shifts	Run Completed (9/28/75) PL-52B,243(1974); PRD15,604(1977); PR-D17,24(1978)
P-367 (12/13/73)	E-367 (1/8/74)	D. K. Robinson, (Case Western Reserve U. Carnegie-Mellon U.. U. of Michigan)	Study of 6 and 12 GeV/c p-p Interactions Using the ZGS Polarized Beam and the 12-Foot Chamber	EPB II 9 or 10 Pol. Protons	12' HBC	336,340 pictures	Run Completed (2/25/76) NP-B123,361(1977);PRD20,596(1979)
P-368 (12/21/73)	E-368 (1/8/74)	T. Kitagaki (Tohoku U., Tohoku Gakuin U., Nara Women's U.)	Study of Antiproton-Proton Annihilations at Rest by Using the 12-Foot TST Bubble Chamber	EPB II (10)	12' HBC (TST)	310,775 pictures	Run Completed (4/10/77)
P-369 (12/26/73)		L. J. Koester, Jr. (U. of Illinois)	Measure the Production of ρ^- Mesons in the Reaction $\pi^- \; C12 \to \rho^- \; C12^*$ (4.44 MeV) at 6 GeV/c	EPB II (8)	SC, NaI, Lead Glass	100 shifts	Withdrawn (8/1/77)
P-370 (12/26/73)	E-370 (3/7/74)	V. P. Kenney (Notre Dame U.)	Study of 8 GeV/c π^- p Interactions in the 12-Foot Bubble Chamber Using a Track Sensitive Target	EPB II (9 or 10)	TST 12' HBC	100,000 pictures	Changed to 8 GeV/c π^+p (6/3/76) Withdrawn (1/27/77)
P-371 (1/4/74)	E-371 (3/7/74)	C. Fu (Illinois Institute of Technology, ANL)	Study pp Interactions in the 12-Foot Bubble Chamber Using a Track Sensitive Target	EPB II (9 or 10)	TST 12' HBC	25,000 pictures	Run Completed (12/17/75)
P-372 (2/2/74)	E-372 (3/7/74)	D. Miller (Northwestern U., ANL)	Proton-Proton Elastic Scattering with Polarized Proton Beams and Determination of p-p Scattering Amplitudes	EPB I (22B) Pol. Protons	PC, Ctrs PPT III	45.1 shifts	Run completed (10/29/74) PR-D12, 2594(1975);PRL-36,763(1976) PR-D16,2016(1977)
P-373 (2/21/74)	E-373 (3/7/74)	R. Segel (ANL, U. of Chicago, Northwestern U., U. of Pittsburgh)	Nuclear Reactions with Fast Hadrons	EPB II (42)	Ctrs	60 shifts	Withdrawn (3/11/75)
P-374 (2/22/74)	E-374 (3/7/74)	E. P. Colton (ANL)	Proton Radiography Studies in Beam 42	EPB II (42)	Ctrs Film	19.8 shifts	Run completed (4/10/74)

ZGS EXPERIMENTS

PROP. NO. (DATE)	EXPER. NO. (DATE)	SPOKESMAN (USER GROUPS)	TITLE	BEAM	DETECTOR	AMOUNT OF RUN	REMARKS (PUBLICATIONS)
P-375 (4/19/74)		R. M. Heinz (Indiana U.)	Search for Polarization Zeros in the Reaction pp → d π+	EPB I (1)	MWPC Ctrs	46 shifts	Withdrawn (1/10/78)
P-376 (5/29/74)	E-376 (6/5/74)	R. E. Diebold (ANL)	Measurement of the Polarization Parameter for pn Elastic Scattering from 2 to 6 GeV/c	EPB IA (21) Pol. Protons	WSC, Ctrs EMS	35.3 shifts	Run completed (10/29/74) PRL-35,632(1975)
P-377 (6/3/74)	E-377 (9/12/74)	H. A. Neal (Indiana U.)	High Precision Studies of the Large $\mid t \mid$ Elastic Proton-Proton Polarization	EPB I (1)	MWPC Ctrs	60.1 shifts	Run Completed (1/30/75) PR-D13,I(1976)
P-378 (7/30/74)	E-378 (9/12/74)	A. W. Key (U. of Toronto, McGill U.)	He^4 Test Film from the ANL 1.5 M Streamer Chamber	EPB I (22 A)	Streamer Chamber with He^4	Not Specified	Withdrawn (1/13/75)
P-379 (8/30/74)	E-379 (9/12/74)	E. I. Shibata (Purdue U., ANL)	Measure the Differential Cross Section for Several Baryon Exchange Reactions	17°	WSC, Ctrs (LH$_2$ Tgt)	21.8 shifts	Run Completed (1/30/75)
P-380 (8/30/74)	E-380 (9/12/74)	N. W. Reay/N. R. Stanton (Ohio State U. Michigan State U.,U. of Toronto, Carleton U.)	A High Statistics Measurement of ρ-ω Interference in πN → π+ π- π0 N	EPB II (8)	Ctrs (LH$_2$ Tgt.) WSC, PC	79.0 Shifts	Run Completed (5/30/75)
P-381 (11/6/74)	E-381 (11/6/74)	A. D. Krisch (U. of Michigan, ANL, St. Louis U.)	Elastic p-p Cross Sections Using a Polarized Target and a Polarized Beam	EPB I (1) Pol. Protons	Ctrs (PPT V)	145 shifts	Run Completed (4/17/75) PRL32,77(1974)
P-382 (11/26/74)	E-382 (4/10/75)	K. Bizzarri (U. of Rome, U. of Naples, U. of Padova, U. of Trieste, CERN)	p̄p Interactions Around 1 GeV/c Using the TST	EPB II (10)	12' HBC TST	100,000 pictures	Withdrawn (1/27/77)
P-383 (12/3/74)	E-383 (1/14/75)	W. Venus (RHEL, ANL, Carnegie-Mellon U. Melbourne U.)	Study of p̄p Interactions at 5.7 and 2.6 GeV/c Using the TST Facility.	EPB II (10)	12' HBC TST	150,000 pictures	Withdrawn (1/27/77)

ZGS EXPERIMENTS

PROP. NO. (DATE)	EXPER. NO. (DATE)	SPOKESMAN (USER GROUPS)	TITLE	BEAM	DETECTOR	AMOUNT OF RUN	REMARKS (PUBLICATIONS)
P-384 (10/1/74)	E-384 (1/14/75)	E. P. Steinberg (ANL, U. of Chicago, Purdue U., Carnegie-Mellon U., U. of Illinois)	Use ZGS Internal Beam for Study of Reactions of Complex Nuclei	ZGS Internal Target	Chemical Analysis	55 hours	Run Completed (4/22/76) PR-C14,1121(1976);PR-C14,1534(1976) PRL-38,269(1977);PR-C18,1349(1978)
P-385 (12/23/74)	E-385 (1/14/75)	B. Sandler (ANL, Northwestern U.)	Proton-Proton Scattering with S-Type Polarized-Proton Beams, N-Type Polarized Target and a Spin Analyzer for Recoil Protons	EPB I (22B) Pol. Protons	PPT III MWPC Ctrs	72.2 Shifts	Run Completed (9/28/75) PRD20,21(1979)
P-386 (1/6/75)	E-386 (1/10/75)	K. C. Stanfield (Purdue U., ANL)	Backward π^-p Scattering Near 5.12 GeV/c	17°	WSC, Ctrs (LH$_2$ Tgt)	41.8 shifts	Run Completed (1/20/75)
P-387 (1/7/75)		A. Yokosawa (ANL, Northwestern U.)	Two-Gamma Decay New Particle Search in the Mass Range from 1500 to 3150 MeV	EPB I (22B)	MWPC Ctrs Lead Glass Ctrs	100 shifts	Withdrawn (4/15/75)
P-388 (1/14/75)	E-388 (1/14/75)	D. H. Miller (Northwestern U., ANL)	A Search for γ-rays Emitted in $\bar{p}$p Annihilations at Rest	EPB II (42)	NaI Ctrs	30 shifts	Withdrawn (3/17/75)
P-389 (2/3/75)	E-389	R. L. Martin (ANL, U. of Chicago)	Proton Radiography	Booster	Integ. Ctrs	67.9 shifts	Run Completed (3/15/75)
P-390 (2/3/75)	E-390	C. Pelizzari (ANL)	ZING Prototype Studies	Booster (low energy neutrons)	Neutron Detect.	158.9 shifts	Run Completed (3/17/75) J. Appl. Cryst. 10,79(1977); NIM-145,91(1977)
P-391 (3/31/75)	E-391 (4/10/75)	S. Kramer (ANL)	Measurement of Polarization Effects Using the Effective Mass Spectrometer and Polarized Beam at 9 and 12 GeV/c	EPB I (21S)	EMS	142.5 shifts	Run Completed (12/21/76) PR-D17,1709(1978)
P-392 (4/8/75)		W. Z. Osborne (U. of Houston)	Search for $K^+_{\pi 2}$ Production-Decay Chain Correlations	EPB I Test Beam 18	Ctrs	40 shifts	Withdrawn (8/5/75)

ZGS EXPERIMENTS

PROP. NO. (DATE)	EXPER. NO. (DATE)	SPOKESMAN (USER GROUPS)	TITLE	BEAM	DETECTOR	AMOUNT OF RUN	REMARKS (PUBLICATIONS)
P-393 (6/4/75)	E-393 (8/1/75)	M. Marshak (U. of Minnesota, ANL, Columbia U.)	Measurements of the Asymmetry in Inclusive PP Scattering	EPB I	Scint. and Cerenkov Ctrs (LH$_2$ Tgt)	60.9 Shifts	Run Completed (9/28/75) PRL-36,929(1976); PRD15,602(1977)
P-394 (6/30/75)		A. Lesnik (Ohio State, ANL, U. of Chicago)	Measurement of Spin Rotation Parameters in Λ^0 Production at 12 GeV/c Using Polarized Proton Beam and the EMS	EPB I (21S)	EMS	60 shifts	Withdrawn (2/11/77)
P-395 (7/14/75)	E-395 (8/1/75)	G. C. Phillips (Rice University)	Measurement of the Total Cross Section for Proton-Proton Scattering in Pure Initial Transverse Spin States in the 1-3 GeV/c Region	EPB I (1) Pol. Proton	PPT V MWPC Ctrs.	69.9 shifts	Run Completed (9/8/76) PL-73B,235(1978)
P-396 (7/24/75)		L. Price (U. of California, Irvine)	Final State Interactions and Pion Correlations in p̄p Interactions at 1 GeV/c in the TST	EPB II (10)	TST 12' HBC	75,000 pictures	Withdrawn (8/1/77)
P-397 (7/24/75)	E-397 (8/1/75)	K. W. Edwards (Carleton U., U. of Toronto, McGill U., Ohio State)	Search for A_1^0, H, D, δ, · · ·	EPB II (8)	SC, Ctrs (LH$_2$ Tgt)	191.2 shifts	Run Completed (4/30/76) PRL38,930(1977);PRL42,346(1979)
P-398 (7/24/75)		D. R. Rust (Indiana U.)	Measure Polarization in the Coulomb Interference Region in PP Elastic Scattering	EPB I (2) Pol. Protons	MWPC, Ctrs (LH$_2$ Tgt)	40 shifts	Withdrawn (8/5/75)
P-399 (7/24/75)	E-399 (8/1/75)	S. W. Gray (Indiana U.)	Study of Inclusive Reactions Using the ZGS Polarized Beam	EPB I (2) Pol. Protons	MWPC, Ctrs (LH$_2$ Tgt)	203.9 Shifts	Run Completed (6/25/78)
P-400 (7/24/75)	E-400 (9/12/75)	N. M. Cason (U. of Notre Dame, ANL)	Study of the $\pi^0\pi^0$ System in the Reaction $\pi^+p \rightarrow \Delta^{++}\pi^0\pi^0$	EPB I (22A)	Str. Chmbr Pb Glass	165.1 Shifts	Run Completed (6/22/77)
P-401 (7/24/75)	E-401 (8/1/75)	A. Beretvas (Northwestern U., ANL)	Measurement of Observable (N,S; O,S), (O,S; O,S), and (N,O; O,N) at 6 GeV/c	EPB I (22A) Pol. Protons	PPT III (R&A) MWPC	79.6 Shifts	Run Completed (5/15/77)

ZGS EXPERIMENTS

PROP. NO. (DATE)	EXPER. NO. (DATE)	SPOKESMAN (USER GROUPS)	TITLE	BEAM	DETECTOR	AMOUNT OF RUN	REMARKS (PUBLICATIONS)
P-402 (7/24/75)	E-402 (5/13/76)	K. Nield (ANL, Northwestern U.)	Measurement of Observable (S,S;O,O) in Proton-Proton Elastic Scattering at 2, 3, 4, and 6 GeV/c	EPB-I (22B)	PPT-III (R&A), Spin Rot'd Sole-noid MWPC	24.8 shifts	Run Completed (6/1/76) PRL-37, 1727 (1976)
P-403 (7/29/75)	E-403 (8/1/75)	D. Nagle (LASL, U. of Chicago, U. of Illinois)	Search for Parity Violation in p-Nucleus Scattering	EPB-I (2) Pol. Protons	Ctrs, IC	57.6 shifts	Run Completed (6/1/76)
P-404 (8/13/74)		W. M. Bugg (U. of Tennessee, ORNL)	Interaction of 8 GeV/c π^+ Mesons in a D_2 TST in the 12' Bubble Chamber	EPB-II (10)	12' BC TST	100,000 Pictures	Withdrawn (8/1/77)
P-405 (8/13/75)		W. M. Bugg (U. of Tennessee, ORNL)	Interaction of Stopping Antiprotons in the 12' Deuterium Chamber Fitted with Thin Elemental Plates	EPB-II (10)	12' DBC w/plates	50,000 Pictures	Withdrawn (8/1/77)
P-406 (9/17/75)	E-406 (11/19/75)	A. Kanofsky (Lehigh U., BNL)	Study of Multiplicity from Particle Nuclei Interactions	EPB-II (22A)	Streamer Chamber	105,000 Pictures 29.9 Shifts	Run Completed (12/28/75) NIM140,433(1977)
P-407 (11/3/75)	E-407 (11/19/75)	M. Marshak (U. of Minnesota, ANL, Rice U.)	Measurement of Depolarization in the Process p + p → p + N*	EPB-I (5)	MWPC (2 LH Tgt.)	60.3 Shifts	Run Completed (6/1/76)
P-408 (11/3/75)	E-408 (11/19/75)	J. Roberts (Rice U., ANL, U. of Minnesota)	Inclusive Pion Asymmetries in Polarized Proton-Proton Collisions	EPB-I (5)	Ctrs.	57.4 shifts	Run Completed (3/1/76) PRD18,3939(1978)
P-409 (12/27/75)	E-409 (1/6/76)	T. Kitagaki (Tohoku U., Gakuin U., Nara Women's U.)	Low Energy Antiproton Interactions in the Neon-Hydrogen Mixture	EPB-II (10)	12' BC Ne-H_2 Mixture	39,000 Pictures	Run Completed (1/2/76)
P-410 (1/5/76)		J. G. Rushbrooke (U. of Cambridge)	Hyperon Production from Interactions of 9 GeV/c Polarized Protons in a Hydrogen Target	EPB-I (22A)	Streamer Chamber LH Target	60 Shifts	Withdrawn (6/14/76)
P-411 (1/9/76)	E-411 (2/19/76)	K. Ruddick (U. of Minn., Columbia U., ANL)	Psi Meson Production Near Threshold in π^-p Collisions	EPB-I (5)	MWPC (LH Tgt)	33.5 Shifts	Run Completed (4/30/76) PRD17,52(1978)

ZGS EXPERIMENTS

PROP. NO. (DATE)	EXPER. NO. (DATE)	SPOKESMAN (USER GROUPS)	TITLE	BEAM	DETECTOR	AMOUNT OF RUN	REMARKS (PUBLICATIONS)
P-412 (1/9/76)	E-412 (2/19/76)	L. G. Hyman (ANL, Carnegie-Mellon Purdue)	Studies of ν Interactions in Deuterium	EPB-II (9)	12' DBC (plates)	1,629,391 Pictures	Run Completed (2/13/78)
P-413 (3/24/76)	E-413 (6/3/76)	R. A. Singer (ANL, California-Irvine, Tennessee, ORNL)	Stopping $\bar{p}$ in D_2	EPB-II (10)	12' DBC	240,460 Pictures	Run Completed (4/24/76)
P-414 (5/26/76)	E-414 (6/3/76)	M. Marshak (Minnesota, UCLA, ANL, Columbia, Texas)	Measurement of Polarization in p-He^4 Elastic Scattering	EPB-I (5)	Cntrs. LH Tgt.	78.6 Shifts	Run Completed (9/8/76) PRL38,1272(1977)
P-415 (5/26/76)	E-415 (6/3/76)	L.E. Price (ANL, Columbia, Minnesota, Rice)	Direct Electron Production in pp Collisions	EPB-I (5)	Cntrs.	201.5 Shifts	Run Completed (8/7/77) PRL41,367(1978)
P-416 (5/28/76)	E-416 (6/3/76)	K. Nield (ANL, Northwestern)	Measurements of Total Cross Section Difference in pp Scattering in Pure Longitudinal Spin States at Incident Momenta of 1.2 to 6.0 GeV/c	EPB-I (22B)	PPT-III (R&A) MWPC	73.5 Shifts	Run Completed (12/19/76) PL67B,113(1977),PL70B,475(1977), PRL41,1436(1978)
P-417 (5/28/76)	E-417	K. Nield (ANL, Northwestern)	Measurement of Observable $C_{LL},(L,L;0,0)$ in Proton-Proton Elastic Scattering at Incident Momenta 1.2 to 6.0 GeV/c	EPB-I (22B)	PPT-III (R&A) MWPC	Not Specified	Combined with E-416
P-418 (5/28/76)	E-418 (6/3/76)	E. Peterson (Minnesota, ANL, Rice)	Asymmetries in Large Angle Nucleon-Nucleon Scattering	EPB-I (5)	MWPC Cntrs. LH Tgt.	99 Shifts	Run Completed (12/21/76)
P-419 (9/10/76)	E-419 (9/23/76)	Y. Watanabe (ANL)	Initial Determination of pp Scattering Amplitudes at 6 GeV/c.	EPB-I (22B)	PPT-III (R&A) MWPC	62.1 Shifts	Run Completed (12/13/76)
P-420 (9/13/76)	E-420 (9/23/76)	J. Gandsman (Carleton, McGill, Ohio State, Toronto, York)	Study of Neutral Mesons Decaying Into $\pi^+\pi^-$ 3γ and $\pi^+\pi^-$ 4γ	EPB-II (8)	SC,Ctrs. (LH$_2$ Tgt)	126.6 Shifts	Run Completed (4/11/77)
P-421 (7/8/76)	E-421 (9/15/76)	A. D. Krisch (Michigan, ANL, AUA, Abadan Inst. of Tech.)	$p(\uparrow) + p(\uparrow) \rightarrow p\uparrow p$ at Large Momentum Transfer	EPB-I (1)	Ctrs PPT-V LH$_2$ Tgt.	257.4 Shifts	Run Completed (6/1/77) PL63B,239(1976),PRL39,733(1977); PRD16, 549(1977)

ZGS EXPERIMENTS

PROP. NO. (DATE)	EXPER. NO. (DATE)	SPOKESMAN (USER GROUPS)	TITLE	BEAM	DETECTOR	AMOUNT OF RUN	REMARKS (PUBLICATIONS)
P-422 (9/16/76)	E-422 (9/23/76)	E. Steinberg (ANL, Chicago, Purdue, Carnegie-Mellon, Illinois)	Nuclear Chemistry Program at the ZGS	ZGS Internal Target	Chemical Analysis	67 Hours	Run Completed (1/20/78)
P-423 (10/20/76)		W. Fickinger (Case Western Reserve)	Study of Reaction $P{\uparrow}P \to n\, \Delta^{++}$ at 3 GeV/c	EPB-II (10)	12-Foot Bubble Chamber	200,000 Pictures	Withdrawn (6/8/77)
P-424 (1/11/77)	E-424 (1/27/77)	E. Steinberg (ANL Chemistry Division)	Correlation of Excitation Energy and Linear Momentum Transfer in the Interaction of High-Energy Protons with Complex Nuclei	EPB-I (12)	Semiconductor Detectors	80.4 Shifts	Run Completed (4/11/77)
P-425 (1/14/77)	E-425 (2/2/78)	J. Roberts (Rice, ANL)	Measurement of C in Large Angle Neutron-Proton Scattering	IPB-I (5)	PPT-VI MWPC Pol. Deut. Beam	80.0 Shifts	Run Completed (12/20/78)
P-426 (1/14/77)	E-426 (1/27/77)	A. Krisch (Michigan, ANL, Oxford, Nordita, Abadan Inst. Tech)	Proposal to Measure 90° c.m. Proton-Proton Elastic Scattering in Pure Initial Spin States from 2-6 GeV/c	EPB-I (1)	Ctrs. PPT-V	110.4 Shifts	Run Completed (7/3/78) PL-74B,273(1978)
P-427 (1/17/77)	E-427 (1/27/77)	D. Ayres (ANL, Elmhurst College)	Study Exclusive Λ-Production Reactions with the ZGS Polarized Proton Beam	EPB-I (21S)	EMS MWPC	204.6 Shifts	Run Completed (7/30/79)
P-428 (1/17/77)	E-428 (1/27/77)	T. Yoon (Carleton, McGill, Ohio State, Toronto)	Study of Mesons in $\omega\pi^-$, $\omega\pi^+\pi^-$ and (4π) Channels	EPB-II (8)	WSC, Ctrs.	96.6 Shifts	Run Completed (7/25/77)
P-429 (1/13/77)	E-429 (1/27/77)	V. Kenney (Notre Dame, ANL, LBL, Tennesce, ORNL, Melbourne)	4.1 GeV/c $\bar{p}$ in the 12-foot Bubble Chamber with TST	EPB-II (10)	12' BC TST 35% Ne-H$_2$	226,200 Pictures	Run Completed (7/28/77)
P-430 (1/27/77)		T. Kitagaki (Tohoku, Notre Dame, U.C. Davis)	Investigation of Hadron Bremsstrahlung in the 12-foot Bubble Chamber with TST, using 7.3 GeV/c π^+,π^-, and p	EPB-II (10)	12' BC TST 35% Ne-H$_2$	150,000 Pictures	Withdrawn (1/28/77)
P-431 (2/14/77)	E-431 (4/28/77)	H. Spinka (ANL)	Measurement of Observable (L,S:O,N) at 6 GeV/c.	EPB-I (22B)	PPT-III (R&A) MWPC	38.5 Shifts	Run Completed (6/1/77)

ZGS EXPERIMENTS

PROP. NO. (DATE)	EXPER.NO (DATE)	SPOKESMAN (USER GROUPS)	TITLE	BEAM	DETECTOR	AMOUNT OF RUN	REMARKS (PUBLICATIONS)
P-432 (4/18/77)	E-432 (5/12/77)	A. Krisch (Michigan, ANL, AUA, Niels Bohr Inst., Abadan Inst. of Technology)	P-p Elastic Scattering at 90°CM from 6 to 12 GeV/c in Pure Initial Spin States	EPB-I (1)	Counters PPT-V LH$_2$ Tgt.	147.6 Shifts	Run Completed (2/19/79)
P-433 (4/26/77)	E-433 (5/12/77)	A. Yokosawa (ANL)	P-p Scattering Experiments with a Longitudinally Polarized Beam and Target Between 1 and 6 GeV/c	EPB-I (22B)	MWPC PPT-III (R&A)	33.6 Shifts	Run Completed (9/4/77) PRL41,354(1978)
P-434 (5/2/77)	E-434 (5/12/77)	J. Roberts (Rice, ANL, Illinois)	Proton-Proton Total Cross Sections and Small Angle Elastic Scattering with a Polarized Beam and Target	EPB-I (23)	MWPC PPT-VI	217.7 Shifts	Run Completed (10/7/78)
P-435 (5/2/77)	E-435 (5/12/77)	I. P. Auer (ANL)	Measurement of $\Delta\sigma_L$ and the Single Scattering Parameters CLL, CLS, CSS in pp Elastic Scattering at 12 GeV/c	EPB-I (22)	MWPC PPT-III (R&A)	293.2 Shifts	Run Completed (5/8/79)
P-436 (5/3/77)	E-436 (6/27/77)	N. Cason (Notre Dame, ANL)	Study of Multiple Neutral Final States in π^-p Interactions at 8 GeV/c	EPB-I (22A)	Streamer Chamber	17.3 Shifts	Run Completed (7/4/77)
P-437 (5/3/77)	E-437 (5/12/77)	M. Marshak (ANL, Berkeley, UCLA, Minnesota, Rice)	Measurement of the Polarization in pd Elastic Scattering Near 1 GeV	EPB-I (5)	Counters LD$_2$ Target	30.7 Shifts	Run Completed (9/3/77) PRL41,1098(1978); PRC18,331(1978)
P-438 (6/3/77)	E-438 (10/13/77)	A. Krisch (Michigan, ANL, AUA, Oxford, Niels Bohr Inst., Abadan Inst. of Technology)	Measurement of $\vec{p} + \vec{p} \rightarrow p + p$ at Very Large P²	EPB-I (1)	Counters PPT-V LH$_2$ Tgt.	133.3 Shifts	Run Completed (6/1/78) PRL41,1257(1978)
P-439 (6/13/77)	E-439 (10/13/77)	A. Krisch (Michigan, ANL, AUA, Oxford, Niels Bohr Inst., Abadan Inst. of Technology)	Feasibility Study to Measure the Spin-Spin Interaction in Proton-Proton Inclusive Cross Sections at 11.75 GeV/c	EPB-I (1)	Counters PPT-V LH$_2$ Tgt.	Not Specified	Experiment parasitic with E-438 and E-426. 6 GeV/c operation approved 6/8/78. Run Completed (7/3/78)
P-440 (10/3/77)		J. Roberts (Rice, Minnesota, ANL)	Polarization Asymmetries in Proton-Proton Inelastic Scattering at Low Energies	EPB-I (23)	MWPC LH$_2$ Tgt.	120 Shifts	Withdrawn (3/1/78)
P-441 (10/13/77)	E-441 (2/2/78)	A. B. Wicklund (ANL)	Study of the Reaction pp $\rightarrow$ p$\pi^+\pi^-$ p with the ZGS Polarized Beam	EPB-I (21S)	EMS MWPC LH$_2$ Tgt.	181.3 Shifts	Run Completed (4/17/79)

ZGS EXPERIMENTS

PROP. NO. (DATE)	EXPER. NO. (DATE)	SPOKESMAN (USER GROUPS)	TITLE	BEAM	DETECTOR	AMOUNT OF RUN	REMARKS (PUBLICATIONS)
P-442 (12/12/77)	E-442 (2/2/78)	G. Igo (UCLA, LBL, ANL, Minnesota)	Deuteron Spin Alignment in High Energy Elastic Proton-Deuteron Back-Scattering	EPB-I (5)	Counters LH$_2$ Tgt.	9.4 Shifts	Run Completed (11/19/78) PRL43,425(1979)
P-443 (12/12/77)	E-443 (2/2/78)	G. Igo (UCLA, LBL, ANL, Minnesota)	Measurement of d-p Elastic Scattering at Small Momentum Transfers with Beams of Aligned Deuterons at Momenta Equivalent to 0.6 - 1.0 GeV Protons	EPB-I (5)	Counters LH$_2$ Tgt	44.1 Shifts	Run Completed (11/17/78)
P-444 (1/6/78)	E-444 (2/2/78)	A. Krisch (Michigan, ANL, AUA, Abadan Inst. Tech.)	A Measurement of A$_{nn}$ in n-p Elastic Scattering	EPB-I (1)	Counters, PPT-V LH$_2$ Tgt.	54.9 Shifts	Run Completed (12/20/78) PRL43,983(1979)
P-445 (1/23/78)	E-445 (2/2/78)	H.E. Miettinen (Rice, Illinois, ANL)	Asymmetries in Inclusive Pion Production with a Polarized Beam and Target	EPB-I (23)	MWPC PPT-VI	96.0 Shifts	Run Completed (4/1/79)
P-446 (1/23/78)	E-446 (2/2/78)	E. Swallow (Elmhurst, Illinois, LASL, ANL, Ohio State, LBL)	Search for Parity Violation in Polarized Proton Scattering at 6 GeV/c	EPB-I (2)	Counters H$_2$O Tgt.	126.8 Shifts	Run Completed (7/30/79)
P-447 (5/31/78)	E-447 (6/8/78)	A. Yokosawa (ANL)	$\Delta\sigma_L$ Measurements Between 1 to 6 GeV/c	EPB-I (22)	Counters MWPC PPT-III (R&A)	51.9 Shifts	Run Completed (7/3/78)
P-448 (5/31/78)		R. Wagner (ANL)	Measurement of the Total Cross Section Difference in pp Scattering for Longitudinal Spin States at Incident Momenta of 6-12 GeV/c	EPB-I (22)	Counters MWPC PPT-III (R & A)	105 Shifts	Withdrawn (6/8/78)
P-449 (5/31/78)	E-449 (6/8/78)	D. Underwood (ANL)	Measurement of $\Delta\sigma_L$(pm)	EPB-I (22)	Counters MWPC PPT-III (R&A)	197.7 Shifts	Run Completed (7/26/79)
P-450 (5/31/78)	E-450 (6/8/78)	G. Theodosiou (ANL)	Measurement of $\Delta\sigma_L$(pp), the C$_{LL}$, C$_{LS}$, CSS, Parameters of pp Elastic Scattering & Search for Multibody Decays of Dibaryon Resonances	EPB-I (22)	Counters MWPC PPT-III (RGA)	127.1 Shifts	Run Completed (10/1/79)
P-451 (9/25/78)	E-451 (2/28/79)	A.B. Wicklund (ANL)	Study of Longitudinal Spin Correlations in the Reaction pp$\to$pπ^+n at 6 and 12 GeV/c	EPB-I (21S)	EMS MWPC LH$_2$ Tgt.	41.4 Shifts	Run Completed (7/9/79)

ZGS EXPERIMENTS

PROP. NO. (DATE)	EXPER. NO. (DATE)	SPOKESMAN (USER GROUPS)	TITLE	BEAM	DETECTOR	AMOUNT OF RUN	REMARKS (PUBLICATIONS)
P-452 (9/25/78)	E-452 (10/5/78)	A. Krisch (Michigan, AUA, ANL, U. of Miami)	Measurement of $p\uparrow p\uparrow \to pp$ at 90° CM at P_{lab} = 11.75 and 12.75 GeV/c	EPB-I (1)	Counters PPT-V LH$_2$ Tgt.	249.9 Shifts	Run Completed (6/4/79)
P-453 (2/7/79)	E-453 (2/28/79)	A. Krisch (Michigan, AUA, ANL, U. of Miami, U. of Kiel)	Measurement of $p\uparrow p\uparrow \to pp$ at P_{lab} = 6.0 GeV/c at Intermediate P_T^2	EPB-I (1)	Counters PPT-V LH$_2$ Tgt.	72.0 Shifts	Run Completed (7/30/79)
P-454 (2/8/79)	E-454 (2/28/79)	T. Mulera (Rice)	Double-Spin Asymmetries in Inclusive π^- Production near 90° CMS	EPB-I (23)	MWPC Counters PPT-VI	60 Shifts	Withdrawn (3/20/79)
P-455 (2/9/79)		A. Krisch (Michigan, AUA, ANL, U. of Miami, U. of Kiel)	Measurement of Pure 3-spin $p\uparrow p\uparrow \to p\uparrow p$ Cross Sections at 11.75 GeV/c and Large P_T^2	EPB-I (23)	Counters PPT-V LH$_2$ Tgt.	250 Shifts	Withdrawn (2/28/79)
P-456 (2/9/79)		A. Krisch (Michigan, AUA, ANL, U. of Miami, U. of Kiel)	Measurement of $n\uparrow p\uparrow \to np$ at 6 GeV/c at Large P_T^2	EPB-I (23)	Counters PPT-V Pol. Deut. Beam	90 Shifts	Withdrawn (5/16/79)
P-457 (2/12/79)		G. Igo (UCLA, LBL)	D-p Elastic Scattering as a Source of Information about both the Deuteron D-Wave and the Spin Structure of the NN Amplitudes	EPB-I (23)	MWPC PPT-VI Pol. Deut. Beam	30 Shifts	Withdrawn (5/16/79)
P-458 (2/15/79)	E-458 (2/28/79)	M. Arenton (ANL)	Study of p K$^+$K$^-$p and pppp Final States Using the Effective Mass Spectrometer	EPB-I (21S)	EMS MWPC LH$_2$ Tgt.	48.0 Shifts	Run Completed (5/8/79)
P-459 (5/7/79)		J. Roberts (Rice)	Continuation of the Measurement of A_{NN} in Neutron-Proton Elastic Scattering	EPB-I (23)	MWPC PPT-VI Pol. Deut. Beam	60 Shifts	Withdrawn (5/16/79)
P-460 (5/7/79)	E-460 (5/18/79)	J. Roberts (Rice, ANL)	Measurement of $\Delta\sigma_T$ in pn Scattering at Low Energies	EPB-I (23)	MWPC PPT-VI	135.9 Shifts	Run Completed (10/1/79)
P-461 (5/7/79)	E-461 (5/18/79)	J. Roberts (Rice)	Polarization Asymmetries in Proton-Proton Inelastic Scattering at Low Energies	EPB-I (23)	MWPC PPT-VI	45 Shifts	Withdrawn (8/1/79)
P-462 (7/11/79)	E-462 (7/23/79)	A. B. Wicklund (ANL, Elmhurst College, Rice)	Proposal to Study $pp \to pn\pi^+$ Between 1.5 and 2 GeV/c	EPB-I (21)	EMS MWPC LH$_2$ Tgt.	23.6 Shifts	Run Completed (8/25/79)

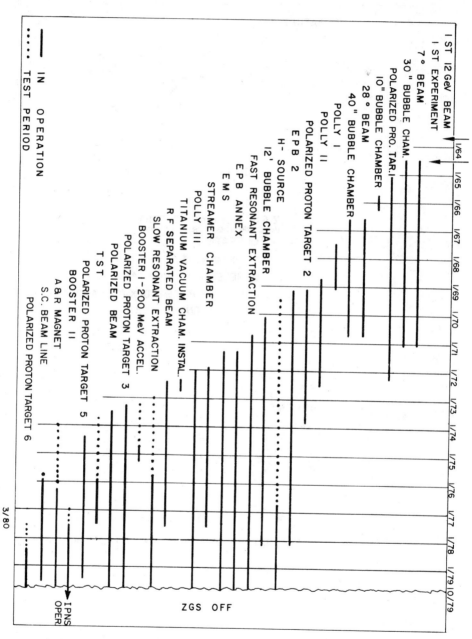

ISSN:0094-243X/80/600371-10$1.50 Copyright 1980 American Institute of Physics

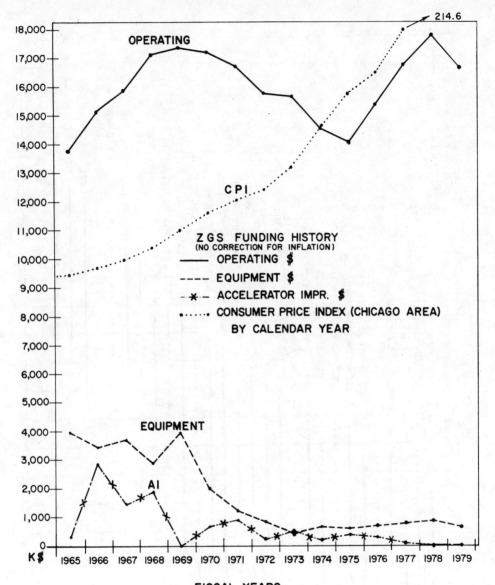

K$ 1965 1966 1967 1968 1969 1970 1971 1972 1973 1974 1975 1976 1977 1978 1979

FISCAL YEARS

3/80

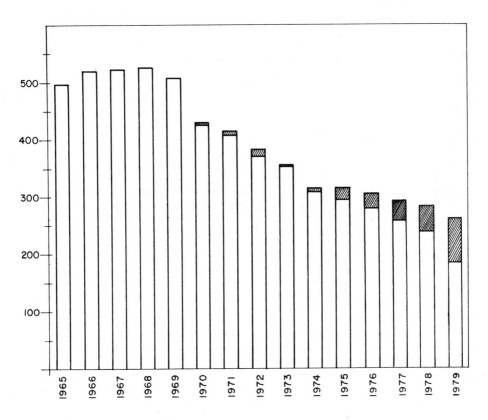

ZGS MANPOWER LEVELS AT FY END   — (REGULAR EMPL. ONLY)

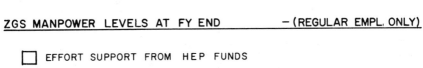

☐ EFFORT SUPPORT FROM HEP FUNDS

▨ MAN YEARS SUPPORTED FROM FUNDS OTHER THAN HEP

3/80

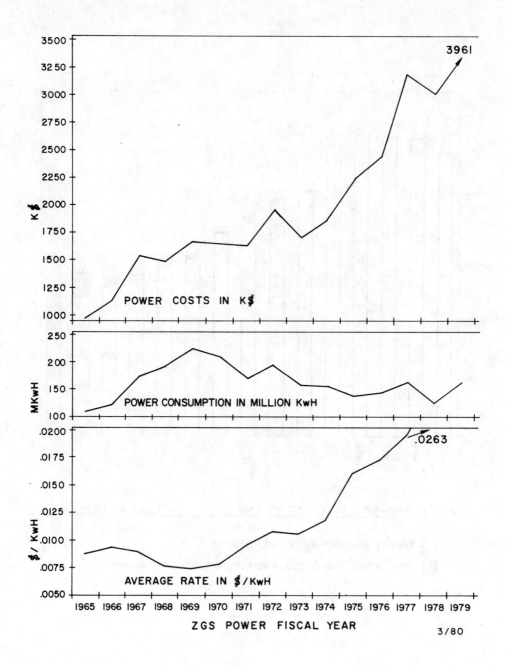

3961

3/80

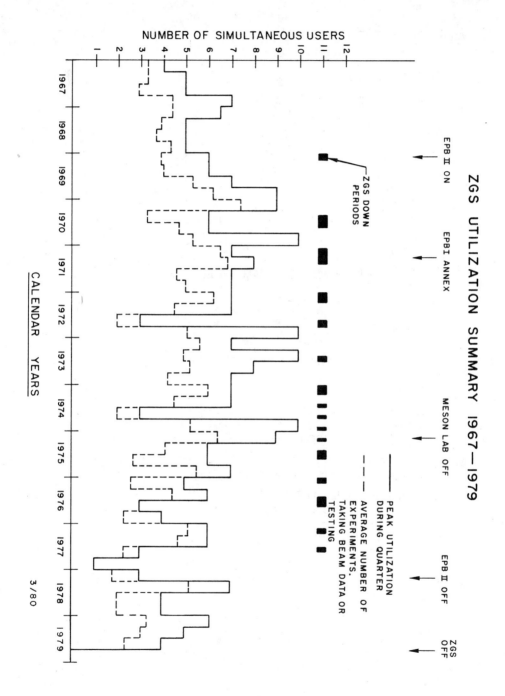

ZGS UTILIZATION SUMMARY 1967—1979

NUMBER OF SIMULTANEOUS USERS

CALENDAR YEARS

3/80

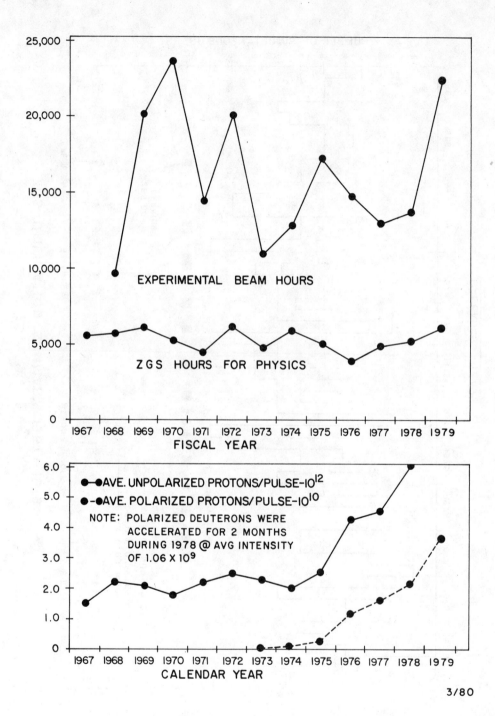

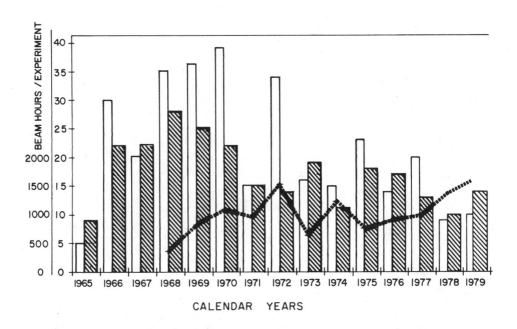

EXPERIMENTS PROPOSED AT ZGS

EXPERIMENTS COMPLETED AT ZGS

AVERAGE BEAM HOURS PER COMPLETED EXPERIMENT

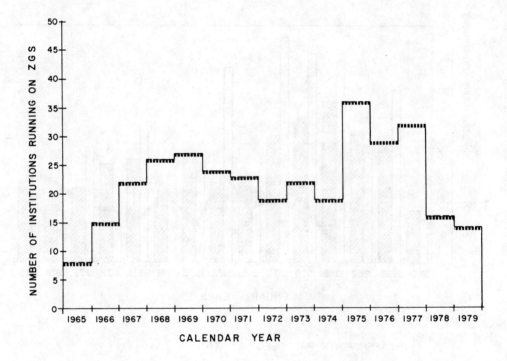

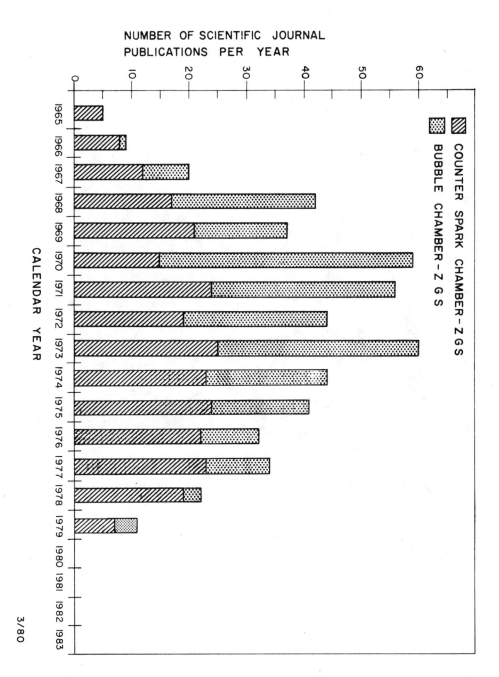

NUMBER OF SCIENTIFIC JOURNAL
PUBLICATIONS PER YEAR

COUNTER SPARK CHAMBER-ZGS

BUBBLE CHAMBER-ZGS

CALENDAR YEAR

3/80

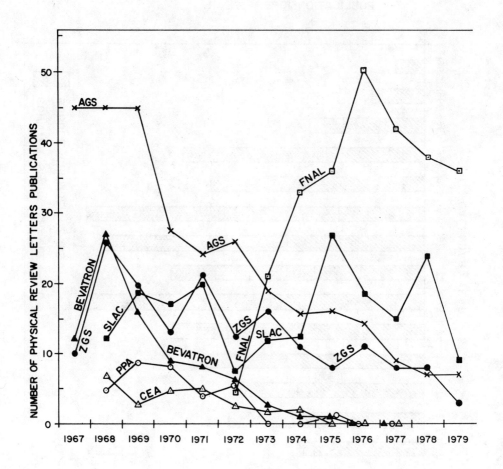

CALENDAR YEAR

THE HISTORY OF THE ZGS - AN ESSAY ON SOURCES

Allan Needell
American Institute of Physics, New York, NY 10017

Is to be hoped that historians tempted to further research by their reading of the accounts and reminiscences assembled in these Proceedings will have access to the primary documentation required for productive research. At present there is a wealth · of important unpublished material relating to almost every conceivable aspect of the history of the ZGS. But that record is widely scattered and poorly organized. For that and for other reasons it is far from certain that such material will be preserved or that it will ever become available to scholars.

At Argonne National Laboratory, records documenting the history of the ZGS are found in many individual and office files. For example, there is a complete set of minutes from the committee that was established to approve experiments and allocate machine time, and there is a complete set of operating logs from the main control room of the accelerator. The former are on file in the office of the Associate Laboratory Director for High-Energy Physics; the latter are stored in the administrative office of the ZGS Operations Group of the Laboratory's Accelerator Research Facilities Division. Neither office is likely to survive the reorganization that has been made necessary by the termination of the ZGS research program. Until now no formal arrangements have been made for the long-term care of those documents.

Recently the Department of Energy and several of the DOE sponsored research laboratories have begun to address the problem of identifying and preserving historically significant records in such a way that they might eventually be used for historical research. Archival programs or history programs have been initiated at a few laboratories and the DOE Office of Energy Research has supported the AIP Center for History of Physics in its efforts to encourage the documentation of modern physics.

As part of the AIP effort I was engaged in a study of the records and records preservation practices at Argonne during the final months of the active life of the ZGS. The occurance of this Symposium during that same period provided the opportunity for me to extend my search for ZGS records beyond the confines of the Laboratory itself.

The following is an account of some of the records brought to my attention over the last several months. It is presented as a postscript to these Proceedings in the belief that however incomplete it is and however quickly it becomes obsolete, such an account may prove useful to scholars. The account is also presented with the expectation that pointing out the existence of this material will increase the likelihood of its preservation and care.

THE ORIGINS OF THE ZGS

The first major high-energy particle accelerators funded by the Atomic Energy Commission were the Brookhaven Cosmotron and the Berkeley Bevatron. (They were funded in 1948 and completed in 1952 and 1954 respectively.) The Argonne Zero Gradient Synchrotron was the first accelerator located in the Midwest. As a number of these Symposium papers have indicated, the debate on the type of accelerator and the way it should be managed was long and complex. University-based physicists of the region as well as physicists associated with Argonne began as early as 1952 to plan for an accelerator research facility comparable to the ones available to their east and west coast colleagues. The Midwest Universities Research Association (MURA) was formed in 1954 by a number of regional universities for that purpose. At the same time Argonne began a serious separate effort to obtain AEC support for a vigorous accelerator program.

The result of these efforts was the decision late in 1955 that Argonne should proceed with the construction "on an urgent basis and for early completion" of an accelerator of more or less conventional design "with a top energy in the 10-15 bev range." Meanwhile, the AEC encouraged MURA to proceed with their design studies for "a machine of advanced design" for construction at a site to be determined and at a date not specified. All attempts to encourage the MURA group to work toward building a machine at Argonne had failed.

The Argonne-MURA controversy is the aspect of the history of the ZGS that has received the greatest amount of attention from professional historians. Even so, much of the primary documentation on that subject has not yet been fully exploited. D.S. Greenberg in The Politics of Pure Science (1967) devoted two chapters to struggles over the siting, selection, and management of high-energy accelerators; he included in those chapters a discussion of the struggle to obtain an accelerator in the Midwest. For information Greenberg relied almost entirely on published accounts (in newspapers and in Science magazine, many of which he had written himself) and interviews with many of the scientists, administrators, and government officials involved; he cites no unpublished records.

Lenard Greenbaum in A Special Interest (1971), a book written under the auspices of Argonne Universities Associated (AUA), describes the history of the relationship between Argonne National Laboratory and the Universities of the Midwest. His book focuses on matters of administration and policy brought to the fore by the high-energy physics program at Argonne and by the conflict with university scientists of the region. His sources include many records that had not previously been available to outsiders.

Records at Argonne

In the course of my investigations into records at Argonne
I was not able to locate some of the most important record collections cited by Greenbaum and described in his own discussion of
sources. Specifically, the files of Associated Midwest Universities (AMU) and of the Policy Advisory Board of the University
of Chicago were essential for Greenbaum's study but could not be
found during my six-month stay at Argonne. There is no record
of this material having been transferred from the possession of
AUA or of its having being destroyed. The AUA files do show that
those records were at Argonne when they were made available to
Greenbaum and therefore there is some hope that they will turn up
with additional searching.

Greenbaum was not given unlimited access to the files of
Argonne National Laboratory itself, or for that matter to the
files of the U.S. Atomic Energy Commission. He was permitted to
see some of the accelerator related files from the Laboratory
Director's Office and all of the files of J.C. Boyce, the Argonne
Associate Laboratory Director for Educational Programs and University Relations (1950-55). All Argonne files were reviewed by
representatives of AUA and the AEC prior to their release.

Much of the Argonne material released to Greenbaum was from
a special collection of documents assembled in 1956 from the
Director's Office file. That collection was assembled in response
to a request for information from the Joint Committee on Atomic
Energy of the U.S. Congress which was itself preparing for an
investigation of the accelerator setting policies of the AEC.
Currently that collection of documents and the J.C. Boyce files
are retained by the Argonne Technical Publications and Classification Office. That Office also maintains a number of other
collections of historical interest, including the Laboratory
Director's Office files from the tenure of Walter H. Zinn,
Argonne National Laboratory's first director (1946-56). The
collections maintained by the Classification Office are indeed an
important and unique source of information on all aspects of
Argonne's history prior to 1956.

Records Outside Argonne

Recently a large amount of record material from persons and
organizations outside of Argonne but related to the early history
of the ZGS have been identified. Most have to do with the
activities of MURA and many are gradually becoming available to
researchers. For example, a large fraction of the corporate
records of MURA have been deposited in the University of Minnesota
Archives along with the papers assembled by Lawrence R. Lunden
(University of Minnesota Vice-President for Business Administra-
tion--1958-71) as a member of the AUA Board of Trustees. In
addition, the papers of several of MURA's founders have been
deposited or promised to a number of MURA university archives.

Most notable are the papers of D. F. Kerst, K. Symon, and R. Rollefson. Kerst's papers are located in the Archives of the University of Illinois, Urbana. The papers of Symon and Rollefson are currently being reviewed by D. F. Moyer for a study he is conducting on the history of MURA. Following his study the papers will be deposited in the University of Wisconsin Archives at Madison. Already in that Archives are the papers of the three university chancellors from the MURA period and the papers of A. W. Peterson, the University of Wisconsin business officer. Attempts are being made by Moyer and others to secure the papers of other MURA scientists and administrators for appropriate university archives.

References to the subject of the Argonne accelerator program and to MURA can be found in the Atomic Energy Commission files maintained by the Department of Energy archivist in Germantown, Maryland. No attempt was made during this study to investigate that material in detail or to locate records from the Atomic Energy Commission's Chicago Operations Office. A check was made of the files maintained by the High-Energy Physics Office at the Department of Energy's headquarters. No primary source material related to the early history of the ZGS are preserved there.

ZGS CONSTRUCTION AND OPERATION

Director's Office

Records of the central administration of the Laboratory from 1956 to the present are maintained by the Director's Office (formally the Office of the Director or OTD). A comprehensive numerical indexing system for those records was established in 1956 and has been in effect with modifications and expansions since that time. Among the filing units is one entitled "ZGS Project." It currently consists of 23 volumes and takes up approximately 2.5 cu. ft. It contains copies of all correspondence initiated or received by the Director's Office related to the ZGS as well as copies of documents sent to the Director's Office for reference purposes. In addition to the "ZGS Project" filing unit, the OTD file contains records on many specific aspects of the ZGS program including the design, construction, and procurement of various ZGS components and experimental facilities. For example, there are folders on the various committees and organizational units established to manage phases of the ZGS program and there is a folder on the "200 MeV Linac injector" and one on "bubble chambers and spark chambers." As far as could be determined, none of the records in the OTD file have been available to anyone writing on the history of the ZGS.

Physics Division

Less formal filing systems have been used by most of the divisions and subordinate offices of the Laboratory. That is true for the Physics Division, the division within which Zinn

established an "Accelerator Group" early in 1954 asking the group
members to conduct preliminary work toward an Argonne high-energy
accelerator. The director of the Physics Division at that time
was Louis A. Turner; J. J. Livingood was chosen to lead the
Accelerator Group. Among the small number of files that survive
from the period of Turner's directorship of the Physics Division
there are two folders labeled "Accelerator Group." They contain
correspondence and memoranda documenting the participation of
Physics Division personnel in accelerator-related activities.
These include the early discussions among physicists from all over
the Midwest of the possibility of obtaining a "Midwest Cosmotron"
and the first phases of the work of the Argonne accelerator design
group. The file contains documents related to the recruitment of
personnel as well as the relationship of the Argonne effort to the
activities of MURA. Many, but by no means all, of the documents
in those folders are duplicates of documents contained in the OTD
file described above.

In 1956 the Accelerator Group was elevated in status to
become the "Particle Accelerator Division" (PAD). That was the
first of many organizational changes associated with the growth of
the ZGS and high-energy physics research at Argonne. (See
figure 1). J. J. Livingood, who had directed both the Group and
the new Division, recently assembled collections of documents from
his own files. One collection consists of copies of personal
letters, memoranda and meeting notes related to MURA and another
contains copies of all the design studies and cost estimates
prepared by the Accelerator Group between 1954 and 1956.

Those collections, together with other copies of correspondence and memoranda from Livingood's files dated between 1956 and
1958, have been deposited with a special collection of potentially
important ZGS records that were set aside during my study. Henceforth the special collection of ZGS records will be referred to as
the "ZGS Record Collection." That collection is temporarily kept
in the high-energy physics building at Argonne. It is to be hoped
that provisions for the long-term preservation of the collection
will be made.

The office files of the Particle Accelerator Division (1956-67)
were divided into an administrative file and a technical subject
file. The administrative file was turned over to the Accelerator
Division (AD) when it was formed from parts of the Particle
Accelerator Division in 1967, and then went to the Accelerator
Research Facilities Division (ARF) in 1973. (See figure 1). At
present there are no plans to review or to transfer any of these
files, although several contain documents of potential historical
value. For example, separate files have been kept of the correspondence between the Particle Accelerator Division and other
laboratories and organizations, including Lawrence Berkeley Laboratory, Brookhaven National Laboratory, Los Alamos Scientific
Laboratory, and MURA. Much of this correspondence documents the
way in which the laboratories shared information and advice. Personnel records and correspondence files of staff members are also

among the administrative records that have been retained.

One specific collection of documents from the PAD administrative files has been identified and transferred to the ZGS
Record Collection. That is the complete set of divisional staff
meeting minutes. The minutes are in two series: 340 meetings of
PAD staff dated from 4/16/56 to 3/14/66 and 50 meetings dating
from 3/21/66 to 11/17/67. At the last few meetings the staffs
of both the Accelerator and High Energy Facilities Divisons were
included. A subject index exists for meetings #289-340 of the
first series and for the entire second series.

The technical subject file of the Particle Accelerator
Division (11 cu. ft.) has been preserved in its entirety and
transferred to the ZGS Record Collection. It is arranged
following an alpha-numerically coded subject index. Subjects
range from "Accelerator theory" to "Targetry" and contain
specific entries for all of the many ZGS components. The file
contains correspondence, calculations, photographs, and drawings
as well as reports and technical notes. Some of the reports are
duplicates of those in the master report file described below.
Approximately 3 cu. ft. of this material is related to the design
and manufacture of the titanium vacuum chamber installed in the
ZGS ring magnet in 1969.

Particle Accelerator Division

To keep track of the many informal design studies and
reports, technical notes, and calculations, each report was
assigned the initials of its author(s) and a number. A complete
master report file of these materials (approx. 5 cu. ft.) has
been preserved and deposited with the ZGS Record Collection. The
reports are arranged alphabetically by author and there is an
index of all reports written prior to 1966. Reports concern
almost every aspect of machine and component design including
the design of the ion source, the injector, the beam lines, and
the experimental area.

Records of the many early subcontracts for manufacture of
ZGS components have not been specifically retained. Records show
that a collection of correspondence with the magnet manufacturer
was retired to the Argonne Records Center, but I was not able to
locate it. Records of the consulting contract let to William
Brobeck and records of the magnet coil contract with the National
Electric Coil Company were found in the Records Center.

Technical records that have not been transferred to the ZGS
Record Collection, but that remain in the office files of the
Accelerator Research Facilities Division, include files of the
polarized proton ion source and polarized proton targets developed
for use at the ZGS. Original tracings of ZGS component drawings
are kept in a storage area near the ARF Division drafting room.

ZGS Operations Groups

Records documenting the actual operation of the ZGS begin in 1963 and are currently being maintained by the offices that have been responsible for that operation during the final phase of the ZGS program. The main control room log books record specific operating procedures, machine parameters and all unusual incidents for the more than 16 years of ZGS operation. The logs are stored in the offices of the ZGS Operations Group and will eventually be transferred to the ZGS Record Collection. The Operations Group Office also maintains a complete "machine research" file documenting all of the machine studies and improvements carried out on the ZGS. That file will also be transferred to the ZGS Record Collection.

Other operations records from the Experiment Planning and Operations Group have already been included in the ZGS Record Collection. These include a complete set of ZGS "long term schedules," with 10-month projections of experiments, experimental set-ups, and beam and facility usage. Also preserved is a complete set of "ZGS Weekly Operations Schedules" and "As-Run Schedules" (both weekly and monthly). The "As-Run schedules" are charts showing the actual performance of the accelerator, beams and experimental facilities.

COMMITTEE RECORDS AND ZGS USERS GROUP RECORDS

Office of the Associate Laboratory Directory for High Energy Physics

The earliest formal ZGS committee was the "ZGS Complex Review Committee." It was established by the University of Chicago in 1958 and was composed of seven outstanding physicists and accelerator experts. The committee was asked to evaluate and make recommendations on all aspects of the accelerator program at Argonne. In 1960 the purview of the review committee was expanded to include the work of the High Energy Physics Division of the Laboratory as well as the Particle Accelerator Division. Since 1971 the committee has been appointed by AUA rather than by the University of Chicago. Although review committee reports have always been treated as privileged information, copies of the reports are on file in the office of the Associate Laboratory Director for High Energy Physics at Argonne.

An important aspect of the history of the ZGS is the attempt that was made from the outset to overcome the legacy of mistrust that had grown out of the debate over the siting and management of a Midwest accelerator. A major part of that attempt involved the establishment of an organizational framework for the participation of university-based physicists in the planning of experimental facilities and the scheduling of ZGS operations.

The Associate Laboratory Director (ALD) office maintains several user related files. These include a complete set of minutes of the Scheduling Committee (later called the Program Committee). It was that committee that decided which of the proposed experiments should be run at the ZGS and what priorities should be assigned to the approved experiments. The committee consisted of representatives of the Users Group selected by the Users Advisory Committee with the consent of the Associate Laboratory Director. In addition to minutes of the meetings, the ALD office maintains a complete record of appointments to that committee.

ZGS Users Group Office

The first meeting of a ZGS Users Group was held at Argonne on January 16 and 17 of 1959; all interested physicists from Midwest universities were invited. The Users Group appointed a number of ad hoc and standing committees to serve in an advisory capacity; these included the so-called Users Advisory Committee. Records of many of these commitees were passed from chairman to chairman and have not been preserved. Written reports were prepared following the annual or semi-annual Users Group meetings, and these reports are on file in the User's Group office at Argonne. (In fact, several of the meetings held in the middle and late 1960s were tape-recorded. The tapes are stored along with the Users Group records.) All of the Users Group meeting records in the office at Argonne will eventually be transferred to the ZGS Record Collection. Another related collection currently being maintained is the official applications and renewals of formal User status granted by ANL. These documents provide a complete record of the physicists who actually registered as ZGS Users.

Experiment Planning and Operations Group Office

Once the Program committee approved an experiment it became the responsibility of an internal ZGS Operations Committee to oversee the installation of the necessary apparatus, assign experimental areas and equipment, and provide for the implementation of the priorities assigned by the program committee. To these ends regular (weekly) meetings of the ZGS Operations Committee were held and minutes of the meeting were recorded. A complete set of these minutes (approx. 3 cu. ft.) maintained by Experiment Planning and Operations Group Office has been included in the ZGS Record Collection.

To assist experimenters in planning and proposing ZGS experiments, the Experiment Planning and Operations Group prepared a ZGS Users Handbook. That handbook was first issued in 1965 and was periodically revised. It described in detail the specifications of available beams, experimental apparatus, shielding etc., and the procedures by which experiments should be proposed. The handbook in several editions is included in the ZGS Record

Collection.

Another extremely important set of documents in the ZGS Record
Collection is a set of ZGS News Bulletins (later the ZGS News
Bulletin and ZGS Monthly Research Report). The Bulletins were
regularly prepared by the Experimental Planning and Operations
Group and distributed to all active users. They contain summaries
of the actions of the Program Committee and the Users Advisory
Committee, progress reports on ongoing experiments, summaries of
the ZGS and related facility performance, and news about all
planned modifications and improvements in the ZGS and experimental
areas.

EXPERIMENT RECORDS

Formal proposals for experiments automatically resulted in
the assignment of a proposal number ("P" number) by the Associate
Laboratory Director. If approved by the Program Committee, the
experiment was reassigned an Experiment ("E") number. Folders
for each proposal and experiment, containing a copy of the proposal
and all related correspondence, are maintained by the Associate
Director's Office.

Similar files were maintained by the Experiment Planning and
Operations Group. These files contain occasional sketches of
experimental set-ups and specific correspondence concerning the
installation of experimental apparatus. Pending a review of the
ALD experiment files and a determination of the extent to which
the information contained in the Experiment Planning and Operations
Group files usefully supplements the information in the primary ALD
experiment files, the Operations Group file is preserved with the
ZGS Record Collection.

Detailed records of individual experiments, if they have been
preserved, are in the possession of the experimental group themselves. In the cases when experiments were performed by an
experimental group from the Argonne High Energy Physics Division
these records, if preserved, are at Argonne. For example, the
experimental group led by A. Yokosawa has maintained an experimental area log book for the recording of all actions taken by the
experimenters of that group while "running" at the ZGS. These
logs provide insight into the actual procedures used as well as
information on the way in which experiments were carried out.
These log books are stored in the experimental group offices.
Similar log books and experiment files undoubtedly exist for
experiments performed by outside users, but I was not able to
gather any detailed information on them or to actively seek to
insure their preservation.

CONCLUSION

At this time, plans for the records included in this report may be summarized as follows:

In the ZGS Records Collection:
  J. S. Livingood: copies of ZGS related documents assembled from his files as leader of the Accelerator Group (1954-56) and as director of the Particle Accelerator Division (1956-58).
  Particle Accelerator Division: complete set of unpublished notes, reports, design studies, etc.,1956-76, complete set of divisional staff meeting minutes, 1956-67; and a complete set of the division's technical subject file, 1956-1972.
  ZGS Experiment Planning and Operations Group: complete set of ZGS long term schedules, ZGS Weekly Operations Schedules, and As-Run Schedules; files of proposals and experiments including occasional sketches of experimental set-ups and correspondence on the installation of experimental apparatus; several editions of the ZGS Users Handbook; and a complete set of the ZGS News Bulletin (later the ZGS News Bulletin and ZGS Monthly Research Report).
  ZGS Operations Committee: complete set of minutes of weekly meetings.

To be transferred to the ZGS Records Collection:
  ZGS Operations Group: operating log books, 1963-79, and complete file of machine studies and improvements.
  ZGS Users Group: incomplete set of minutes of various committees as well as written reports and several tape recordings of Users Group meetings.

ZGS records not tagged for transfer:
  ZGS Users Group: file of official applications and renewals of formal User status.
  High-Energy Physics Division: experimental area log book documenting actions of A. Yokosawa. Other records of individual experiments may be in the possession of other experimental groups.
  Accelerator Research Facilties Division: records of the development of the polarized proton ion source and polarized proton targets; and original tracings of ZGS component drawings.
  Contractor records: consulting contract let to W. Brobeck; and magnet coil contract with the National Electric Coil Co. (Records now in Records Holding Area.)

391

ZGS documentation in other collections at Argonne:
Director's Office: special collection of documents assembled in 1956. (Now in the hands of the Classification Officer.)
Director's Office: records of W. H. Zinn, 1946-56. (Now in the hands of the Classification Officer.)
Director's Office: records, 1956 to date.
Associate Laboratory Director for Educational Programs and University Relations: records of J. C. Boyce (1950-55). (Now in the hands of the Classification Officer.)
Associate Laboratory Director for High-Energy Physics: complete set of minutes of the Scheduling Committee (later called the Program Committee) and the record of appointments to the Commitee; and files of folders for each proposal and experiment using the ZGS including related correspondence.
Particle Accelerator Division (1956-57), Accelerator Division (1967-73), and Accelerator Research Facilities Division (1973 to date): administrative records.

ZGS related documents in collections at other repositories:
University of Minnesota Archives: MURA corporate records.
University of Illinois Archives: papers of D. F. Kerst.
University of Wisconsin Archives: records of university chancellors from the MURA period; papers of A. W. Peterson, university business officer; and future deposits of the papers of K. Symon and R. Rollefson.

This account, lengthy as it is, in no way exhausts the subject. In it I have not been able to include discussions of records documenting the bubble chambers and spark chambers developed for use at the ZGS even though the design of detectors is one of the most significant aspects of the ZGS program. I have also not discussed the records being maintained by Argonne's Applied Mathematics Division documenting the development of "Polly," a computer assisted device for the analysis of bubble chamber film, nor have I discussed the records in the active files of the Associate Laboratory Director's Office for High Energy Physics. Emphasis has been placed on locating and describing records documenting the early history of the ZGS itself.

Naturally the records of the latter phases of the history of the ZGS, including the decision to cease operations, may prove to be just as significant. Almost all of these records are currently in active files, and there is some concern at Argonne that with the press of new projects and responsibilities, records of the ZGS will be neglected. The existing records management at Argonne does not assume any responsibility for identifying and preserving records of potential historical value, and there is a danger that such records will be lost unless some further action is taken.

It was only by chance that the AIP Center for History of
Physics was already conducting its study of records preservation
at government laboratories, and moreover that the project had
scheduled field work at Argonne, at the time the ZGS was being
decommissioned. We were pleased to have had the opportunity to
assist in carrying out some intitial steps toward providing a
documentary record of the history of the ZGS. We hope that with
the help of this report further necessary steps will be taken.
That will not occur, however, unless interested parties, scientists and historians, make known how important it is to have such
a record and to make it available to scholars.

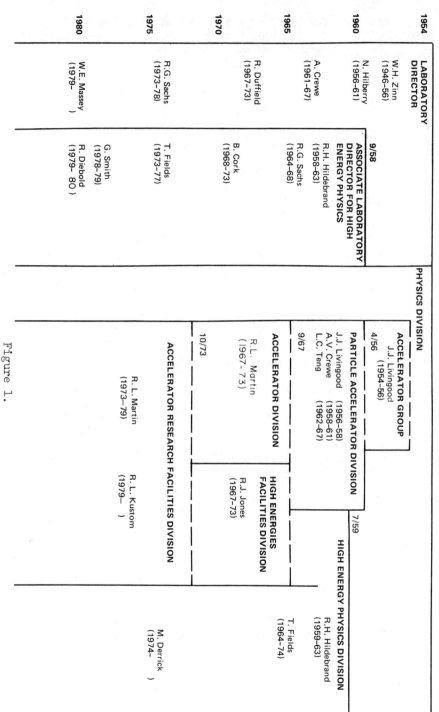

ZGS ADMINISTRATIVE CHRONOLOGY

ZGS RELATED DIVISIONS

Figure 1.

THE LIFE AND DEATH OF A PARTICLE ACCELERATOR
L. G. Ratner
Argonne National Laboratory

Contents

Prologue

Chapter I - Accelerators and Atoms

Chapter II - An Accelerator is Conceived

Chapter III - The Agony of Birth

Chapter IV - Adolescence

Chapter V - Maturity and the Death Knell

Epilogue

List of Special Contributors

Chapter II - J. J. Livingood, (1952-1957)

Chapter III - L. C. Teng, R. L. Martin, (1958-1963)

Chapter IV - D. Getz, L. C. Teng, R. L. Martin, (1963-1965)
        Control Room Logs

Chapter V - B. Cork, T. Fields, R. L. Martin, (1972-1979)
        Control Room Logs

Epilogue - R. Diebold, R. L. Martin

ISSN:0094-243X/80/600394-60$1.50 Copyright 1980 American Institute of Physics

Prologue

You may well wonder why the life and death of a particle accelerator is worth writing about and you may not be too sure just what a particle accelerator is. We will devote Chapter I to telling you what an accelerator is, but here we would like to make a few remarks on why this is being written at all. First, we want to inform those people who have been around the Zero Gradient Synchrotron (ZGS) and their friends and families about how the ZGS came into being and what its purpose and accomplishments were. We would also like to let the general public know what we have done at the ZGS. To begin, let us say that a particle accelerator is a very expensive tool for doing scientific research into the nature of matter and it costs millions of dollars to build and millions of dollars a year to operate. Such costs can only be borne by the Federal Government and have been so borne since the early 1940s, when physics research became so important to national defense and the production of new energy sources. There has been a progression of machines each bigger, more powerful, and more costly than its predecessors, and as each new machine became operational, it was necessary to phase out some of the older machines in the interest of economy.

The Zero Gradient Synchrotron (ZGS) at the Argonne National Laboratory was such a machine and we feel that its history provides an interesting and informative picture of some of the sociological and scientific factors influencing the progress of research. The ZGS has a more interesting history than other installations, since it was born amidst controversy between the government-funded Argonne National Laboratory and the midwest university community which would eventually become its prime user. Its demise was also marked by some controversy between the scientists using the facility and the Federal Government. Although the government admitted that the research which was being terminated was of the highest quality and interest, there could be no change in the decision that a facility had to be turned off in order to operate newer facilities more efficiently.

In the beginning the problems associated with starting the ZGS were symptomatic of the adjustments which were being made by the government and the universities as they came to grips with the problem of forming a mutually satisfactory organization for operating and using the national laboratories. When they were first started, the national laboratories had specific missions dealing with various aspects of atomic energy; design and build bombs, propulsion reactors, power reactors, etc. The laboratories were begun using people from the universities. In the decade following World War II, as the above missions began to be accomplished, investigations moved into more basic research and the laboratories assumed some university functions. Along with this came increased federal funding for research to the universities themselves. Thus the era of "Big Science" came into being, where federal support for science was recognized as a national purpose needed for the furtherance of national policy, both domestic and foreign.

The government's concern with mission-oriented research tied to national goals could conflict with the scientists' and the universities' concern for promoting basic research. Basic research is the "abstract" and "theoretical" knowledge which has no immediate direct application but is always necessary before the mission-oriented or applied research can begin. Big Science funding, representing hundreds of millions of dollars, can lead to conflicting views such as

a. Support well-known "centers of excellence."

b. Support less-developed centers in order to broaden the base of "excellence."

c. Support regional centers regardless of quality.

d. Support only research applied to national goals.

e. Support basic research.

Such conflicts had to be resolved in some way during the funding, construction, and operation of the ZGS with one additional complication. The Argonne National Laboratory had been started and operated by one university and throughout its history, even before the ZGS, there had been conflicts between this university and the other universities in the Midwest on the use of the facilities and in planning for future facilities. Both the University of Chicago, which operated Argonne for the Atomic Energy Commission (AEC), and the Midwest University Research Association (MURA) were proposing basic research facilities. A sometimes bitter confrontation occurred on the questions of where and when such facilities should be built, who should use and operate them, and who set policy for them.

These conflicts have been described elsewhere,[*] so that we will only refer to them as they affected the scientific and economic decisions that were made in creating the ZGS. Our major interest will be to describe how the ZGS was born, lived, and died, what it accomplished, and how it did it against the backdrop of the milieu in which it was conceived.

---

[*]Leonard Greenbaum. A Special Interest, The University of Michigan Press, Ann Arbor 1971.

I. Accelerators and Atoms

To those scientists who investigate the nature of the physical universe around us and try to understand how it was created, how it operates, how matter interacts with itself and with energy, and how the universe will end, the accelerator is one of the very important tools. It is the tool with which the scientist can study the very, very small. Together with its complement, the telescope which studies the very, very large, it has brought us an almost unbelievable knowledge about how our universe was born, how it works today and its possible future.

Everyone is familiar with telescopes and the objects which they study. But the accelerator is very young compared to telescopes and the objects of its study have been known about for less than a hundred years. To many people, accelerator is a meaningless word when applied to something that doesn't move. To understand what an accelerator is and what it does, we will make a brief detour and learn a little about the material of which our world is made. As with many ideas in Western Civilization, the story begins in Ancient Greece.

In about 400 B.C., Leucippus and his student Democritus were proponents of a theory that stated that everything was made up of tiny bits of matter called atoms that swirled around in empty space and combined to form the universe and all that it contains, including human beings. Democritus also believed that atoms were indestructible, and when a tree died and decayed, its atoms were not destroyed but combined again to make new things. This appears to be the first concept of an indivisible, indestructible unit which makes up all things.

Two thousand years were to pass before any further insights into the nature of matter occurred. In the 17th and 18th centuries the experimental work of many chemists and physicists led to the kinetic theory of heat and the laws of chemical combination. These investigations made it clear that the matter we are familiar with is made up of molecules and atoms and that the smallest constituent which retains the property of the element is the atom. By 1869 some 60 known elements could be arranged in families in the so-called "Periodic Table." All of matter was reduced to the elements of the table and though there were still some 30 blank spaces, scientists felt that these would be discovered. But even some 90 different objects making up all matter seemed like a lot. In fact in 1815 the hypothesis was advanced that all elements were built up out of hydrogen and that all matter could be thought of as being combinations of hydrogen atoms. So by 1870, there were thought to be a number of different elements each composed of its own atoms and a hypothesis that the internal structure of these atoms was composed of hydrogen atoms. Further information had to wait for new techniques, but these were not long in coming. Before the end of the 19th century, the discovery of radioactivity, X-rays, and the phenomena of electrical discharges in evacuated tubes made possible further knowledge of the nature of matter. One of the fundamental experimental techniques was conceived of by Lord Rutherford when he realized that one might learn about the structure of atoms if one could bombard matter with smaller objects and determine how they bounce off or pass through matter. A well-known analogy is bouncing tennis balls off of a barn door – it's obvious that when the ball hits, it bounces back; when it misses, it doesn't. By carefully throwing many balls we can determine the size and shape of the barn door. It had been discovered that radium, when it decayed, shot out alpha particles at very great speeds and Lord Rutherford thought of using this source of tiny objects to try to determine the nature of an atom. The concept of bombarding matter with fast particles was born and thus the radium nucleus became the world's first particle accelerator. After all, a particle accelerator is only a machine that can produce many very, very fast bombarding particles and differs from the natural source only in velocity and type of particle used. The first experiment with projectiles from a particle accelerator – the radium nucleus – showed a remarkable result when thin metal foils were bombarded. One out of thousands of particles would not go straight through but would be deflected through a large angle, as though it had bounced off of something hard and heavy. This turned out to be the heavy core or nucleus of the atom. It is indeed strange that matter is mostly empty space with all the mass concentrated in the tiny nucleus.

By 1911 an atom was known to have a tiny, massive, positive charge core surrounded by various numbers of negative electrons at very large distances from the core. By 1920 a great number of isotopes were known. These isotopes are various forms of the same element, chemically identical but differing primarily in atomic weight. It appeared that the weights of all atoms were almost exactly whole-number multiples of the weight of one hydrogen atom. This strongly suggested that all elements were built up from the hydrogen atom, or more exactly from the hydrogen nucleus. A hundred years of effort seemed to have confirmed the speculation of 1815. The hydrogen nucleus as a basic building block of matter was given the Greek name "proton" [the first one]. However, things still did not quite add up. For example, copper with an atomic weight of 64 had only 29 electrons. If it had 64 protons it could not have a balanced charge. Again the bombardment of matter by alpha particles brought a solution to this problem. When beryllium was bombarded by alpha particles it gave off a very penetrating radiation. This was identified by 1932 as a particle about as massive as the proton, but with no electric charge. This was called the neutron. So now our copper nucleus could contain 29 protons to balance the electron charge and 35 neutrons to bring its mass to 64. The proton and the neutron are thus the constituents of the nucleus. From radioactive decay studies one found that electrons and positrons (positively charged particles with the mass of electrons) are emitted in processes in which a proton changes into a neutron or a neutron into a proton.   This has led to the present idea that the proton and neutron are different states of the "nucleon."

Our story has led from the matter we can touch to molecules and then to atoms and then to the nucleus and then nucleons. As we look at matter in smaller and smaller pieces, we find that the technique of scattering projectile particles is one of our most valuable tools.   Today the atom is reasonably well understood and its properties can be calculated from quantum theory.   The nucleus is not as well understood and models explain some, but not all, of the phenomena.   Experiments are now probing the nucleon and we have a quark model which explains some of the phenomena, but again not all.   It is now thought that quarks make up the nucleon in much the same way as nucleons make up nuclei and nuclei make up atoms.   These results have been made possible by the experiments done using particle accelerators which accelerate proton and electron projectiles to bombard matter.

The year 1932 was not only marked by the discovery of the neutron, but also by the advent of machines to produce more energetic probes than those available from natural sources.   In this year Cockcroft and Walton in England produced a particle accelerator that reached 800,000 electron-volts, which was enough energy to penetrate light nuclei and gave us new knowledge about nuclear structure.   But this and the Van de Graaff accelerator which got up to 10 million electron-volts were still not enough.   Technical difficulties prevent building larger machines of these types, but a new idea - that of multiple acceleration - was invented also in 1932 by E. O. Lawrence in a device called the cyclotron.   In this device a large electromagnet bends particles into circular orbits and they gain energy in every turn when they pass through an accelerating cavity.   By 1945, the use of frequency modulation led to synchro-cyclotrons and energies were pushed to the 500-million electron-volt range (500 MeV).   In 1945, E. M. McMillan in the United States and V. Veksler in the USSR published the idea of the "synchrotron principle", which allowed still higher energies at reasonable costs.   Today such machines reach energies of 500-billion electron-volts (500 GeV) and still higher energy machines are being planned for the future.

The need for higher and higher energies is simply related to the fact that the size of the bombarding particle becomes smaller as the energy increases, allowing us to see smaller structures.   The relationship between energy and size is given by $E = hc/\lambda$ where h is a number called Plancks' constant, c is the velocity of light, E is the energy, and λ is related to the size.   The significance of λ can be understood by considering the ordinary microscope that uses light which is reflected or transmitted by the specimen.   Since the wavelength (λ) is about one hundred-thousandth of an inch, details this small can be seen.   Electron microscopes can show details about a thousand times smaller and indeed have shown pictures of large atoms.   Electron microscopes use beams of electrons with energies in the kilovolt range.   To see finer structures such as nucleons and their constituents, λ must be about a millionth of a billionth of an inch, which requires GeV energies. Hence, the desire for higher energies is really a desire to see smaller and smaller structures. There is another feature which the push to higher energies revealed. As energies reached higher and higher thresholds, new particles began to be produced. Just as the discovery of electrons, protons, and neutrons led to new insights so did the discovery of pions, kaons, lambdas, the J-PSI, the upsilon and many more. This process is still continuing at the highest energies achievable today and scientists predict more new particles when still higher energies are reached.

So, an accelerator is a device which produces energetic particles which can be directed at targets allowing the scientist to explore the nature of the target down to smaller and smaller distances. There has been a continuous increase not only in energy but also in intensity (number of particles per second) from 1932 to the present day. The ZGS was about half way along in this energy domain – it was about 100 times higher in energy than the early cyclotrons and about 100 times smaller in energy than the Tevatron (one million MeV) that is presently under construction at the Fermi National Accelerator Laboratory. With its peak intensity of almost 10^{13} protons per pulse, the ZGS was within a factor of 3 of the world record. We will try to explore in the following chapter why the intensity and energy for the ZGS were chosen and what criteria, both scientific and political, were involved in that decision.

II. An Accelerator is Conceived

The genesis of high energy accelerators occured in 1947 and 1948 when plans were made for the 1-GeV machine at Birmingham, the 3-GeV Cosmotron at Brookhaven, and the 6-GeV Bevatron in Berkeley. Guidance for the protons in these devices was accomplished by the so-called "weak focusing" principle, the magnets all having a constant radial gradient; i.e., the magnetic guide field is constant as a function of radius.

In the fall of 1952 physicists in the Midwest started becoming vocal about the need for an accelerator in the Midwest. To keep high quality people and do competitive physics, they needed to match the work being done on the East Coast where the 3-GeV machine was operating and on the West Coast where the 6-GeV machine would be operational within a year. The university community wanted the government to support a new facility and a new laboratory which would be independent of Argonne, or if at Argonne, to be controlled by the using physicists. The Atomic Energy Commission, if it supported an accelerator at all, wanted it sited at Argonne. In the event it were sited at Argonne, the Laboratory management wanted no external control by the university physicists. It was also about this time (1952) that the alternating gradient scheme (AG) (ultimately of enormous importance to accelerators) was first considered at Brookhaven. Because the inevitable oscillations in the orbits are thereby reduced in amplitude, the magnets can be much smaller and less power is required to drive them than with the constant gradient design. Theoretical studies at the Brookhaven National Laboratory (BNL) and at the European Center for Nuclear Research (CERN) unearthed many potential hazards of operation. Many of these were laid to rest when in 1955 a small working model for electrons was completed employing electrostatic fields and in 1957 a 1-GeV electron synchrotron was put in operation at Cornell. Several years of study and experimentation with magnetic fields then followed, the CERN 28-GeV and the Brookhaven 33-GeV machines becoming operative in 1959 and 1960.

In early 1953 Argonne possessed two Van de Graaff machines of a few million volts and a brand new 21-MeV cyclotron for deuteron acceleration. Several members of the Physics Division (then directed by Louis Turner) felt that the Laboratory should have a more potent armament. To arouse such interest, Jack Livingood gave a series of lectures in the fall covering the mathematics of weak focusing cyclotrons, betatrons, synchrocyclotrons, synchrotrons and linear accelerators. Morton Hamermesh continued with a detailed discussion of the "strong focusing" alternating gradient idea then being studied at Brookhaven and CERN. Ed Crosbie and Mel Ferentz became sufficiently intrigued to join these lecturers in forming a group of four that concentrated on understanding and advancing (where possible) the theory of orbits in magnetic fields of various sorts including the recently studied fixed field alternating gradient (FFAG) idea.

Meanwhile, physicists at the Midwest universities continued studies on their own and formally incorporated in an organization known as the Midwest Universities Research Association (MURA) on April 15, 1954. The Atomic Energy Commission did not provide any funds for MURA and indeed on June 17, 1954, the AEC authorized and funded ANL to begin studies on an accelerator. The informal Argonne group thus became an Accelerator Study group under the direction of J. J. Livingood in July of 1954. This group was augmented in 1955 by Martyn Foss, John Martin and Lee Teng.

On May 6, 1955, MURA proposed to the AEC the construction of a 20-GeV FFAG accelerator. They were told by the AEC to work through Argonne.

On July 25, 1955, a formal proposal was made to the AEC by ANL for a 25-GeV proton accelerator of a novel type, embodying a two-stage machine. Stage 1 consisted of a synchrocyclotron with a single dee, wherein ions starting from rest would reach 2 GeV at a radius of 12 ft. Because of the increasing mass of the particles due to relativity, the oscillator frequency was to be lowered from 16 to 11 MHz 20 times per second, so as always to match the rotation frequency. The magnetic field of about 25 kG would be static in time and contoured according to the FFAG principle so as to provide radial and axial stability. Particle extraction would be brought about by an elaboration of the Tuck-Teng-LeCouteur regenerative scheme, already employed at the Liverpool synchrocyclotron under the guidance of Albert Crewe, and then being incorporated in the Chicago machine. Using this technique backwards, the ions were then to be introduced into Stage 2, which was an FFAG ring magnet of 20 kG with 150 ft radius. Acceleration was by a single gap cavity, the frequency rising from 32.7 to 33.6 and then falling to 33.0 MHz, 5 times per second. The final energy was 25 GeV with an output of 2 x 10^{10} protons per pulse.

This ambitious proposal died on the vine, because of the projected three years of model magnet study and six years of construction. The FFAG idea, originally proposed by Thomas in 1938, had not been understood until 1950, when two small electron cyclotrons embodying it had been built at Berkeley under strict security restrictions. The scheme was independently rediscovered in 1955 by Symon and by a trio of Russians, with a much more thorough understanding and analysis; a working model of a sophisticated electron machine was made in 1957 at Cornell University by R. R. Wilson and his group.

On November 23, 1955, a compromise between ANL and MURA was initiated by the AEC wherein Argonne would build a 12-GeV machine on a crash program in three or four years, and MURA would be given funding for studying the design of the "world's finest accelerator" to be built at some unspecified place and time. The MURA seemed to have won its fight for an eventual "dream" machine under university physicists' control. Argonne and its Director, Walter Zinn, who had fought very hard for a larger machine, were unhappy with building a machine that was so far below the energy of the 25-GeV Brookhaven machine. A visit to Washington by Livingood and Zinn to argue for a 15- to 25-GeV machine met with no success. The AEC felt that Argonne should build a 10- to 15-GeV machine in four years and that such a machine would give the U.S. "program superiority" over the Russians, who were constructing a 10-GeV machine, until the 25-GeV machine at Brookhaven became operational.

Since at this time neither the AG nor the FFAG ideas had been experimentally demonstrated and with orders to proceed at once, the only alternatives were to build a constant gradient machine (well proven in three laboratories), or the zero gradient device (which had been used in the static fields of mass spectrometers since 1951). On January 24, 1956, Laboratory Director Walter Zinn, at the request of the AEC, asked the design group to prepare, in 24 hours, a proposal and cost estimate for a 10- to 15-GeV proton accelerator to be completed in four to six years, this time scale and energy range having been set by the AEC. The instant-design-team percolator was warmed up, and a six-page document was delivered the next day (with the aid of a long night) describing a 12.5 GeV machine of constant gradient at a cost of $30 million. Zinn, unhappy with the AEC decisions, soon resigned. On April 2, 1956, Livingood was appointed Director of the Particle Accelerator Division with orders to proceed with all deliberate speed on a well-considered plan to outclass the Soviet 10-GeV constant gradient machine then under construction, and to keep costs low. In view of the restrictions on the type of machine, the design group chose 12.5 GeV as being larger than the Russian machine and sufficiently above threshold for the production of all the then-known particles. To meet the criteria of speed and low cost they chose the zero gradient machine as the one with the simplest design and easiest construction.

In this type of machine the magnet edges are set at an angle to the orbits, so that the radial component of the fringing field supplies the axial focusing force. The top and bottom surfaces of the magnet are parallel rather than being sloped with respect to each other. This insures easier fabrication and assembly; model magnet studies should take less time and, most important, multiturn injection is assured, making possible the design goal of 10^{12} protons per pulse.

Thus the ZGS was conceived. During the balance of 1956, the staff grew to 42 warm bodies. Buildings 805 and 806 were remodeled as offices and laboratories. Building 818 was acquired in January 1958 and experiments were started on model magnets.

As expected, the magnetic field was adequately flat across the chamber up to 14 kG, but the gradient rose disastrously at 20 kG. Models were then made with longitudinal holes drilled through the iron in the neighborhood of the coils on the basis of cut-and-try methods salted by the intuition of Foss. Finally, patterns were found which produced acceptably flat fields up to 21.5 kG. Maximum fields in the constant gradient Cosmotron and Bevatron were limited to 13.8 and 16 kG.

Pulsed magnet models then followed, and excellent progress was made under the leadership of John Martin on the design and construction of a prototype accelerating gap tuned by biassed ferrites. Rolland Perry and Philip Livdahl undertook the design of the 50-MeV Alvarez-type linear accelerator for injection (to be preceded by a 750-kV Cockcroft-Walton pre-injector). E. Frisby, K. Burba, and G. Calabrese worked on the design of the magnet power supply; D. McMillan battled with an early form of computer employing multitudinous mechanical relays for rapid measurement of changing magnetic fields; R. Lykken was busy with mechanical design; and J. Moenich, R. Trcka and K. Weberg worried over vacuum pumps.

A host of other problems were attacked by the group of seventy-some enthusiasts, augmented by A. Crewe, J. Marshall, R. Hildebrand, U. Kruse and S. Wright, borrowed from the University of Chicago to advise on experimental areas and beam steering magnets therein, while J. Fitzpatrick concerned himself with the design of buildings.

In August 1957 Congress authorized $27 million for construction, and $1.5 million was released in December to acquire an architect-engineer. Fitzpatrick reduced a preliminary list of 65 companies to 21, which were invited to submit proposals. Ten of these were interviewed and on April 9, 1958, the choice was made in favor of Sverdrup and Parcel of St. Louis. With the basic design now settled and a competent group at work on construction, Livingood resigned as Division Director in the fall of 1958 to write a text book on accelerators. Albert Crewe took over as Division Director and the final design and construction stage began.

III. The Agony of Birth

It was still to take five years to build the ZGS and the three-year crash program of 1955 was apparently long forgotten. By the time the ZGS was completed many of the reasons for its construction were no longer valid. The Russian machine never worked well and had few physics results; the 25- to 30-GeV machines at Brookhaven and CERN worked better than anticipated, and came into operation with high intensities four years before the ZGS. The ZGS, however, did provide a high energy facility in the Midwest. It did draw people to the universities, it did do a large number of good physics experiments, and it did finally provide a very good User-Argonne relationship.

In this chapter we will take a look at the construction and design, with emphasis on some of the more interesting and innovative features of the machine. The other important activity during this period was to formulate a mutually satisfactory arrangement between the users and Argonne. Up to this time accelerators had no formal user groups and a format was invented that since has become a model for use at other facilities.

With the selection of Sverdrup and Parcel, design of the facility began to take a more concrete form, but approval for construction had not yet been given by the AEC. During 1958 and 1959 design continued on the many accelerator components and on the general plant facility. It was perhaps the first time in the history of accelerator building that the architect-engineering firm was located so far from the construction site. For well over a year several people led nomadic lives, commuting once or twice a week to St. Louis. Finally on February 17, 1959, the AEC approved full authority for construction, and on June 27 the ground-breaking ceremony took place. Grading of the area, including 190,000 cu yd of earth-moving, and excavation for the machines and buildings began immediately.

One of the interesting features of the facility is the magnet foundations, which sit on concrete pillars. These pillars are known as caissons and extend all the way down to bedrock. The caisson-making machine was a Benoto deep-caisson excavator developed in France, and was a fascinating piece of machinery. The unit is operational at a job site by utilizing self-contained hydraulic controls and cable crane lift, and is capable of moving about under its own power. It can move forward, backward, and sideways; to the right and to the left by means of a system of rail wheels, skids propelled by hydraulic rams, and four corner jacks acting as moveable hinged knee-like supports. Once in position it acted like a huge drill. A steel cylinder about 1 m in diameter, with cutting teeth at the front edge, formed the drill bit. It was thrust into the earth by a to and fro rotary motion. As it sank out of sight new lengths of steel cylinder were added. After it reached bedrock, the dirt in the middle was dug out with a clam shell. A man with a steam hammer was then lowered to the bottom to level out the bedrock. Concrete was poured and the entire length of steel cylinder was pulled out before the concrete hardened. The caissons and magnet foundations were completed in early 1960.

Another interesting system was the double vacuum chamber for the ring magnet. The outer chamber depended on the magnet for structural strength and gave a rough vacuum. This allowed the inner chamber to be of thin-wall construction to avoid eddy current effects, while being strong enough to sustain its own weight over the width of the gap and such additional atmospheric loading as was imposed by the outer chamber. By 1959 only two specific construction methods were still being considered after many others had been eliminated. One was an epoxy chamber, which would have severe disadvantages from radiation damage, and the other was a "Spacemetal" chamber. Spacemetal was a material developed by North American Aviation under an Air Force contract. The Spacemetal chamber incorporates the use of a stainless steel sandwich made up of two outer skins with a corrugated core metal between. This became the final choice and was superior to the epoxy chamber in being stronger, much more radiation resistant, having a lower outgassing, and having less eddy current effects. Construction and testing of a prototype chamber continued throughout 1960. Using the techniques developed during this period, the 54-ft long, 32-in. wide by 5 3/4 in. high vacuum chambers for the ring were all constructed at ANL.

The ZGS ring magnet itself is also worthy of special mention. Careful model magnet studies were made to determine corrections that would allow the magnet to operate at high magnetic fields without distortion. It was found that holes drilled through the laminations close to the pole faces and near the magnet coils would preserve the flat field (zero gradient) even when the magnet was driven far into saturation. The ZGS design field of 21.5 kG was the highest of any accelerator ever built. By 1959 most of the model studies were completed and the major design criteria of the magnet were decided upon.

In general, other systems, such as the preaccelerator, linear accelerator, and radio-frequency system, were similar to the hardware at other laboratories. However, I would like to mention the ring magnet power supply which was built by Allis-Chalmers. The motor-generator set for the power supply consists of two 3-phase generators rated at 34,620 rms kVA each, a wound rotor motor rated at 14,500 hp, and a 68-ton flywheel interposed between the two generators. The total weight of the set is 841,000 lbs. The generators are connected to four rectifier transformers rated at about 17,000 kVA each. The final stages give about 11,000 V and 11,000 A across the magnet coils. The soundness of this supply is demonstrated by the fact that it has pulsed over 90,000,000 times during its use with the ZGS.

One other innovative feature of the ZGS was the design of external proton beam areas as an integral part of the facility. Other accelerators had extracted proton beams to experiments, but the ZGS was the first to design special areas where several experiments could be set up in series along the beam line and all take data simultaneously. Initially the design used transmission targets in the beam so that only a small fraction of the protons were used by each experiment. The last one in line could use a heavy target and use up all the beam if so desired. Later the proton beam could be split by special magnets and still more users could be accommodated. The use of proton areas and beam splitting became a standard design feature in later accelerators such as the Fermi National Laboratory synchrotron.

During 1959 and 1960 orders were placed for components, construction continued, and refinements to the design and plans for using the ZGS were formulated. By May 1961 the preaccelerator power supply had arrived from Switzerland, all the linear accelerator tanks had been delivered, all the drift tubes and linac quadrupoles had been fabricated and were being installed in the tanks. Final tests were being performed on the linac radio-frequency (rf) system, the synchrotron rf system was proceeding on schedule and the ring magnet supply was in an advanced stage of construction. Most of the vacuum system had been delivered and the inner vacuum chamber was being fabricated on site. However, one fly in the ointment was making itself felt - the ring magnet fabrication was running into difficulties. The schedule would be delayed, but at this point no one yet knew the magnitude of the difficulties.

By December of 1961, the fabricator had made ten center blocks for the ring magnet and one had arrived at ANL. If another fabricator could be used to make the end blocks, it might be possible to have all the magnets by fall of 1962. However, renegotiation of the contract was not possible. More seriously, the ten blocks had significant shorts between laminations and were unacceptable. At this crucial time, Laboratory Director Norman Hilberry resigned, and Al Crewe, Director of the Particle Accelerator Division, became Laboratory Director. Lee Teng became the Division Director and the on-going problem of negotiations with Baldwin-Lima-Hamilton (BLH) became his headache. In the summer of 1962, Ronald Martin joined the Division as Associate Division Director. The ring magnet problems and the tremendous pressure put upon the Division to finish the machine by August 1963, despite the late delivery of magnets, is well given by his recollections. "Acquiring the ZGS magnet is a burning impression in my memory. A significant part of the drama had been played out before I arrived at Argonne. Very detailed model magnet measurements were made, specifications had been written, and a contract let to Baldwin-Lima-Hamilton in Pittsburgh. But the first ten magnet blocks they made (80 center blocks and 48 end blocks are required for the ring) all had serious shorts between laminations and were unacceptable. Negotiations had been going on for some time to get BLH to modify their techniques and place higher priority on this sensitive job. Bob Lykken was head of the mechanical engineering group and had visited BLH a number of times. The breakthrough came when Charlie Laverick was sent out to negotiate. In characteristic style Charlie negotiated only with top level BLH management. He convinced them to make this job the prime responsibility of a vice president, to give the project high priority, and to assign a project manager acceptable to us.

"That agreement broke the ice but it still was not the end of our problems. Eventually much higher quality control was required. Emphasis was placed on cleanliness so stacking the laminations for the 30-ton magnets was done in a clean room with new fixtures for heating the magnets to bond the laminations. The most important role here was played by Herb Ross. He agreed to live with the magnet construction in Pittsburgh until it was completed. I was on the phone to him nearly every day, or at least three times a week. His suggestions to BLH and me and his daily monitoring of the situation were most important contributions to the eventual success of the magnet.

"Martyn Foss had designed the magnet on the basis of most detailed magnet studies of an 1/7 scale model. One of the critical features was holes in the laminations close to the pole face and near the limits, inside and outside of the region in which beam was to be contained. The purpose of these holes was to preserve the flat field (so called zero gradient after which the ZGS was named) to as wide a radial aperture as possible when the magnet iron was driven into saturation. The field shape depended very critically on the size and location of these holes. Since the calculations had been based on scaling measurements from a small model it had been Martyn's plan to modify the holes in many (perhaps as many as 32) of the center blocks based on magnet measurements of many of the first full scale magnet blocks. But no magnet measurements had yet been made when we passed this point in the magnet construction. So what to do? I ordered eight additional center pole magnets without holes. (The ZGS has eight octants and it is necessary to maintain symmetry). This seemed adequate since Herb Ross had demonstrated in Argonne's central shop that one could drill these holes on completed blocks without introducing shorts between the laminations and on a curve (the magnet blocks are tapered radially). Martyn, who had left ANL to go to Carnegie Tech, was hired as a consultant to help me understand how to make the hole correction. However, the information on which the shape and size of the hole was based had all been empirical and the data of the model measurements was all preserved in the computer output. I took one look at the volume of computer printout - measuring about ten feet wide by eight feet high and I estimated that it might take me three years to go through it all! If I recall properly, the order for the additional magnets was cancelled and we lived with the original design.

"That decision proved not to be too bad because the magnetic field proved to be quite good rather far into the saturation of the iron up to about 21.5 kG. We later did observe with beam a narrowing of the good field region at about 20.5 kG. Since the beam is narrower at full energy this restriction did not hinder acceleration past it, but if we wanted to target at this field the beam was somewhat unstable. The maximum energy did not make all that much difference anyway, and since it was significantly cheaper in electrical power not to push the magnetic field so high, the ZGS has been run for most of its life at 20 kG.

"I was later to learn what an efficient magnet design the ZGS was. This occurred when we designed a superconducting stretcher ring to fit into the same tunnel. It would operate with the same peak energy of the ZGS (increasing the energy by stronger superconducting fields would not get the ZGS into a new area of physics since the AGS operated up to 33 GeV). The stretcher ring was a strong focusing ring with independent dipoles and quadrupoles. Even with 30-kG dipoles the radius of the ring had to be 10% larger than that of the ZGS to contain 12-GeV protons. The ZGS design demonstrates a remarkably high average magnetic field (including straight sections). Thus maximal efficiency in the use of the magnet was obtained

"As time passed and BLH began producing good magnet blocks with regularity, the magnet measurements became a major crisis. Ed Berrill, who was a very slow and careful worker, made the search coils. In the end, the precision with which he did the job paid off. Still, delays threatened our schedule, and in April 1963 not a single magnet block had been installed in the ring! We had anticipated measuring many blocks before deciding how to place them around the ring. In April there was no time for that. The magnets were being measured on a six day/week schedule with two 10-hour shifts a day, and they had to be installed as soon as they were measured. Tat Khoe assigned the blocks to different octants, first arbitrarily, and later to minimize field errors up to the fourth harmonic as results of the magnet measurements came in. In this process I was warned that you shouldn't push people as hard as I was, and that their efficiency and productivity would drop off with extended duration of 60-hour weeks. Commercial data indicated a decline in productivity after three weeks on such a schedule. It is to the credit of that crew that their productivity under this schedule held up for about ten weeks, allowing the magnets to be all measured in time to meet the August 1 turn on."

The ring magnet was obviously the most formidable problem, but there were other crises that occurred before we were able to put beam into the ZGS. One was discovering a high resistance in one of the 54-ft long magnet coils after it was delivered to ANL. Precise electrical measurements located the problem to be in one of the cross-over joints in one end, which was not at all accessible.

Nevertheless, the insulation was carefully dug out using dental tools, protection inserted, the joint rebrazed and reinsulated. It was amazing that one could do such delicate work on such an enormous structure, and it was a tribute to the very high level of capability of a large number of people. Years later, this ability to handle serious problems was even more spectacularly demonstrated in replacing portions of several turns of a coil in place, as a result of shorts during operation. Such coil faults occurred on two separate occasions in the 16-year operating history of the ZGS. Another crisis arose when it was discovered that the very ambitious program to be the first accelerator to use computer control was not going to be ready in time. There was a crash program to install 40-50 scalers for timing signals and the barest essentials of a conventional control system. This initial system saved the day and subsequently the computer control system came on line and proved to be very valuable in operating the accelerator. Still another problem, which eventually proved to be a blessing in disguise, was the inadequate diagnostics to monitor beam conditions in the ZGS. This led to the gradual development of a great deal of new equipment and included some new innovations such as nondestructive instrumentation, which were copied by accelerators in Western Europe and the USSR and were then reinjected back into the USA at Brookhaven and Fermilab. The final problem that had to be overcome was the ring magnet power supply with its 144 rectifier tubes. This crisis and the drama of getting the first beam in the ZGS is well described by giving Ron Martin's recollections.

"Mercury vapor rectifier tubes can easily flash over (carry current when they are not supposed to). They have to be conditioned, brought up to operating voltage and power levels gradually, with faults occurring each time one raises either the voltage or current. They eventually operate reliably at a given level. They wear off sharp points, absorb impurities, or whatever. It's something of a black art, like the conditioning of a high voltage column. Since there were 144 rectifier tubes involved, it soon became very clear that conditioning them was a major problem and we didn't have a proper load to do it. Our last major crisis in this year was that the ring magnet power supply would take months to operate reliably, and we couldn't even begin until we had the magnet together!

"To gain some ground on this situation, we decided to put power from two rectifier banks into the first two ZGS octants completed. These were octants 4 and 5, and I believe we rotated the rectifier banks feeding them, so that all of the rectifier tubes could begin the conditioning process. Hence, my main battle station for the months of June and July was the control room for the RMPS in the Center Building. I learned more about 12 phase rectifiers, effects of different firing angles, the necessity of advancing firing angles proportional to the current, etc., than I ever cared to know.

"This was my location on the night of July 30, 1963. We were about to make the last connections on the magnet. We were trying to get vacuum on the entire ring for the first time and complete the electrical connections from the ring magnet power supply to the ZGS magnet. The tension built up. The proof that we had indeed completed the assembly of the ZGS was to be when we injected and circulated one complete turn of protons at 50 MeV. But from 6 p.m. until nearly midnight, there was a series of problems. Vacuum leaks, which were inevitable, had to be located and eliminated. Cooling water was leaking from the large bus lines to the magnet. The connections had to be rebrazed and welders were called in. Finally, when we thought we were nearly ready to go, something in the injection straight section box sprung a water leak. All of the techs had gone home and it seemed like curtains for the evening. But John Martin thought we could fix it ourselves. I recall taking wrenches in hand, opening the box (which required operating the crane to lift the box cover), fixing the leak and reclosing the box. In a half hour we had vacuum again, and power on the magnet. Lee Teng was in the main control room with another group and we said that we were ready for injection. Within a short time they announced that one turn had been circulated! We shut down and all went to the main control room to celebrate. A dream had come true. We had beaten by two days a schedule set more than a year before for assembling the ZGS. We accomplished something that AEC people as late as April had claimed was just not possible to do. It represented a lot of hard work and long hours for a large number of people.

"An important precedent was set that night which was to affect the attitudes of the Division throughout the years: if the machine is down, it's an emergency. The highest priority is to fix it as quickly as possible, and even senior staff are not immune.

"Injection was resumed the next day but we still could not get more than one turn to circulate. The signals from the scintillating screens in the ring were quite confusing even when they could be seen because of the low injected current. Beam positions did not fit any smooth curves. This was confusing to me because the ZGS magnet has a very large aperture, and it should be easy to get the beam to go around the ring several times. It would have to go around one million times in one second for acceleration to 12 GeV. I decided there was some rather major problem and resisted suggestions over the next few days to use the "kicker coils," which were for fine tuning the equilibrium orbit, until we had located the real problem. It might be an obstruction or a number of other things but the orbits couldn't possibly be that bad.

"Finally, I relaxed the restriction on the use of the kicker coils and several turns of coasting beam were obtained almost immediately, with rather surprising orbit correction currents required in octants 4 and 5. Later, when we were able to sort everything out we did discover very large orbit warps because octants 4 and 5 were quite different than the others! Finally, it became clear that the magnets weren't different, only the residual fields because we had used these two octants as a load to precondition the rectifier tubes, and they had been driven to higher fields than any of the others. We tried demagnetizing but this did little good until we could drive the entire magnet to higher excitation fields than we had previously driven octants 4 and 5. Eventually, when we could drive the whole magnet to full power, we several times removed the invert pulse so that the magnet was driven into an overcurrent trip set at 13,000 A (normal peak current is 10,000 A), enough to drive the magnets strongly into saturation. Only then did the differences between octants 4 and 5 and the other octants of the ZGS disappear! This experience, plus later experience on the sensitivity of the beam to the power level of the previous pulse, gave me a healthy respect for the complexities of magnetic fields even in iron – and the fact that it is not precisely calculable because, at least for small effects, it depends on the previous history of the iron.

"August and September of 1963 were a time of conditioning the rectifiers of the ring magnet power supply energizing the total ZGS magnet. Rectifier faults after a few pulses were quite common, especially if we tried to increase the excitation level. In the middle of August the ZGS magnets had been driven to a field equivalent to 3-GeV protons. Many of the senior accelerator people had been assigned to serve on shift during this time and we carried out the tune-up seven days a week. These included John Martin, Larry Ratner, Phil Livdahl, Rolland Perry, Bob Daniels, and myself. Acceleration was first achieved on August 18, a Sunday afternoon, with John Martin at the helm. The beam was carried to the full level of excitation at the time, 3-GeV.

"I recall being sorry I missed that event. I was out house-hunting at the time because my family had just moved here from Long Island a month before. We lived in several adjoining apartments in the guest facilities. People tell me they still remember the six little girls all in a line (not one was called Madeline) dashing over to the Argonne pool. I had been able to spend so much time on the ZGS that first year because I was alone. In August I had to cut back to about 80 hours a week and pay attention to other matters than the ZGS.

"As conditioning of the RMPS rectifiers proceeded, acceleration followed the increasing excitation of the magnet and tension mounted. Finally we were prepared to take the last step to reach full magnet excitation on September 18. I think that everyone again was in the control room in anticipation. Several times we had tried pulsing after the last increase and each time had a trip due to a rectifier fault. Finally the full cycle was achieved without a fault – but beam had not been injected. A second pulse – still no beam. Some acceleration on the third. Full acceleration to 12-GeV on the fourth full-power pulse. Fortunately we got the now-famous picture of the oscilloscope trace to prove it, since on the next pulse the power supply tripped on invert. The ZGS had worked! So we shut down and had a party in the control room. I wired Lee Teng about the event the next day and he announced it at an accelerator conference he was attending in Dubna."

Although the ZGS was on, there would still be a long way to go before it could be used for experiments. During the period of 1958-1963, interest among the potential users was increasing and the modus operandi for user-ZGS interactions had to be established. As we have mentioned in the prologue, there had been many years of mutual distrust between MURA and ANL. In gearing up for a high energy physics program it became necessary to overcome this distrust and to establish a mutually beneficial relationship. In 1958, Roger Hildebrand was named Associate Director of Argonne for High Energy Physics. Hildebrand was very much aware of the whole background and the problem of acute disappointment on the part of many Midwest physicists that the machine had gone to Argonne and not to MURA, and indeed of the general distrust of Argonne and the University of Chicago. One of the first things he did at Argonne was to start carrying out a strong program of getting the university scientists as actively involved in Argonne plans as they were willing to commit themselves. Toward this end he had many discussions with Ned Goldwasser of the University of Illinois, who had been quite active in MURA, but who recognized that reality called for working with Argonne and helping Argonne to provide the best possible facility for high energy physics research there.

At the time, the exodus of good people from the Midwest had reached alarming proportions. It was clear that if Midwestern universities were to continue to contribute effectively to the field of particle physics, it would be necessary to have an attractive high energy facility readily available to them. It further was apparent that in order for a high energy program to be set into operation with optimum efficiency and maximum productivity at Argonne it would be essential to obtain the fullest possible measure of cooperation and participation from the pool of high energy physicists already existing at the Midwestern universities. The need for collaboration was mutual. It was only necessary to bring together the interested people. The local Argonne staff, the universities, and the AEC were the three groups most intimately involved.

Hildebrand asked Goldwasser to provide all the help he could in forming an Argonne Accelerator Users Group which would be very active in all plans and programs in connection with the ZGS. Hildebrand also hired D. R. Getz as his assistant, with explicit instructions to spend as much of his time as was necessary on problems of the Users. Such a group was formed and Goldwasser himself was elected Chairman. A set of bylaws was adopted and various advisory committees appointed. There was a committee on bubble chambers, a committee on external beams, a committee on electronics, and a more general Users Advisory Committee to make recommendations to the Associate Director. Hildebrand also felt keenly the difficulty of trying to develop high energy research facilities in the absence of a high energy physics staff at Argonne, and he really desperately needed help from the users in the university community. Two things were needed. An in-house staff of experimenters who could specify what equipment to build, and then use that equipment. Because it was not realistic to expect such a staff to be available in a reasonably short time period in the number necessary to help with all the immediate tasks, a mechanism was needed to call on outside people for help in designing and building the facilities. Hildebrand was working hard on trying to recruit good physicists for the High Energy Physics Division research staff. At the same time, both he and Goldwasser were trying to find outside people who were willing to take on responsibilities for facility design at Argonne. The User Committees were asked for advice and assistance along these lines. In conjunction with the users and as a further attempt to satisfy MURA, an organization called Associated Midwest Universities (AMU) was formed, and John Roberson was named Executive Director. The AMU drew its funds from the AEC and was ostensibly to provide financial support for Users as well as give them an unofficial voice in the affairs of Argonne, including areas other than high energy physics. In fact it did provide funds for User activities, and help and advice on administrative, financial, and organizational problems. However, the office was regarded by many of the Users as a sop provided by the AEC and the University of Chicago as a token way of making Users think they had some entry into Argonne policy-making.

Within the AEC itself there was a careful evaluation of ideas concerning the part to be played by universities at AEC Laboratories. In a 1960 Report of the Joint Committee on Atomic Energy on the Future Role of the AEC Laboratories, the following statement was made:

"Progress has been made in recent years in arrangements for the utilization of these facilities by universities, both in the conduct of research programs and in affording opportunities to graduate students to do advanced work at the laboratories. Such arrangements have required modification of the normal practices of both the laboratories and the universities.

"Enough experience has been gained . . . to demonstrate that such arrangements can succeed . . . . In a number of specialized fields there are relatively few universities in the country that can offer first rank facilities . . . . Large accelerators and other unusual and expensive research tools, whether provided on university campuses or on Commission-owned sites, are intended to serve qualified researchers from other institutes as well as those on the staff of the operating institutions. The Commission will insist that these opportunities for broad use of its facilities in the interest of science be maintained."

This statement and others of similar nature influenced Midwestern university physicists to look with renewed confidence toward the possibilities for carrying on active research programs in high energy physics.

As the Users' Group gained in strength, many of the doubts that originally existed about the feasibility of participation in a research program at Argonne were allayed.

One area which Hildebrand felt from the beginning would be a fruitful one for User participation was that of design and construction of bubble chambers. He was, on the other hand, firm in his belief both that a very large bubble chamber was of utmost importance to the ZGS research program and that the large bubble chamber was such an important financial and scientific undertaking that it had to be designed and built in-house. He spent much time recruiting for a person to head this project, and eventually was very pleased when R. W. Thompson of Indiana University agreed to take a joint appointment at Argonne/University of Chicago and be in charge. Thompson had a reputation for being a super-meticulous physicist who did his experiments with great precision. Hildebrand felt that although a large bubble chamber designed and built at Argonne by Thompson would be somewhat later, it would be considerably superior to the Berkeley or Brookhaven large chambers. Many people interested in a large bubble chamber wanted to have it as fast as possible after the ZGS began operating, and they were irked that both the person and the arrangement (University of Chicago/Argonne) were not of their choice, nor did they have any official voice in the matter. In fact, the AEC was very short on the kinds of funds Argonne needed to build the large chamber, and thus Hildebrand himself could not establish a schedule for beginning construction of the chamber.

Some of the Midwest physicists who had ties to MURA felt strongly enough about all of the uncertainties that they proposed to Argonne and the AEC that a more modest-size bubble chamber (30 in.) be designed and built at MURA to be moved to Argonne by the time the ZGS was turned on. This was very appealing to everyone. It gave MURA scientists some active participation in the Argonne program and it gave some money to MURA and thus immediate jobs for MURA engineers. It also gave Argonne a reasonably large bubble chamber for turn-on and took some of the heat off the very large bubble chamber program. The AEC came up with appropriate funds and design began at MURA by physicists from MURA, Wisconsin, Purdue, Indiana, and Michigan State. In addition, a small bubble chamber was being built at Illinois by Goldwasser. He agreed that if Argonne would provide a magnet for this chamber, it could come and be under the aegis of Argonne as a general facility. All of the above chambers were liquid hydrogen chambers. Two other chambers were proposed by Users. A University of Michigan group under the leadership of B. Roe, D. Sinclair and J. VanderVelde had developed a preliminary design for a large heavy-liquid bubble chamber and they felt this would be a good complement to the Argonne program. Also, M. Block of Northwestern proposed to build a liquid helium bubble chamber for use as a general facility at Argonne. The entire proposed program was laid out before the Users and they were asked for advice. The result was that Argonne's bubble chamber program was:

a) build the large hydrogen bubble chamber at Argonne by Argonne,
b) build the 30-in. chamber at MURA/Argonne by MURA,
c) build the magnet for the Goldwasser chamber at Argonne,
d) build the large heavy liquid-chamber at Michigan/Argonne by Michigan.

During this same period, and closely associated with the above decisions, heated debates were going on about Argonne's policies with regard to the use of general facilities. If a university group spent months or years designing and building a facility should they not have initial priority on its use? But what about a single person, such as Thompson and the large hydrogen bubble chamber? He would have devoted years of his life to the chamber. Was he to get a guarantee of certain amounts of running on the chamber for himself?

These were issues where emotions ran strong. Hildebrand, with help from Goldwasser and the Users, arrived at the policy that builders of facilities would have first crack at using the facility but would have no priority after that. This policy was so distasteful to Thompson that he resigned, and again Hildebrand was left with no large hydrogen bubble chamber at Argonne. But the relationship with Users had improved.

Ned Goldwasser served as Chairman of the Users' Group for two years, 1958-1960. Keith Symon of MURA was Chairman from 1960-1962, and Kent Terwilliger of Michigan served from 1962-1964. During this time meetings were held twice each year to report on progress in construction and facilities at Argonne and to have somewhat informal presentations to the Users' Group by top physicists from the U.S. and Europe. Plans and developments at other accelerator centers were also presented.

Communications between the Argonne administration and the Users during this time became well established. Keith Symons' concluding remarks at the last Users' Group Meeting he chaired (May 11-12, 1962) included the remarks: "[we] have found the Associate Director [Hildebrand] cooperative almost beyond the call of duty. We have not always been in agreement with him but we have thrashed out our difference of opinions and our meetings have usually ended in agreement. When the Committee's deliberations have resulted in recommendations contrary to previous Laboratory policy, I believe I can say that in virtually every case our recommendations have been accepted."

It would thus seem that all of the problems were at an end. But such was not really the case. In fact, the Users were pleased with the situation except for one important thing. The situation still apparently depended on the personalities of individuals such as Hildebrand to be successful; there was still nothing official or contractual which gave any assurance that some change of individuals at the Associate Director level or above would not cause the whole situation to revert to that of a few years in the past. One of the Midwest physicists said more or less the following at that time: "We appreciate very much that Argonne has made us feel like especially welcome guests here. We feel wanted and necessary. But we want an <u>official</u> voice which will <u>guarantee</u> that our desires and needs are and will continue to be considered contractually by the AEC and the Operating Contractor of Argonne." From such thoughts as this came the eventual formation of the Argonne Universities Association (AUA) and the tripartite contract — AUA/University of Chicago/AEC — wherein policy decisions resided with the AUA.

By the time the ZGS was completed, User-ANL cooperation was good and the Users had a great deal of input to design decisions for the use of the synchrotron. However, during the next couple of years the growing pains of trying to make the accelerator and all the experimental equipment work well led to User unhappiness again.

IV. Early Childhood

The next few months were devoted to improving the intensity, debugging and improving systems, and learning how the accelerator operated. By October the intensity had gotten over 10^{11} protons per pulse, but it wouldn't improve by more than a factor of 2-3 for a long, long time. The machine was limited by an unexpected instability, a "resistive wall effect," which caused the beam size to increase to the point where it could not be contained in the aperture. Nevertheless, the machine dedication ceremony was held in December 1963, with G. Seaborg, Chairman of the AEC, and G. Beadle, President of the University of Chicago, in attendance. The principal speaker at the evening banquet was G. Bernardini of CERN and he gave a very optimistic outlook on the physics that could be done with the ZGS. Despite this optimism, it would be difficult for the ZGS to do frontier physics in competition with Brookhaven and CERN whose machines had twice the energy and had already been running for two years. To compensate for this the ZGS tried to provide better and more sophisticated experimental equipment. The 30-in. chamber had superb resolution and a very versatile separated beam, and the 40-in. heavy-liquid bubble chamber was the largest of its kind at the time of its first operation. Also, the pion focussing horn for the first neutrino experiment was a world-class instrument. The operation of a multitude of sophisticated beam equipment, in addition to commissioning and learning to operate the ZGS, made the transition from construction to operation a rather complex and demanding endeavor.

To add to the initial difficulties a hardware disaster occurred in early January 1964. The outer vacuum chamber of one octant collapsed. Because of the very wide aperture, the relatively thin inner vacuum chamber was not designed to withstand the full atmospheric pressure. In case a leak developed in the rough outer vacuum, a set of electrically operated safety valves was supposed to open and allow the equalization of pressure in the two systems. This electrical circuit failed to function and full atmospheric load collapsed the inner chamber. A complete repair was not made for about six months, but instead the chambers were pushed open as best as one could from the inside and the ZGS ran with a reduced aperture of about 3-1/2 in. instead of the normal 4-1/2 in. In September the octant magnets were taken apart and the chambers fully repaired. A simple puncture diaphragm was put in between the inner and outer chambers - these rupture automatically when the outer vacuum becomes too large and are fail-safe devices.

In March 1964, the machine was operating at an intensity level of 1-2 x 10^{11} protons per pulse, and we were beginning to insert targets to produce secondary particles for the three meson area beam lines. By the end of March one was able to target with the rf off and to spill beam out with the magnetic field. A group led by S. Marcowitz was busily engaged in the beam lines trying to measure the number and type of particles which were coming out of the machine. This "Beam Survey" would provide the users with information that would be useful in designing their experiments. Meanwhile studies continued on trying to improve the beam intensity and more of the experimental facilities became operational. On April 14, cosmic ray tracks were observed in the 30-in. bubble chamber and on April 22 dynamic polarization was achieved in a polarized target. On April 24, the first extracted proton beam came out of the machine with about a 10% extraction efficiency. Besides the extracted line, another experimental area (Meson Area) was set up with three beam lines which transported secondary particles. These particles were produced when the circulating beam hit a target inside the ZGS. The beam lines were set up at 7°, 17°, and 30° with respect to the target and circulating beam. During this time users were setting up equipment and tuning in the 30° beam line and in the neutrino area. By May 29, the beam survey was finished in the 7° beam line and the installation of a separated beam for the 30-in. bubble chamber commenced. On June 22 the beam survey in the 17° beam was finished and the installation of experiment E-3 ($\pi - P$ Elastic Differential Cross-sections from 2.5-6.0 GeV/c) by a University of Michigan group led by K. M. Terwilliger began. During July the separated beam was tuned and the ZGS targeting program began to use two targets simultaneously to send beam to three different experiments. In August and early September the vacuum chambers were repaired and it was hoped that the enlarged aperture would allow for higher intensities. Another milestone occurred in September when the computer programmer was tried in machine operations. The first tests went well and it appeared that within a few months the machine would be operable by the programmer. The 30-in. bubble chamber operated in October taking data and the first liquid hydrogen target became operational in the 17° beam line. In November the ZGS intensity had reached 3.10^{11} protons per pulse and the machine was on for 82% of its scheduled time. Operations seemed to be becoming more reliable and indeed the first two experiments E-5 (Excited States in He^4) by a Northwestern-ANL group led by R. A. Schluter and E-8 (Excited States in L_i and K^- x-rays) by the same group were completed.

Progress was indeed being made, but the continuous effort of installing experiments, getting new equipment to work, getting sufficient intensity and long spill times, and increasing the reliability of many systems was requiring an all-out effort that was as demanding as the hectic days of construction. The switch-over from construction to operation was difficult, and good procedures and lines of communication were still being developed.

In January 1964, R. Hildebrand resigned as Associate Laboratory Director for High Energy Physics and was succeeded by R. Sachs. Professor Sachs was the man whose task over the next few years was to be the establishment of a smoothly running operation which would satisfy the Users and allow the ZGS to take its place as one of the world's finest accelerators.

Also in 1964 the management of the Laboratory was changed so that the Midwestern universities would have an input into the Laboratory's policies and programs. The report of an ad hoc committee chaired by Professor John H. Williams of Minnesota was agreed to by all the parties concerned. This new proposal seemed to meet the desires of the Midwest university community and at long last would give them a voice in Argonne's operation. The following AEC news release of October 21, 1964, sums up the new organization.

"The AEC today approved the seven-man committee's recommendations – which had been approved ealier this month by AMU, MURA, and the University of Chicago. The committee's recommendations follow:

"1) The committee recommends that the prime contract for the operation and management of the Argonne National Laboratory for the Atomic Energy Commission be changed to a tripartite agreement, the parties to be the AEC, a not-for-profit corporation to be organized by a group of Midwest universities and the University of Chicago. The activities of the parties pursuant to this plan shall at all times be subject to the provisions of the tripartite contract.

"2) The function of the new not-for-profit corporation of Midwestern universities shall be to formulate, approve and review Laboratory policies and programs.

"3) The function of the University of Chicago shall be to operate the Argonne National Laboratory in a manner responsive to the policies established and approved by the new corporation.

"4) The terms and conditions of the new tripartite contract shall be such as to assure the new not-for-profit corporation of Midwestern universities that their decisions shall be carried out, and that policies approved by the corporation shall be put into effect.

"5) The terms and conditions of the new tripartite contract shall be such as to assure the Univerity of Chicago that it will be able to effectively operate the Laboratory in a manner responsive to the policies established by the new corporation.

"6) The committee recommends further that the parties recognize (a) that they share a mutual responsibility for promoting the maximum scientific progress and engineering development made possible by the funds and facilities provided by the Government, (b) that they must cooperate in order to stimulate scientific and technological advancement in the Midwest community and the Nation, and (c) that these purposes can be attained only by continued emphasis on recruiting and retaining on the staff of the Laboratory the most competent and creative scientists and engineers available and by affording them full support."

Glenn T. Seaborg, Chairman of the U.S. Atomic Energy Commission, commenting upon the AEC approval of the plan, said:

"The AEC is pleased to accept the principle proposed by the AMU-Argonne-Chicago-MURA Ad Hoc Committee for the tripartite arrangement for the management and operation of the Argonne National Laboratory. The AEC believes that this arrangement will promote the increased participation of the academic institutions of the Midwest in the work of the Argonne National Laoratory. It is another important event in the life of an outstanding laboratory whose significant research and development accomplishments are recognized around the world. The Argonne National Laboratory will continue to play a vital role in the engineering and development aspects of the reactor development program as well as continuing to conduct important research in the basic physical and life sciences. The morale and excellence of Argonne's staff, under the able direction of Albert Crewe, are a cause for pride among all of us. We know that the future will bring even greater service from them to science and to the Nation.

"This new plan will help to stimulate continued growth of Argonne's programs and facilities. It will enable the many important institutions in the Middle West to develop their own programs more efficiently through direct familiarity with proposals for new programs and facilities at Argonne. At the same time, the plan proposes to retain the competent and experienced management provided for so long at Argonne by the University of Chicago.

"The committee's plan is unique and creative. It offers promise of meeting two important AEC objectives in the administration of our national laboratories. The first objective is to foster strong, unified management of the total research and development program assigned to each national laboratory. The second objective is to permit major universities in each region to take part in planning the scientific program and facilities. We are confident that the tripartite concept will permit the accomplishment of both objectives at Argonne."

With the ZGS beginning to operate more reliably, and with the Users getting the official voice that they had desired, 1965 appeared to be the dawn of a new era for high energy physics in the Midwest, but high energy physics is a tough and expensive game and despite some great successes in 1965 there were disappointments which again led to User bitterness.

The ZGS was operating consistently with over 80% reliability and accelerating $2-3 \times 10^{11}$ protons per pulse. Studies had been made of the causes of the intensity limits and equipment was being designed to overcome them. An idea of the problems and solutions is given by the following description from Lee Teng's report to the Users in May 1965.

"We have also tried to understand our intensity limitations and I think we are now in the state where we can explain these limitations and are doing something about them. In this, we have been greatly helped by the people from MURA in both understanding the intensity limitations and providing equipment for overcoming the limitations. The intensity limitation is as follows. The Cockcroft Walton and 50-MeV linac injector have been operating for more than a year without difficulty. They are very reliable and the injected curent has steadily gone up. We can count on injection of more than 15 mA, with peaks of 30 mA. This current allows us to inject into the ring a coasting beam of $5 \times 10^{12} - 10^{13}$ particles per pulse. The theoretical value for the rf pickup is about 30% and we indeed pick up just about that much. In other words, immediately after the rf is turned on, we can accelerate $1-2 \times 10^{12}$ particles for a few milliseconds. The first limitation comes at about 30 msec after acceleration, at a kinetic energy of about 200 MeV. At that time, we have a limitation which we understand to be a resonance between the vertical and radial betatron oscillations, $\nu_x = \nu_y$. Our present mode of tuning the accelerator with dc magnets is inadequate for tuning out this resonance. What we really need are poleface windings to shape the field in the octants. Our long-term program includes new vacuum chambers with poleface windings incorporated in them. But this will take a minimum of a year and a half. More immediately, we are simulating poleface windings by putting tuning coils on the end guards of octants. They will be added by July or August and will, we hope, be able to cope with the 30-msec resonance limitations.

"At about 100 msec (at more than 1 GeV in energy) we encounter another intensity limitation, the so-called coherent resistive wall effect that was discovered last year. This causes the beam to undergo coherent vertical oscillations. Eventually, we lose the beam on the chamber top and bottom. It has been shown both at MURA and at the Cosmotron that the effect can be damped by essentially inverse feedback. Equipment to do this is being built with the help of MURA. We expect to install the equipment in July."

Indeed the installation of the damping equipment immediately led to beams of $4-6 \times 10^{11}$ in August and with the addition of the end guard coils this was upped to $5-6 \times 10^{11}$ by December. This first trial was successful, but indicated the need for better amplifiers and also pointed out that many machine parameters would have to be optimized before a total correction could be achieved.

At the time of the May 1965 meeting both the Users and the AEC were satisfied with the progress being made as evidenced by the fact that the AEC was funding the large bubble chamber (the 12 ft) and a second proton area at a cost of $20 million, and by the following statement by Professor W. D. Walker of the University of Wisconsin.

"I would like to make a few remarks in my role as chairman of the Users' Group. First of all I might try to give a bird's eye view of the operations at the ZGS for the past year. My view is that it has been a pretty tough year. We had various difficulties with the bubble chamber that extended up until about the first of November. I think the accelerator operation was also fairly sporadic until that time. Since then, I think we have done exceedingly well. I don't think this reflects on anyone. The year was spent partly in learning how to run these various pieces of apparatus. It is extremely difficult to start a large operation like this from scratch, and I think the people who have done it are to be complimented for their great efforts."

Despite the progress and the compliments, there was frustration behind the scenes. The experimental program was going slowly and the Users lashed out at the way things were operating and at the quality of the staff. The following letter of July 29, 1965, to Associate Director Sachs from a User gives the flavor of this unrest which lay below the complimentary surface.

"On two occasions yesterday you asked me to document statements.

(1) In the case of recuiting engineers.

(2) On overdesign of beam transport equipment.

"The first example is almost self evident from the lack of competent, creative engineers around PAD. It might not be just a matter of recruitment, but also of the rather bewildering organization that exists in PAD. This is going to be an extremely difficult thing to remedy. Both you and Lee are theoretically oriented and I am not sure you can see what is obvious to an experienced, hardware-oriented experimentalist.

"In the case of (2) I am sure I can get facts and costs from Brookhaven, but do you _really_ want this. It will not help particularly to rub Teng's nose in past mistakes. What is desperately needed is a reorientation. Things are done in an incredibly expensive way when you compare with the often makeshift operations at Brookhaven. You should have talked to some of our bubble chamber technicians after they visited Brookhaven.

"One or two really good internal groups could possibly change the _style_ of operation at the ZGS. One or two able physicists could make an enormous difference. Lots of your energy has gone into trying to recruit physicists for administrative purposes. This is important, but simply having one or two more competent experimentalists could make enormous differences."

These allegations were rebutted and the Argonne position outlined in the following answer from Dr. Sachs.

"I am writing in response to your letter of July 29, 1965 which concerns the recruiting of engineers in PAD and remarks on "overdesign" of beam transport equipment.

"It is my impression from the discussion with the Users' Advisory Committee that this is not intended to be interpreted as an official statement from that committee, but only as an expression of some personal impressions.

"I am sure that there have been, and always will be, isolated experiences which tend to annoy you and other users of the ZGS from time to time. However, I have usually found, after careful investigation, that many of the strong criticisms, especially the unfavorable comparisons with other laboratories, are the result of ignorance of the facts. For example, we have prepared a very careful comparison of our operating costs relative to those of the Brookhaven high energy physics program and find that the situation is not nearly so bad as most people make it out to be. Furthermore, I find that at least two careful examinations of the so-called "overdesign" of our beam transport equipment have been made, again with results that do not bear out the casual comments.

"I have found that Lee Teng keeps operating conditions under continual review and that he makes changes in policy and style where the evidence shows that they are clearly needed. These changes are not made easily or quickly because many factors are involved, but they do take place in a systematic fashion. That is not to say that he cannnot learn anything from others, but it does mean that serious suggestions for changes should be made on the basis of undeniable facts. Furthermore, before going into an investigation of the facts, an effort should be made to determine whether Teng is aware of the problem, whether he has already gathered the facts, and whether he has or has not instituted changes in policy as a consequence of his investigations.

"In reference to your remarks on recruiting, I hope that you realize that we do have "one or two able groups" here. However, we all agree that more strength is needed and we have made continual efforts to add strong people. In this connection, you might realize that one of our difficulties in recruiting strong physicists is the unfavorable image of our operations that is being casually and irresponsibly perpetuated in the world of physics, often by our Users' Group. This may serve the purposes of Brookhaven and Berkeley, but it certainly does not help high energy physics in the Midwest. I can tell you that these stories do get back to the AEC and can lead to severe reductions in the support required for a viable Midwestern high energy physics program. Therefore, I again recommend that you do everything you can to educate the community to the need for <u>responsible</u> criticism based on thoughful study of the facts.

"If we are to be successful in running a high energy physics program dedicated to the needs of the university physicists, we must develop a sense of mutual loyalty and e'sprit de corps in the community. Efforts in this direction must be made from both sides. From the Argonne side, Lee Teng knows that the only ultimate objective of his work is to give the experimentalists an opportunity to do physics, so his loyalty to the community manifests itself as a fierce determination to accomplish this objective as best he knows how. From the users' side this loyalty has been shown through the willingness of many of you to take on responsibility for specific tasks, but that must be supplemented by a deeper and continuing sense of responsibility in regard to the entire effort if we are to succeed."

Despite the frustration and the painful progress of early childhood, nine experiments were completed in 1965 and operations appeared promising for the future. Many more experiments were to be completed in 1966 and the crawling stage of childhood entered into the walking stage of adolescence.

V.   <u>Adolescence</u>

The year 1966 opened with the Particle Accelerator Division going all out to try to improve operations in line with some of the Users' criticisms. One important effort was the manning of shifts in the experimental area by senior staff members. The initial operations of the Experimental Area Planning and Operations Group (EPOG) had been organized by M. Hildred Blewett and she had done a tremendous job in getting completely inexperienced people to learn how to operate all the equipment, interact with the users, and maintain equipment, as well as build and install new equipment. However, the tremendous amount of new facilities and new equipment that was being built and operated required more engineering expertise than was available to the operating crews. For this reason, and to also try to improve communications from shift to shift during operations, a group consisting of Royce Jones, Assistant Division Director for Experimental Areas Operations, Phil Livdahl, Deputy Group Leader of EPOG, Wayne Nestander, Associate Group Leader of EPOG, R. C. Juergens, Mechanical Engineering Group Leader, R. Trcka, Mechanical Systems Section Leader, and F. C. Beyer, Cryogenics Systems and Mechanisms Section Leader, began lending their talents to the multiple-shift operations. Their expertise, as well as the fact that one was likely to find the Assistant Division Director working the owl shift, not only aided the technical efforts but also emphasized to all the importance of the operations. Indeed the fruits of these labors were clearly indicated by the fact that the number of completed experiments jumped to 21 during 1966.

The 30-in. bubble chamber with its complex separated beam line completed ten experiments including E-6, which was an exposure of 5.5-GeV/c K^- mesons and an experiment that was competitive with the work on the higher energy machines at Brookhaven and CERN. In this year the bubble chamber and its physics gained an important position in the world's physics output. Another world first was achieved when the first superconducting bubble chamber (a 10-in. liquid helium chamber) completed a successful run with a K^- meson exposure.

Another important feature was the operation of the external proton beam (EPB) with two sequential foci, from each of which more than one experiment could operate. The EPB was able to complete four experiments during this year. Physics is, of course, not only concerned with the number of experiments but also with their quality; i.e., their importance in understanding nature and discovering new phenomena. The quality of the work being done on the ZGS might be indicated by the fact that two experiments attracted the attention of the national news media as well as being mentioned in the Congressional Record as shown by the following excerpt from the Congressional Record-Appendix of April 20, 1966.

Extension of Remarks
of
Hon. Weston E. Vivian
of Michigan
In the House of Representatives

Mr. Vivian: "Mr. Speaker, it is with great pride that
I call to the attention of my colleagues recent
accomplishments by a team of physicists from the
University of Michigan, Ann Arbor, in the Second
Congressional District of Michigan; the district I am
privileged to serve in the U.S. Congress.

Within recent days, Drs. Alan D. Krisch, John
R. O'Fallon, Keith Ruddick, and graduate student
Steven W. Kormonyos, all members of the Department
of Physics at the University, and Lazarus G. Ratner
of the Argonne National Laboratory, in Argonne, Ill.,
have reported on a significant discovery which they
made of a hitherto unobserved sub-nuclear particle.

The second experiment on proton-proton elastic scattering at
90° in the center-of-mass did not make the Congressional Record, but
as shown by the following press release of November 22, 1966, did
receive national coverage in the news media and postulated that
there was an internal structure to the proton.

Physicists have found evidence for several layers
within the proton, the heavy, positively-charged
particle within the atom's nucleus. The proton was
thought to have structure, but there was no strong
evidence that it did. This finding supports the
idea of structure in the proton.

Discovery of the new layers in the proton was made
by a team of scientists using the 12.5 billion
electron volt (GeV) Zero Gradient Synchrotron (ZGS)
at Argonne National Laboratory near Chicago.

The physicists were Carl W. Akerlof, Alan D. Krisch,
and graduate student Ross H. Hieber of the University
Michigan; Kenneth W. Edwards, University of Iowa;
Keith Ruddick, University of Minnesota; and Lazarus
G. Ratner of Argonne. They reported their findings
this week in Physical Review Letters, under the title,
"Proton-Proton Elastic Scattering at 90° and Structure
Within the Proton."

Other noteworthy accomplishments during 1966 were the
completion of E-1 (Quasielastic Neutrino Scattering) by T. Novey's
group and the completion of a 400,000 picture exposure of
5.5 GeV/c K^- in H_2 in the 30-in. bubble chamber. This latter
experiment, by a group led by T. Fields, produced an enormous amount

of data which was analyzed and published over the following seven years. Besides the high energy results, there were also results from a very strong program in nuclear chemistry by a group led by Ellis Steinberg of ANL. Indeed, this program continued throughout the life of the ZGS and even provided techniques for measuring the proton beam intensity, an important factor in several high energy physics experiments.

The completion of this many experiments obviously indicated that the ZGS was also running well. There was an increase in reliability and total running time despite the fact that the machine was down for eight weeks to rewind the rotor of the ring magnet power supply. Routine inspection had shown that the rotor windings were beginning to bulge out and it was only due to excellent maintenance procedures that this was caught before a major catastrophe could occur. Within two days after the reinstallation of the rotor, the ZGS was in operation with a beam intensity of 3.5×10^{11} protons per pulse. Also, the hard work of the last two years on studying and trying to correct the problems causing the intensity limitation began to bear fruit. The analog damper for the resistive wall instability became operational and with some tuning of the end guard correction coils the ZGS finally broke the 10^{12} proton per pulse barrier in September 1966. The ZGS had reached its design goal three years after completion of the machine. A peak pulse intensity of 1.7×10^{12} was recorded and the machine reliability for November and December 1966 reached 91%.

Another important capability of the ZGS which played an important role in the carrying out of the proton-proton 90° scattering experiment was the use of the "front porch" spill. This allowed the extraction of some of the beam at any intermediate energy for one experiment, and allowed the rest of the beam to be carried up to full energy for the use of other experiments on the same machine pulse. This is not done even today on the Brookhaven accelerator. For the proton scattering experiment this front porch was set for 51 different momenta between 4.0 GeV/c and 13.4 GeV/c. It only took ten minutes to make each change and the extraction efficiency over almost this entire range stayed at a very good level of 20%.

As the year 1967 began, new equipment and new modes of operation were initiated as the development work of the previous years began to be complete. The 30-in. bubble chamber went to a two picture per machine pulse operation and in the first quarter of 1967 some 888,000 pictures were taken. Also at this time a scheme devised by Lee Teng to inject beam with the rf on, at a low level, and prebunch the beam before acceleration led to a new intensity high of 2.4×10^{12} protons per pulse. New monitoring equipment was installed in the ZGS and this Beam Profile and Position Monitor, developed by W. H. DeLuca and F. Hornstra, gave a visual picture of the beam size and position during the acceleration cycle. This would prove very useful in determining where beam was being distorted and becoming less intense and less suitable for targeting.

In June the bubble chamber went to a three picture per pulse operation and a slow spill for other beams was used between bubble chamber pulses. By the end of the year beam control and a sophisticated targetry system allowed seven individual spills per pulse. There were three fast spills to the 30-in. bubble chamber and one fast spill to the 40-in. bubble chamber at the full energy flat-top. There were also two slow spills interlaced with these to the 17° beam with a 600-msec length. In addition there was a 300-msec spill to the EPB at a lower energy front porch. With the ZGS intensity at 2.8×10^{12} there was now enough beam for this ambitious targeting program. The target position requirements were staggering: positional precision in radial position and height of 10 mils; highly reproducible, longitudinal positon variable over six feet within the magnetic field of the octant four magnet, angular variation in three dimensions; and readouts of all settings. Originally there had been an attempt by Brobeck Associates to develop a hydraulic system, but after some work this did not appear to be feasible and the contract with Brobeck was cancelled. Instead the job was turned over to the ZGS Conventional Controls Group led by Andy Gorka, who fortunately turned out to be one of the world's cleverest gadgeteers. He might put Rube Goldberg to shame, but Andy's devices really work, even though some of the parts come from the Sears, Roebuck catalog. The sophisticated target manipulations that were required for the seven-spill program gave ample proof of Andy's success.

The successful high intensity running, however, was bringing new problems. Blisters were appearing from radiation, especially in the areas where there were targets. From the very beginning it was recognized that the Spacemetal chambers bonded with epoxy would be damaged by radiation, would outgas, would become brittle, and eventually fail. Will Hanson, who was responsible for their design, was even then searching for a better solution. He did discover an ingenious solution to the problem: diffusion bonding of titanium, a process which North American Rockwell was developing. A prototype chamber arrived at Argonne in 1966 and soon North American Rockwell was producing chambers. The first 54-ft long octant arrived at Argonne on October 12, 1968. These chambers had no organic material and were impervious to radiation problems. Another important feature was incorporated in them. This was the addition of "pole face windings" which would permit modification of the magnet field of the ring magnet. They would do a much better job of correction than the end guard coils and would be able to correct the ZGS tune through the acceleration cycle. They would also provide the necessary field changes to do resonant extraction of the proton beam - a much more efficient method than targeting. These were not finished until 1970 and not installed until 1972 and there were to be continual vacuum chamber problems for the years between. Despite this, the year 1967 saw an operating efficiency of 90%. Between January 1966 and September 1967 the operating efficiency had increased from 80% to 90%, the time for physics had increased from 500 hours per month to 600, and the intensity had increased from 4×10^{11} to 2×10^{12}.

In September 1967, Lee Teng resigned to go to the then-forming National Accelerator Laboratory and work on the 400-GeV machine being. designed there. The Particle Accelerator Division was reorganized and Ron Martin became the Director of the new Accelerator Division and Royce Jones became the Director of the High Energy Facilities Division. Operations continued to go well and experiments were being done on a realistic schedule. The machine seemed to have been nursed through its growing pains and when Bob Sachs resigned in June of 1968, he could look back on a job well done. It would be the task of his successor, Bruce Cork, to oversee the operations of the adult machine and to foster the program of introducing new and innovative features that continued to keep the ZGS in the forefront of high energy physics during its remaining life.

The machine operation continued to improve during 1968 both in reliability and intensity, and new equipment kept making its mark. In April, a beam splitting septum was installed in the EPB and still more users could be put on line. By September of 1968 the control and monitor computer, which had been developed by Lloyd Lewis and his group, was an essential part of machine operations and was capable of also doing background as well as foreground jobs such as tuning for the extracted proton beam lines while still running the accelerator. Another 21 experiments were completed in 1968, bringing the four-year total to 75. The efficiency rose to 91.7% with an average intensity of 2.24×10^{12} per pulse for the entire year. A peak pulse of 3.3×10^{12} was recorded.

The year 1968 was a good one for the ZGS and again saw some important physics output. Two papers were published in 1968 following one paper in 1967 on the results of a series of experiments begun in 1966 by an ANL, Iowa, Michigan, NAL collaboration led by Professor A. D. Krisch of Michigan. These were the first experiments on the so-called "inclusive cross-section," i.e., the reaction $p+p \rightarrow (\pi^{\pm}, K^{\pm}, p^{\pm})$ + anything. This type of physics became an important part of the world's physics programs over the next four years. Also during 1968, the bubble chamber program got into full swing with 15 experiments completed in the 30-in. chamber and two in the 40-in. heavy liquid chamber. Another activity in which ANL was becoming a world leader was in the use of polarized targets. Polarization effects were measured in π P elastic scattering by a group led by N. Booth of the University of Chicago and A. Yokasowa of ANL. With these experiments completed in 1967 and 1968 it was clear that the ZGS had come of age and was competitive with the older laboratories both in quality and quantity of physics.

VI. Maturity and the Death Knell

The ZGS was now in full swing serving the Midwest as a regional facility for high energy physics. Previously, the programs of the Midwestern universities had been limited because the East and West Coast laboratories were regional facilities for their areas and there was very little left for outsiders. The ZGS was capable of running a dozen simultaneous experiments and with its 90% operating efficiency, many experiments were completed each year, despite substantial budget reductions every year from 1969 on. The manpower level dropped from a high of 525 in 1968 to about 250 in 1978. About 50 more people were supported in 1978 by non-high energy physics funds as some other projects started up. With the demise of the ZGS in 1979, the Midwest may again suffer the problem of doing experiments at distant sites. The coming of the Fermi National Accelerator Laboratory ameliorated this condition to some extent, but the operating mode was different. A small group could no longer do an experiment in a few months, but had to join large collaborations, and take years to mount and do experiments. The opportunities for younger people to do things on their own is seriously curtailed in such a situation and this slows their development. All research facilities must eventually end, but in the case of the ZGS it was particularly painful. The ZGS marked both the beginning and the end of the Midwest regional facility, in contrast to the West and East Coasts which still maintain such facilities. Although these are becoming more "national," they are still most fully utilized by universities in their respective areas.

But back in 1969 the ZGS facility was still growing, improvements continued, and new capabilities became operational. On February 12, the second extracted beam (EPB II) installation was complete and by March the experimental program there had started. At the same time a third sequential focus was set up in EPB I and more experiments came on line. In April the 30-in. bubble chamber started quintuple pulsing and the ZGS supplied slow spill to the counter experiments in between the bubble chamber spills. The rapid analysis of the prodigious number of bubble chamber pictures was made possible by the development of POLLY II, an automatic scanning and measuring device. This system, developed by a group led by H. Bruce Pillips, performed so well that many universities built or bought copies for their own use. Another development that was important for tuning and operating the many experiments in the proton beam lines was the addition of a 48-channel integrator and electronic scanner for the Segmented Wire Ion Chambers (SWIC). This gave quantitative, normalized beam profiles at many stations on each ZGS pulse and these SWICs were connected to the CDC-924-A computer in the main control room for on-line monitoring of beam spot size and position. The computer generated a set of data related to the size and position and gave profiles and digital information to the control room operators and the experimenters. This capability was very important for the efficient tuning of the beam lines and also for the operation of the individual experiments. The accelerator was running well and 26 experiments were completed in 1969.

In July of 1969 the Accelerator Group achieved a world's first when they successfully injected H^- instead of H^+ into the ZGS and were able to accelerate the proton beam (H^- was stripped to H^+ at injection) to full energy. This was a very important achievement since H^- injection avoids many normal injection problems and allows the machine to be filled to its space charge limit very easily. The ZGS eventually went to normal operations with H^- and was able to achieve new intensity highs. This method was later adopted by Fermilab and is now being installed at Brookhaven. The Chinese are now building a 50-GeV accelerator and will also use this method. This injection concept also led to new ideas for a Proton Diagnostic Accelerator for medical radiography and perhaps most important it stimulated ideas on a very ingenious scheme for fusion power which is now being actively studied by the Accelerator Groups at Argonne and other laboratories. The H^- program also led to another significant activity. Though H^- was injected directly into the ZGS, it would lead to much higher intensities if it could be injected first into a 500-MeV booster. The program of constructing Booster I and Booster II went forward and although it was not completed in time for use in the ZGS, it led to a very useful accelerator for producing neutrons for use in studies in solid state and material properties. Booster II, now known as the Rapid Cycling Synchrotron (RCS), will remain in operation after the ZGS shutdown. In fact, the Intense Pulsed Neutron Source program is planning the design of an even more intense machine which would be competitive and in some aspects superior to the best that can be done with the reactors which are the present principal neutron sources.

There was yet another first in 1969 when the first tracks were observed in the superconducting 12-ft bubble chamber. The successful operation of this chamber marked an important advance in the use of superconductors and showed the remarkable progress that had been made at Argonne by the people working with Charles Laverick and John Purcell. Back in 1964 Laverick and George Lobell had constructed the coils for the 10-in. liquid-helium bubble chamber. Laverick had continued his experiments to improve the superconducting wires and braided cables and became one of the outstanding contributors to the knowledge of how to make and use superconducting devices. Indeed, he built several record-breaking magnets, each one achieving higher magnetic fields than the previous ones. Argonne's involvement in this field advanced the state of the art for the whole world and gave Argonne a group of experts who have now built a superconducting magnet for Fermilab's 15-ft bubble chamber and a superconducting magnet for a magnetohydrodynamic (MHD) power plant in the Soviet Union. This joint US-Soviet project is designed to develop the use of MHD plants which are able to convert fossil fuel to electricity much more efficiently than conventional plants. Argonne is now building another large coil for a U.S. MHD facility and is recognized as a world leader in the construction of large superconducting magnets.

In January 1970, the beam intensity average reached a new high of 2.4×10^{12} with the installation of a higher power amplifier in the vertical damper system. The addition of a radial damper also seemed to keep down the radial size of the beam. The ZGS went off for two months on April 21, 1970, when there was a main magnet coil failure in octant 2. During this period a spare coil was installed as well as a new high gradient column for the preaccelerator and a resonant extraction magnet for EPB II. Resonant extraction more than doubled the amount of beam that could be extracted compared to the targeted extraction scheme. The ZGS started up again on June 26. On November 13, 1970 the first neutrino event was observed in the 12-ft bubble chamber and this facility was on for production.

The years 1969 and 1970 saw some significant physics being done at Argonne and the annual review committee mentioned the following four experiments as being especially noteworthy.

1. Phase of the CP violating amplitude for $K_L^0 \pi^+ \pi^-$ by a University of Chicago – University of Illinois Circle Campus collaboration.

2. Showing of the high degree of ρ^0 coherence in the process $pp \rightarrow \pi^+ \pi^- (\rho^0 \rightarrow \pi^+ \pi^-)$ by an ANL group.

3. The sensitive search for the exotic (non eight-fold way) hyperon, Z^* by looking at proton polarization in elastic $K^+ P$ scattering by an ANL–University of Maryland group.

4. First possible evidence in hyperon semi-leptonic decay that the "single angle" Cabibbo Theory is inadequate by an ANL–Ohio State–University of Chicago–Washington University Group.

However, things never go smoothly and on January 9, 1971 another main magnet coil failure occurred in octant 3 and there was no spare coil. The repair of the two damaged turns (out of 30 turns) of the main coil of octant 3 presented some challenging problems. The damaged area was 14 ft from one end of the 54-ft long coil and we preferred not to disturb the delicate hand-fitted insulation at the coil ends. It was therefore necessary to insert a copper splice into each of the two damaged turns (the copper cross section is 0.9 in. x 1.4 in.) without affecting good insulation in either direction from the splice. The solution adopted was to free the two turns from the main body of the coil for 7 ft in either direction from the final and most critical joint of a 3-ft splice. A specially designed fixture was then used to bend outward in an "S" curve each part of a turn by a precise amount (2 in.).The thermal expansion due to local heating at $1300^\circ F$ then allowed a silver braze to join the two coil sections with about 200–400 psi of compression. The compression was monitored by strain gauges at either end of the turn being spliced. After the final braze joint was completed the turn was held in this outward position without putting large forces on the epoxy joints 7 ft away by heating a large portion of the free section of the turn to 150°F. This allowed cleanup of the brazed joint. Cooling the turn to the temperature of the main body of the coil moved it back into its proper position in the coil.

Reinsulation also presented some challenging problems which were solved by cutting carefully controlled 0.010 in. slots in the remaining 0.040 in. G10 layer insulation in the damaged regions to make a lap joint with new insulation, and vacuum impregnation with epoxy of the entire region. These repair techniques were successful and the ZGS resumed operations on April 26. Special mention should be made of the ingenuity and all-around technical expertise of R. Wehrle and J. Bywater who led this successful effort.

During this down period, H^- was successfully injected into Booster I and a new experimental area, the EPB-I Annex, was completed and in May experiments were being done there. On July 29 there was a vacuum leak and a blister 10 in. x 3 in. had to be repaired. Plans were firming up for installing the new titanium chambers and it was hoped that the old ones would last until then. In the first week of October the ZGS reached new intensity highs after some noise was eliminated in the rf system and when the acceleration from injection to full field was slowed down. The weekly average reached to 2.8×10^{12} with a peak of 3.4×10^{12} and the weekly efficiency was 95%. The annual review committee again felt that the ZGS was doing significant physics and listed the following research:

1. Determination of (K_S^o, K_L^o) mass difference and the phase of η^{+-}.

2. Search for Time Reversal Violation in $K^o\mu_3$ decays.

3. Experiments on K^+ and $\pi^\pm$ P scattering using a polarized proton target.

4. Many precision measurements of two-body and quasi-two-body differential cross sections both in counter and bubble chamber experiments.

This year also marked the transfer of the 30-in. bubble chamber to Fermilab. The chamber had made a great reputation for its quality and productivity and it was admirably suited to the exploratory kind of physics that would be initially done at Fermilab. The entire 30-in. bubble chamber program at Argonne National Laboratory reflected great credit on those who operated and used it.

On May 1, 1972, the ZGS was shut down to replace the original epoxy-bonded stainless steel vacuum chamber with the new all-metal titanium chamber. This replacement required almost complete disassembly of the accelerator ring. The original vacuum chamber had been in use in the ZGS since 1963 and evidence of radiation damage had been increasingly apparent the last two years. In addition, the lack of precise tune control due to the absence of pole face windings in the old chamber was impeding progress in fully implementing slow spill resonant extraction at the ZGS. The new titanium vacuum chamber has 42 pole face windings incorporated into it, of which 28 were used for the passive correction of the eddy current effects resulting from the all-metal vacuum chamber construction. This correction was vital since this was an all-metal chamber (thinned to 11 mils titanium thickness over most of the area) and eddy currents were much stronger than in the old Spacemetal chamber. The remaining 14 windings are available for making the rapid tune shifts required for resonant extraction.

The titanium chamber consists of eight sections which were fabricated by the North American-Rockwell Corporation and at the time of their assembly were the largest diffusion-bonded structures ever built.

The pressure to maintain operation of the ZGS for the experimental program had been considerable in the past two years and the task of identifying the most suitable five-month shutdown period for the installation of the new chamber proved to be very difficult. However, in the spring of 1972 a combination of financial pressure and the desire to get ahead more rapidly with the resonant extraction program resulted in the decision to proceed with the installation, and the lengthy shutdown began.

The gratifying rapidity with which the ZGS was brought into operation in less than two weeks time reflects the care that went into the design and development of the needed accelerator instrumentation and the thoroughness of those responsible for working out the retuning strategy. These activities were given considerable attention during the spring in the few months that preceded the shutdown. A full "dress rehearsal" was carried out during a machine research period in April when the ZGS was briefly retuned as a 6-GeV accelerator following the proposed retuning plan. This energy was chosen on the basis of previous experience which had suggested that many of the machine characteristics at this energy differed markedly from their familiar behavior at the normal operating energy of 12 GeV.

The retuning of the accelerator during the week of September 18-24, 1972, was carried out almost completely according to plan. Within 24 hours after injection, acceleration to full energy was achieved. Then the currents in the pole face windings were adjusted to "flatten" the tunes to closely resemble what would be expected with an ideal (theoretical) zero gradient accelerator. This phase of the operation was carried out with relative ease and dispatch as a result of the high degree of sophistication attained with the real-time software for the ZGS control computer and its on-line display and with the automatic tune measuring instrumentation. By Monday, September 25, the intensity reached 2.3×10^{12} and by December the ZGS reached a new peak intensity of 3.5×10^{12} protons per pulse.

Other activities during this year included further study of H^- injection into the booster where circulating beams of over 5×10^{11} protons were achieved. During the shutdown, deuterons were accelerated through the linac with reasonable efficiency and one could think of later accelerating them in the ZGS. Also, during the downtime it was necessary to again repair the rotor of the motor-generator set. During 1972, work was also going on to prepare the necessary hardware for the program for accelerating polarized protons, and in January 1973, the ZGS was operated with 3 front porches and a flat-top on the same pulse. This mode would be useful for tuning through the resonances during the initial polarized proton tuneup.

Resonance extraction simultaneously to EPB I and EPB II was accomplished in February 1973. This month also saw the completion of the second preaccelerator (PA II). An H^+ beam from PA I went through the linac and an H^- beam could go to the booster on the same ZGS pulse by time-sharing the linac. This now meant that the booster program could go continuously instead of having to wait for ZGS down periods. By April, 200-MeV beam was extracted from the booster.

However, the most significant event of 1973 was the successful acceleration of polarized protons to high energy. The ZGS was the world's first high energy accelerator with this capability. The sequence of events began to move rapidly when the polarized source arrived at ANL on May 2, 1973. The pulsed quadrupole system for preserving polarization had already been installed in the ZGS and a polarimeter in EPB I had been set up to measure the beam polarization. On July 11 polarized protons were accelerated to 3 GeV/c and on July 22 to 6 GeV/c with 65% polarization. This eleven day period was a hectic and sometimes frustrating time as one searched for the depolarizing resonances. The uncertainty in the knowledge of the tune and magnetic field of the ZGS gave about a 100- to 200-gauss region in which the resonance could exist. Since one had to be within about 10 gauss to find it, a long time was necessary to search this window. The low beam intensity made the operation difficult, since it took a long time to accumulate enough data to get reasonable statistical and systematic errors. One was not sure if the theoretical prediction as to strength and width were correct or if indeed it was even possible to overcome the depolarizing resonances. The real break came when it was found that the depolarization could be enhanced and a sharp dip was found. This finally located the resonance and one was able to adjust the strength and timing of the quadrupoles. Five resonances were corrected and the optimistic design goal of 65% was achieved. One now knew that the theory was correct and that polarized protons could be accelerated to high energies.

The polarized proton project was a whole division effort with many systems coming together and working reliably at the moment of truth. The theory work of T. Khoe and the work of J. Bywater, R. Lari, E. Parker, C. Potts, W. Praeg, N. Sesol, R. Stockley, and

R. Timm produced the designs and the hardware that made it possible. The efforts of the University of Michigan group led by A. Krisch and the efforts of R. Martin, E. Parker, and L. Ratner through eleven days and eleven sleepless nights finally guided the polarized protons through the resonances and into a fruitful era of new results in high energy physics. Even during this brief initial period the experimenters were able to measure the difference in the total cross section between the parallel and anti-parallel alignment of the protons' spin. For the first time one had information on pure spin cross sections at high energy. Later in the year, physics runs started with 1.6×10^8 polarized protons per pulse being extracted by the resonant method. It was gratifying to find that this extraction process did not depolarize the beam as a priori no one knew whether or not this would happen.

Another project that reached an important milestone in 1973 was the Superconducting Stretcher Ring (SSR). This was proposed in 1971 as part of the program to increase the intensity of the ZGS. It had several important goals. It would be used as a "beam stretcher"; i.e., the ZGS could be pulsed and then fill the storage ring and beam could be taken out of the storage ring continuously. This would essentially "stretch" the spill time from a few hundred millisecs to continuous spill. In addition, the ZGS could pulse twice as fast since there would be no need for the flat-top. Alternatively, one could save money on the power bill by pulsing the ZGS much less frequently and still have a continuous beam for the experimenters. Also, there was no experience for operating a superconducting magnet ring and it was felt that the solution of the problems of a "real-life" cryogenic system, of radiation damage, and of field shape would be applicable for future accelerators. In August 1973, the first dipole magnet was successfully tested at 34 kG and construction of several more dipoles and quadrupoles was initiated. These superconducting magnets were to operate at 30 kG and be an operational ring. Therefore, they were designed as simply as possible and not to advance the state-of-the-art in superconducting magnet technology. Although the total project was never funded, the magnets were used in an extracted beam line and this superconducting beam line provided a good proving ground for testing the performance in a realistic operating environment.

The year 1974 saw a great deal of use of the 200-MeV booster beam. In January, tests were made to evaluate the 200-MeV beam for medical diagnostic purposes and for its use to produce neutrons. These tests were promising and in August some million pulses were used for a neutron scattering experiment. In September, 10 million pulses were extracted to the neutron target. Some 10% of the beam passed through a hole in this target to go to a radiography experiment. The proton radiography utilized an automatic stage for moving the specimens so that the collimated 1 mm x 1 mm beam could effectively scan them. The scan information was magnetically taped to form scope pictures of the relative density areas of the specimens. Approximately 50 scans were taken of a skull with and without brain, four brains, a liver, and a pancreas. These results looked promising and it appeared that one could get as good information as X-rays with a significantly lower radiation dose. A proposal to build a special Proton Diagnostic Accelerator was eventually made, but it was not possible to interest the medical community in a new technique, and this was never funded.

The polarized proton operation saw a continual increase in beam intensity due to source improvements made by Everette Parker and his group. Intensity went from its initial 1.6×10^8 to 6×10^8 in May and to 1.2×10^9 in October 1974. It was found that by pulsing this initially dc source, the output could be more than doubled. This and other improvements eventually made more than a factor of 100 improvement in the beam intensity. The ZGS worked well with polarized protons and during the year experiments were done at many momenta between 1.0 and 6.0 GeV/c.

There were also studies to determine if the ZGS could successfully operate with H^- injection, since with the polarized source now in PA II it was no longer possible to run the ZGS and the booster at the same time. First the injection was changed from its normal entrance into the ZGS at 0.7 in. below the midplane and an improvement factor of 1.5 was noted. Then the inflector was removed and a stripper foil to convert H^- to H^+ was placed inside the ZGS. Since no inflector septum existed to intercept the beam, the rate of rise of the magnetic field was reduced to inject for a much longer time; source pulse lengths were increased from 240 μsec to 550 μsec. An improvement factor of five was observed for this mode compared to the off midplane injection. With only 2.4 mA of H^- injected, 2.9×10^{12} protons were accelerated to 700 G and 1.9×10^{12} to full energy. Normal ZGS acceleration with protons requires about 30 mA of injected beam. This successful study of H^- injection was of great significance in that it again allowed simultaneous HEP operation and booster development and tuning.

Besides all these activities, the unpolarized experimental program was going well. There was a wide interest in the physics that could be studied at the ZGS as evidenced by the large number of university groups who were running and who were scheduled for future runs.

Some of the physics of particle resonances, strong and weak interactions, could best be studied in the energy range available to the ZGS. The results of ZGS experiments were important enough to warrant publication in Phys. Rev. Letters at approximately the same rate as the other three major high energy physics laboratories in the country. A very important result published in this year was the confirmation of the existence of the "weak neutral current." This was done in the 12-ft bubble chamber, exposed to a neutrino beam, by a group led by M. Derrick.

By completing 17 experiments in 1974, the ZGS again demonstrated that it was a viable laboratory. In the same period it succeeded in setting up a low-energy separated beam, a K_L^o beam for the University of Chicago, and an rf separated beam.

One very successful counter-spark-chamber facility on the ZGS experimental floor was the effective mass spectrometer (EMS). The spectrometer consisted of a large aperture dipole magnet surrounded by magnetostrictive readout spark charmbers, and measured the angles and momenta of fast forward-charged particles from interactions in a liquid hydrogen or deuterium target. In four years of operation since its turn-on in 1971, the EMS was used by groups from eight institutions to complete fourteen experiments. The EMS program was dominated by investigations of meson spectroscopy and production mechanisms of elastic and inelastic interactions. The facility took full advantage of the fact that cross sections for simple inelastic reactions represent a large fraction of the total cross section at ZGS energies. It was thus easy to perform comprehensive, high statistics studies of related reactions at several energies; precise comparisons were possible because a single apparatus was used for all reactions and energies. Secondary beams of $\pi^{\pm}$, $K^{\pm}$, p and $\bar{p}$ up to 6 Gev/c were available; the use of both hydrogen and deuterium targets allowed the isolation of isospin dependences. The EMS made significant contributions 'to the understanding of even supposedly well-understood reactions, elastic scattering being a good example. By comparing elastic scattering of the six beam particles from protons, the energy dependence of the crossover values was measured precisely, and compared to theoretical expectations.

The systematics of ρ^o production by pions and K^{*o}, $\bar{K}^{*o}$ production by kaons were investigated and compared within the framework of SU(3)-symmetric, strongly absorbed Regge-pole models. Precise measurements of ρ-ω interference were obtained by using π^+ and π^- beams on a deuterium target, and these were combined with measurements of production by the ZGS Charged-Neutral Spectrometer to provide unique constraints for amplitude analyses of vector-meson production. Other high statistics studies of production mechanisms included measurements of exchange-degeneracy breaking in $K^{\pm}N$ charge exchange, of cross sections and Λ polarizations in $\pi^- p \to K^o \Lambda$, ΛK^o, and ΛK^{*o} (890), and of line-reversal breaking in $K^{\pm}N \to K^o \Delta$ (1236). Use of a large downstream Cerenkov counter allowed a unique measurement of $\pi^{\pm}N \to K^+K^-N$, which isolated interferences between isospin-0 and isospin-1 K^+K^- states. Production of S^*, ϕ, f and A_2 mesons and their mutual interferences were clearly observed, and f' production by pions was detected for the first time through the interference of the f and the f'. The same reaction produced precise data on ϕ-meson production by pions, and showed that natural-parity-exchange ϕ and ω production are similar, aside from an overall suppression factor of 300 for the ϕ's, as predicted by the quark model; unnatural-parity-exchange ϕ and ω cross sections were found to differ dramatically, however.

In the streamer chamber the reaction $\pi^- p \to X^- p$ was studied for fast forward protons. The important new result was that no A_1 production was observed. This agreed with expectations from earlier experiments at BNL. The streamer chamber is a second facility which was used for a series of experiments.

Despite the excellence and productivity of the work at the ZGS, the first intimations of its end came when, in its review of the FY 75 budget, the Office of Management and Budget (OMB) requested that the AEC and the Science Advisor develop "a plan for shutting down the ZGS accelerator at the earliest reasonable time." Further discussions suggested an assessment of the ZGS "in the context of an overall assessment of the national needs of the discipline" and indicated an interest in an orderly shutdown.

The ZGS Study Committee and two subpanels were established to prepare a response to the request of the OMB. The Study Committee consisted of representatives of the Science and Technology Policy Office, the Physics Section of the National Science Foundation Research Directorate, and the Division of Physical Research of the AEC. The Physics Subpanel consisted of knowledgeable practitioners of high energy physics selected to give a wide and balanced representation of experts in theoretical and experimental high energy physics. This Subpanel examined the status and capabilities of the ZGS in the context of the overall scientific needs of the nation's high energy physics program and addressed the question of determining the earliest reasonable time for ZGS shutdown. The Management and Procedures Subpanel consisted of experienced accelerator laboratory administrators, some of whom had had recent experience with accelerator shutdowns. This Subpanel studied the proper approach to closing out a major accelerator with specific emphasis on the ZGS.

The following excerpts from the report of the Study Committee speak for themselves as to the value of the ZGS and its programs.

"The ZGS Study Committee has reviewed the ZGS program and the two Subpanel reports. The Committee was much impressed with the present vitality and productivity of the experimental physics program at the ZGS and found no indications of pending weaknesses in that program. Recently completed ZGS experiments show that the ZGS, although lower in energy than the other high energy physics accelerators, is doing research at the frontiers of the field. Specific indications of the strength of the program are:

1. The polarized beam capability at the ZGS is unique. It constitutes an important tool for unraveling the nature of the strong interaction. It appears that technical reasons may preclude accelerating polarized protons to comparable energies at other accelerators.

2. The neutrino program, involving the 12-ft bubble chamber and the complete booster, provides an important independent source of information in the area of low energy neutrino physics. The ZGS program in this area complements the higher energy neutrino programs at the AGS and at the Fermilab.

3. The ZGS program in hadron physics is and gives strong promise of continuing to be outstanding. This program is focused on the low energy region where resonances are best studied. A continuing study in hadron physics in this energy region is likely to be a critical need for the eventual understanding of high energy phenomena.

4. The 12-ft bubble chamber at the ZGS is an important and productive research tool. It is the largest bubble chamber operating in the intermediate energy range. It was the first large chamber to be constructed with super-conducting coils and has operated very successfully for over four years. The Track Sensitive Target, currently being installed in the chamber, will provide an unparalleled capability for observing reactions with neutral particles in the final state.

5. The ZGS has an excellent staff of accelerator physicists whose achievements in accelerator developments are impressive. The ZGS staff has pioneered in the application of H^- injection and in the acceleration of polarized protons.

6. The ZGS program has had considerable beneficial impact on the balance of the ANL program. The exchange of technology and personnel between high energy physics and the other basic and applied program areas at ANL is expected to continue to be important.

"The Committee also notes that the ZGS has a demonstrated capability for contribution to fields of science outside high energy physics. These include nuclear structure research, proton radiography, cancer therapy, materials research, and nuclear chemistry. The materials research, radiography, and cancer therapy programs can be satisfied by the capabilities of the booster and injector."

The following two recommendations of the Committee were the most relevant for the life of the ZGS.

"Recommendation 1: In view of the important contributions of the ZGS experiments to the overall national High Energy Physics program, operation of the ZGS should be continued, possibly at a more intensive level, through FY'76, FY'77, and FY'78. Mid- to late-FY'79 is projected as the earliest reasonable time to shut down the ZGS. The review proposed below would consider whether shutdown of the ZGS at the end of the decade would be in the best interests of the national High Energy Physics program.

"Recommendation 2: A policy of advance notice should be established and announced relative to accelerator shutdown. In the case of the ZGS, the appropriate lead-time appears to be two to three years. The nature of the apparatus and procedures for high energy physics experiments requires such a period for orderly completion of work underway when the shutdown is announced."

This report, issued in September 1974, indicated to the ANL staff that the demise of the ZGS was not imminent and morale and productivity remained high. It also suggested that if the decision to shut down the accelerator was made, special consideration would be given to the retention of the ZGS staff during the final running period and to various programs relating to future job security.

The years 1975 and 1976 continued a smoothly running program with both polarized and unpolarized beams. The intensity of both operational modes continued to improve. Midplane injection and tuning brought the peak beam up to 4.1×10^{12} in January 1975 and to 5.4×10^{12} in March 1976. In October 1976 with the advent of operational H^- injection it reached 6.3×10^{12}. Machine research by Yanglai Cho led to the use of the pole face windings to produce an octupole correction which was very helpful for both intensity increases and extraction efficiency. The extracted beam hit a peak of 3.7×10^{12} in the October running period. Concurrently, the polarized beam intensity kept increasing and in August 1976 reached 1.5×10^{10}, a factor of 100 times its initial level only three years earlier. Also of significance was the first 11.75 GeV/c polarized proton physics run in January and February of 1976. This also saw the use of the superconducting beam transport line which brought these high energy protons to the effective mass spectrometer. This was the first all-superconducting beam line in use for high energy physics and was constructed with the SSR magnets.

Perhaps the best indication of this period is some of the physics results that were noted by the 1976 annual review committee in its report.

"The Michigan-ANL(ARF)-St. Louis collaboration has used a polarized target with the polarized beam to study the parameter C_{NN} at 6 and 12 GeV/c and 0.5 $(GeV/c)^2 \leq p_\perp^2 \leq 2.2$ $(GeV/c)^2$. A dramatic energy dependence is observed, the 6 GeV/c data peaking at $p_\perp^2=0.5$ and decreasing smoothly to zero at $p_\perp^2=0.9$ and rising to a broad maximum at $p_\perp^2 = 1.7-1.8$. It is very tempting to associate this dip at $p_\perp^2 = 0.9$, also seen in the polarization, P, at 12 GeV/c, with the structure in the elastic differential cross section at $-t = 1.3-1.4$ $(GeV/c)^2$, seen as a shoulder at an incident momentum of 20 GeV/c and only developing into a dip at ISR energies. This may be an example of a feature showing up at much lower energies in the spin-correlation data than in the unpolarized differential cross section.

"The other group making extensive measurements on spin-correlations in proton-proton elastic scattering is the Northwestern-ANL(HEP) collaboration. This experiment has a carbon polarimeter in the recoil arm for observation of the recoil polarization. With polarized beam and target, this permits observation of triple-spin-correlation parameters. First results have been obtained on such a parameter.

"The Effective Mass Spectrometer group has made observations on proton-proton and proton-neutron polarization in elastic scattering up to 6 GeV/c using the polarized beam and unpolarized H_2 and D_2 targets. Significant differences are observed for the whole t range $0.2 < t < 1.0$ $(GeV/c)^2$, indicating interesting differences in the exchange amplitudes in the two processes. Such data are important in sorting out the isospin dependence in any amplitude analysis of nucleon-nucleon scattering.

"The EMS group has also had an extensive program on spin-dependent effects in inelastic reactions for both exclusive and inclusive channels. Exclusive inelastic channels provide spin information on resonance production, sometimes interpretable as virtual pion-polarized proton scattering. A beginning has been made on spin effects in inclusive reactions, with much more work planned by the EMS group, and also by the Minnesota-Rice and Indiana groups.

"A vigorous program of counter experiments with unpolarized protons is continuing, primarily concerned with various aspects of meson spectroscopy. Rather sophisticated experimental results have been reported by three functioning spectrometer facilities, namely the Effective Mass Spectrometer, the Charged-Neutral Spectrometer and the Streamer Chamber. The EMS group has reported results on $\pi^- P \to K^+ K^- n$ and $\pi^+ n \to K^+ K^- P$. Results from streamer chamber work in $\pi^- P \to K^0 K^0 n$ of somewhat more limited statistical accuracy have also been reported. Both groups have uncovered evidence for a new, broad S-wave resonance at 1300 MeV in the $K\bar{K}$ channel. This is undoubtedly of great significance for the field of meson spectroscopy. The Charge-Neutral Spectrometer has provided a great volume of new data of high quality. The η, η' and ω^0 components of the various final states studied are clearly and impressively established.

"The Minnesota-Columbia-ANL group has completed a search for the so-called Ericson fluctuations in 90^0 scattering. The indications for large fluctuations over small energy intervals earlier reported by CERN are not confirmed, although there may be some interesting energy-dependent structure in the data. The same group has also carried out a simple yet sensitive search for threshold production of ψ particles with negative results to date.

"The 12-ft bubble chamber facility has operated with excellent efficiency for hadron physics during this year. The taking of 710,000 photographs in a single month, on a double-pulsing mode, would be a notable achievement for even a small chamber, but its accomplishment last March with an instrument as complex and large as the 12-ft bubble chamber is truly remarkable. The new film complete a very large statistics K^-p exposure which should provide substantial new data on the properties of strange baryons such as Ω, Ξ, Σ, and Λ resonances.

"Another significant achievement is the experience gained in running track-sensitive targets with hydrogen on the inside and Ne-H_2 mixtures on the outside to provide high gamma-ray conversion efficiency."

In addition to all of the above activity, members of the High Energy Physics and Accelerator Divisions collaborated with a group from Fermilab, from November 1975 to May 1976, to produce a proposal for constructing a 1000-GeV x 1000-GeV colliding beam facility at Fermilab. This group was headed by R. Diebold of ANL. The project, however, fell by the wayside since Fermilab was completely preoccupied with the Energy Doubler Saver project and subsequently a colliding beam project, Isabelle, was approved at BNL.

Also during this period, plans for high energy physics within the Energy Research and Development Agency (ERDA) and other agencies included the PEP Electron-Positron Ring at Stanford and other new facilities. To win approval for new projects, the national high energy physics program was asked to shut down some other facility. This would affect the ZGS and in March 1976, Dr. James Kane, Deputy Assistant Administrator for Physical Research of ERDA, informed the Laboratory Director that "it is our current expectation that the ZGS will cease operation by the end of 1978." The 1974 AEC-NSF ZGS study report projected mid- to late-FY'79 as the earliest reasonable time to shut down the ZGS and recommended a two to three year lead-time of advanced notice for a shutdown decision. Thus, it appeared that ERDA decided to take the "earliest reasonable time" as the actual shutdown time. In conformity with the report's recommendation, ERDA was giving the Laboratory a 2-1/2 year lead time to permit appropriate planning.

Because of these plans and the previous Study Committee recommendations, a Review Panel was constituted to carry out a scientific review of the ZGS program. The following specific charge for the Panel is contained in a letter to its chairman, Professor R. Walker of the California Institute of Technology, from James S. Kane, Director of Physical Research, U.S. ERDA, dated June 28, 1976:

"You are requested to review the status and role of the ZGS in the U.S. High Energy Physics program and to update the 1974 report of the Physics Subpanel of the AEC-NSF ZGS Study. We are especially interested in obtaining your advice on what approach should be used prior to shutdown to gain the optimum scientific benefit from the ZGS and its special and unique facilities during its remaining lifetime. Present ERDA plans contemplate that the ZGS will be shut down by the end of calendar year 1978."

The recommendations of the review committee were:

"1. Prime emphasis during the next two and one half years should be on the unique program utilizing the ZGS polarized proton beams.

"2. The second generation neutrino experiment using the 12-ft bubble chamber should be carried out, using approximately five months of operating time.

"3. The "conventional program" should be phased out after fulfilling existing commitments with approximately three months devoted to it during FY'77.

"4. A total time of approximately twenty operating months should be devoted to the polarized beam program after October 1976. This will require:

a. continued operation of the ZGS for the polarized proton program for approximately one year beyond CY'78, and

b. a level of funding which permits nine months of operation per year in FY'77, FY'78, and extending through CY'79

"5. Sufficient equipment funds should be provided in FY'77 to construct new apparatus and facilities essential to the polarized beam program."

The extension beyond the FY'78 deadline mentioned in Kane's letter was eventually approved for the ZGS to continue until October 1, 1979. From the reviews of both committees one gets the impression that without the polarized beam and 12-ft bubble chamber, the ZGS would very likely have been turned off in 1976. This would have had a serious impact on two major programs (the Intense Pulsed Neutron Source and Heavy Ion Fusion), which might well have never come to fruition. In the case of Heavy Ion Fusion, as we shall discuss later, this could have perhaps been a serious setback of national and even world-wide significance.

The years 1977, 1978, and 1979 saw a great deal of effort by both the High Energy Physics Division and the Accelerator Research Facilities Division on programs that would be viable after the ZGS was turned off. Despite this, the ZGS operating record continued to improve. The unpolarized beam intensity in the last operating month (February 1978) before the ZGS became totally dedicated to polarized proton operation reached new highs. A peak of 7.5×10^{12} was recorded with a daily high of 6.8×10^{12} and a monthly average of 6.2×10^{12}. The polarized beam intensity hit a peak of 9.33×10^{10} in May 1979 and the average for the month was 5.6×10^{10}. This peak was about 1000 times more beam than was accelerated in the first run in 1973. The intensity was not the only story. The ZGS was able to accelerate polarized deuterons to full energy in December 1977 and the world's first information on pure spin phenomena at high energies was obtained by two experimental groups in the first physics runs in November 1978. Another important improvement was the achieving of high polarization at 11.75 GeV/c. It had initially been about 50 to 55% but in May 1977, 67% was achieved. This improvement resulted from an ingenious and carefully conducted program of investigation and correction of imperfection resonances which are due primarily to slight misalignments of and distortions in the magnets of the ZGS. A total of 22 such resonances were identified and corrected by introducting suitable currents in pole face windings.

Significant physics results kept coming from the ZGS. In particular, it seemed that every polarized beam experiment produced unexpected results which had not been predicted by any theories. In the unpolarized beam, important work was done in the charged neutral spectrometer on radiative meson decays. Additionally, experiments on Σ^- beta decay and on direct electron production were completed. Successful runs in the 12-ft bubble chamber with the Track Sensitive Target (TST) were also completed, as well as a large exposure of neutrinos in deuterium. There were two experiments with polarized protons which received national media coverage as shown by the following excerpts. The first was in Science News of September 24, 1977, and the second from the New York Times of November 24, 1977.

Science News

"Spin was one of the first characteristics of subatomic particles that physicists came across. Ubiquitous and plain, it plays small but important roles in atomic and nuclear structure. Generally, spin is expected to be responsible for effects that appear more or less like fine tuning. After all, the amount of energy involved in altering the orientation of a spin is small compared to other effects involving the same particle. The spins of protons were expected to have only a small effect on the way one proton bounces off another. Surprise. Spin turns out to have a large and philosophically rather curious effect on how protons bounce.

"The experiment that sprung the surprise was done at Argonne National Laboratory near Chicago by J. R. O'Fallon, L. G. Ratner and P. F. Schultz of Argonne; K. Abe, R. C. Fernow, A. D. Krisch, T. A. Mulera, A. J. Salthouse, B. Sandler and K. M. Terwilliger of the University of Michigan; D. G. Crabb of Oxford University; and P. H. Hansen of the Niels Bohr Institute of Copenhagen. Their report is in the September 19, 1977, Physical Review Letters.

"The effects of spin were most pronounced when the bouncing proton came off at a large angle to its original direction. The combination of high energy and high scattering angle indicates that something rather deep inside the target proton is responsible for the observed effect. That is, rather simply, that protons bounce well off each other when their spins are parallel (axes in the same direction, turning the same way). When the spins are antiparallel (clockwise versus counter-clockwise, for instance), the protons don't even seem to notice each other. They appear to pass right through each other as if they were transparent. Bang! Wow! Balloon full of question marks. The physicists and philosophers who have been asking questions about the materiality of matter will have fun with that."

New York Times by Walter Sullivan

"Physicists at Argonne National Laboratory in Illinois believe that by smashing together two protons, the nuclei of hydrogen atoms, they have very briefly created a "diproton" heavier than the combined weight of the two original particles.

"If confirmed, the observation would support speculation that such double particles exist and would culminate years of experiments in search of them.

"According to Dr. Malcolm Derrick, Argonne's Deputy Director for High Energy Physics, (sic) the findings also call into question the widely-held view that all heavier particles must be formed either of two of three quarks. The double particle ("dibaryon" or "diproton") would be formed of six quarks.

"The Argonne experiments have been done with the Zero Gradient Synchrotron, said to be the only machine in the world that produces a high energy beam of polarized protons, that is, one in which most protons are spinning the same way. This beam is fired at a target whose protons are also polarized.

"In the tests, done by Dr. Akihito Yokosawa and nine colleagues, it was found that when the impinging protons were polarized in a manner opposite to that of the target protons, there was little or no effect.

"When the polarizations, or spins, were parallel, however, there was strong evidence that some protons were mating to form a particle with a mass of 2,260 million electron volts (as opposed to 936 for one proton)."

Publicity alone does not make important physics and only the future development of our understanding of nature will say which of the almost 300 experiments completed at the ZGS were the significant ones. All of them increased our knowledge and all of their results will eventually have to be explained by whatever theory that may be devised.

During this "mature" period in the story of the ZGS, there were changes in the Associate Laboratory Director, but both Bruce Cork and his immediate successor Tom Fields, who took over in 1974, not only maintained the environment for producing good physics, but vigorously pushed such projects as the 12-ft bubble chamber, the polarized proton beam, and the booster. These had a strong effect not only on the life span of the ZGS, but also on the production of new and interesting physics results. Tom Fields resigned in 1978 and was succeeded by Gerald Smith for one year and then Robert Diebold became Director in 1979. With a relatively stable operating program and with the imminent demise of the ZGS, the latter two directors were mainly concerned with the longer range planning for post-ZGS activities. Also during this time, Malcolm Derrick served as acting Associate Laboratory Director during two different periods.

Some of these activities, such as the Intense Pulsed Neutron Source, Heavy Ion Fusion, construction of large superconducting magnets, construction of a small tokamak fusion reactor (APEX), and work on Ocean Thermal Energy Conversion, had been going on for several years. Others, such as a High Resolution Spectrometer for an experiment at Stanford, plans for polarized beams at Fermilab and Brookhaven, plans for colliding beam detectors at Fermilab, design for an antiproton-proton colliding beam facility, and plans for a small research accelerator for polarized protons at ANL, were formulated during these last two or three years. With the exception of the small research accelerator, which was to be used to investigate the properties of polarized beams in alternating gradient machines, these programs are now actively being pursued. The small accelerator was superseded by the plan to produce a polarized beam as soon as possible at Brookhaven.

The story is almost over and along the way many technological and physics triumphs have been pointed out. However, before we end, we should say something about Heavy Ion Fusion. This may be very important to all of us someday. The ANL program started out when the success of the H^- stripper injection technique led Ron Martin to the idea of using such a method in an accelerator system to deliver intense beams of particles to implode and ignite the thermonuclear reaction in pellets by inertial confinement. There were programs to do this with electrons and lasers and these programs were well advanced when Ron Martin and Richard C. Arnold, as well as several others at ANL, produced some basic papers during 1975 and in December made a presentation for HEARTHFIRE, an acronym for High Energy Accelerator and Reactor for Thermonuclear Fusion with Ion beams of Relativistic Energies. The vigorous promotion of these ideas led to numerous meetings, summer studies, and workshops. As this work progressed, more and more people began to understand the practicality of the heavy ion approach to pellet fusion and from a budget of $50,000 in 1976, a multi-million dollar program over the next three years is now anticipated. Success of this program could mean an inexhaustible energy source whose hazards are orders of magnitude smaller than fission reactors. There are no possibilities

of "runaways" and the wastes are very significantly smaller. It is
interesting to note that if ANL had not pioneered H⁻ injection, had
not developed the superconducting 12-ft bubble chamber, and had not
had a polarized beam, the ZGS would very likely have been turned off
in 1976 and there may have been no HIF. Although A. Maschke at BNL
independently had similar ideas, the great push for implementation
came from ANL. The technologies developed in high energy physics and
accelerators may indeed provide a solution to the world's energy
problems. There are indeed "practical" payoffs for the support of
the esoteric research of high energy physics. H⁻ injection, also
initiated· by Ron Martin, was a significant technological advance
which not only led him to HEARTHFIRE, but has also been a major
factor in being able to produce intense neutron beams, comparable to
fission reactors, for many important nuclear structure studies and
radiation damage studies. It is primarily due to the vigorous
efforts of the ANL group that HIF is competing now with the older
techniques. Ron Martin resigned as Director of the Accelerator
Research Facility Division in June 1979 to assume the job of Program
Director for HIF. He was succeeded by Robert L. Kustom.

The ZGS died on October 1, 1979, and with it the Midwest's
regional facility for high energy physics, but the expertise
developed will live on and continue to contribute to many fields of
endeavor as has been indicated above and will be expanded upon in
the following Epilogue.

EPILOGUE

With the ZGS close-down scheduled for the end of FY 1979, the Department of Energy (DOE) turned to the question of future involvement of the Argonne groups in the national high energy physics program. Over the years strong, competent groups of both high energy and accelerator physicists had developed around the ZGS. Moving these groups enmasse to another laboratory was impractical, and would have meant a loss of the fruitful interaction between high energy physics and the other Argonne programs. An obvious solution was to have these groups work in a user mode at the remaining accelerator laboratories; indeed, each of the three major in-house Argonne experimental groups had already interleaved experiments at Fermilab while continuing with their ZGS work. Still, this would be a new precedent with major user groups at a national laboratory, neither serving to improve the program on a local machine, nor directly contributing to the formal education of future scientists as do university-based groups. With construction projects underway at the three remaining accelerator laboratories, the tight funding situation led DOE to request a committee chaired by Professor Francis Low of the Massachusetts Institute of Technology to make recommendations on the Argonne post-ZGS activities. The committee met in May 1978 and its recommendations were subsequently reviewed by HEPAP in August. The results are summarized in the following excerpt from the letter of the chairman of HEPAP, Professor Sidney Drell, to Dr. James Kane of the Department of Energy:

"1. The first recommendation is to continue the strong in-house Argonne experimental and theoretical high-energy research program. The experimental effort will henceforth operate in the user mode at accelerators at the other national facilities.

"2. The second recommendation is that the Argonne National Laboratory continue, on a trial basis, to make available its support facilities for university users.

"3. . . .the excellent accelerator R & D group at Argonne should continue to receive support and encouragement to work closely in collaboration with Brookhaven and Fermilab so that the national program not lose their singular talents . . ."

In anticipation of this endorsement of the user mode of operation the HEP Division Director, Malcolm Derrick, had previously organized study groups to consider various possibilities for future programs. Out of these discussions came three major thrusts: 1) an e^+e^- detector for PEP; 2) a colliding $\bar{p}p$ and pp detector for Fermilab; and 3) a polarized beam (using Λ decays) at Fermilab.

The e^+e^- project was developed around the large superconducting magnet previously used for the 12-ft bubble chamber. With an enormous BL^2 this magnet will measure the momenta of charged tracks of factor of ≥ 5 better than the other e^+e^- detectors at PEP and PETRA; for this reason it has been labeled HRS - the high resolution spectrometer. This facility was approved as experiment PEP-12 and is principally a collaboration of Argonne and several Midwest universities. This effort continues the close interaction between Argonne high energy physics and the AUA university groups. With its large shops and engineering capabilities, Argonne plays an important role in the collaboration complementary to the unversity groups. The magnet has been reengineered and turned on its side to give a solenoidal field along the beam direction. It will be trucked to SLAC in the next few weeks, one of the largest loads ever to be moved over the Nation's highway system. The experiment is scheduled to start up in mid-FY 1980.

The Colliding Detector Facility (CDF) at Fermilab is another collaboration with the university community, as well as having strong participation from Argonne and Fermilab. The solenoidal detector is designed for pp collisions up to 150 x 1000 and $\bar{p}p$ up to 1000 x 1000 GeV. Conceptual design of the detector has been made and prototype work on the detector systems is in progress. The detector will study events with electrons and muons, including those from the decay of intermediate vector bosons, $W^{\pm}$ and Z^0, as well as yield information on particles and jets in the region 5^0 to 175^0. First collisions should occur not long after first operation of the Saver-Doubler superconducting accelerator, scheduled for 1982. The project obviously has a close connection with the work being done by the Argonne accelerator group on $\bar{p}$ cooling at Fermilab.

The Argonne polarized-target group has spearheaded an international collaboration in proposing a 300-GeV polarized beam for the Meson Area at Fermilab. This beam will use the proton polarization imparted by the weak decay $\Lambda \rightarrow p\pi^-$. By picking off certain regions of phase space in the Λ decay, different proton polarizations can be obtained. Typical parameters are 10^7 protons per pulse with 45% polarization. Using methods developed by the group for their Argonne experiments, bending magnets in the beam will rotate the spin into the desired direction at the experiment and will be able to flip the spin on alternate machine pulses to reduce systematic errors. The polarized beam will be used together with a polarized target to extend the Argonne measurements to the several-hundred-GeV region, as well as to study new phenomena such as jets at these energies; spin-dependence of quark-quark interactions will be reflected in the resulting jets. Design details and cost estimates are being refined and the project expects official approval in the next few months.

Though the committee report dealt with an ongoing accelerator R & D effort after the closing, the activities of the Accelerator Research Facilities Division (ARF) had to be phased out of ZGS operation by the closing date, but still had to maintain a viable operation until the last minute.

In giving a long lead-time for planning the termination of the operation of the ZGS, DOE hoped that an "orderly transition" could be carried out, and ZGS management was urged to accomplish this goal. At the same time, there was very strong pressure to accomplish the minimum polarized proton program recommended by the review panels. The latter implied continuous operation of the ZGS for the last 12 months of its life. To carry out such continuous running with very little down time for maintenance, even with a full crew, seemed difficult and had not been done in the prior 15 years of operation of the ZGS. To do so with an ever-decreasing staff of people seemed next to impossible. Yet, the operation of the ZGS has been highly successful and is expected to remain so up to the final moment at 0900 on October 1, 1979.

Personnel have been transferred out of operations as openings in other areas occurred. The number of personnel associated directely with the operation of the ZGS has decreased from 140 in 1977 to about 80 at the present time. A part of this loss has been made up by the liberal use of overtime which has more than doubled in the last two year. The sociological impact of the generous use of overtime is of some concern to management because it will stop rather abruptly at some point in time. Nevertheless, it has been an important management tool to accomplish these two seemingly incompatible goals of an orderly transition and full, continuous operation of the ZGS up to the last day.

In the beginning of this transition period there was concern among ARF Division management that there might be a mass exodus of operating people so that the ZGS operation would automatically terminate prematurely and not accomplish the physics desired. It has not occurred. Instead, the decrease in personnel has been gradual though steady.

Finally, there has been a development of job openings in non-HEP activities within the ZGS complex. The latter include the operation and development of Booster II for the intense pulsed neutron program for solid state science and materials damage studies, development of an accelerator demonstration for heavy ion fusion, success with the first, large saddle coil superconducting dipole for magnetohydrodynamic application and subsequent authorization for construction of an even larger superconducting magnet, and construction of a small tokamak for certain very specific experiments in the magnetic confinement fusion program. The technology for all of these programs is closely related (even often identical) to that developed in the HEP accelerator program, and all of them have benefited a great deal from the availability of such highly trained personnel. These and a few other developments represent a very significant spin-off from the country's high energy physics program. Contrary to some statements by competition in other laboratories for these programs, they are all on-going, stand-alone programs justifiable on their own merit. They were developed at Argonne because they were very exciting and the technology had arrived at the stage that made them possible. Had the ZGS operation not been in the process of termination, there would have been strong pressure to expand the accelerator technology effort in the ZGS complex. In FY 1980, these programs collectively are expected to have about 90 people involved.

Thus the skills and new technologies developed by the Argonne staff as they made the ZGS into a highly successful and productive instrument for high energy physics research are not being wasted. They are being utilized both in basic research experiments at other high energy accelerators and in applied research projects aimed at making our society technologically stronger. It might be both pleasant and interesting to have another symposium in 1989 to see what these outstanding hardware capabilities have produced.

AIP Conference Proceedings

		L.C. Number	ISBN
No.1	Feedback and Dynamic Control of Plasmas	70-141596	0-88318-100-2
No.2	Particles and Fields - 1971 (Rochester)	71-184662	0-88318-101-0
No.3	Thermal Expansion - 1971 (Corning)	72-76970	0-88318-102-9
No.4	Superconductivity in d-and f-Band Metals (Rochester, 1971)	74-18879	0-88318-103-7
No.5	Magnetism and Magnetic Materials - 1971 (2 parts) (Chicago)	59-2468	0-88318-104-5
No.6	Particle Physics (Irvine, 1971)	72-81239	0-88318-105-3
No.7	Exploring the History of Nuclear Physics	72-81883	0-88318-106-1
No.8	Experimental Meson Spectroscopy - 1972	72-88226	0-88318-107-X
No.9	Cyclotrons - 1972 (Vancouver)	72-92798	0-88318-108-8
No.10	Magnetism and Magnetic Materials - 1972	72-623469	0-88318-109-6
No.11	Transport Phenomena - 1973 (Brown University Conference)	73-80682	0-88318-110-X
No.12	Experiments on High Energy Particle Collisions - 1973 (Vanderbilt Conference)	73-81705	0-88318-111-8
No.13	π-π Scattering - 1973 (Tallahassee Conference)	73-81704	0-88318-112-6
No.14	Particles and Fields - 1973 (APS/DPF Berkeley)	73-91923	0-88318-113-4
No.15	High Energy Collisions - 1973 (Stony Brook)	73-92324	0-88318-114-2
No.16	Causality and Physical Theories (Wayne State University, 1973)	73-93420	0-88318-115-0
No.17	Thermal Expansion - 1973 (lake of the Ozarks)	73-94415	0-88318-116-9
No.18	Magnetism and Magnetic Materials - 1973 (2 parts) (Boston)	59-2468	0-88318-117-7
No.19	Physics and the Energy Problem - 1974 (APS Chicago)	73-94416	0-88318-118-5
No.20	Tetrahedrally Bonded Amorphous Semiconductors (Yorktown Heights, 1974)	74-80145	0-88318-119-3
No.21	Experimental Meson Spectroscopy - 1974 (Boston)	74-82628	0-88318-120-7
No.22	Neutrinos - 1974 (Philadelphia)	74-82413	0-88318-121-5
No.23	Particles and Fields - 1974 (APS/DPF Williamsburg)	74-27575	0-88318-122-3

Date Due